High Resolution NMR

Theory and Chemical Applications

Second Edition

EDWIN D. BECKER

Laboratory of Chemical Physics
National Institute of Arthritis, Metabolism, and Digestive Diseases
National Institutes of Health
Bethesda, Maryland

1980

ACADEMIC PRESS

A Subsidiary of Harcourt Brace Jovanovich, Publishers

New York   London   Toronto   Sydney   San Francisco

ACADEMIC PRESS, INC.
111 Fifth Avenue, New York, New York 10003

United Kingdom Edition published by
ACADEMIC PRESS, INC. (LONDON) LTD.
24/28 Oval Road, London NW1   7DX

Library of Congress Cataloging in Publication Data

Becker, Edwin D
    High resolution NMR, Second edition

    Includes bibliographical references and index.
    1. Nuclear magnetic resonance spectroscopy.
I. Title.
QD96.N8B43 1980   538'.3   79−26540
ISBN 0−12−084660−8

PRINTED IN THE UNITED STATES OF AMERICA

80 81 82 83   9 8 7 6 5 4 3 2 1

Contents

4.   Chemical Shifts

5.   Electron-Coupled Spin–Spin Interactions

6.   The Use of NMR in Structure Elucidation

7. Analysis of Complex Spectra

8. Relaxation

9. Theory and Application of Double Resonance

10. Pulse Fourier Transform Methods

11. Exchange Processes: Dynamic NMR

12. Solvent Effects and Hydrogen Bonding

13. Use of NMR in Quantitative Analysis

14. Contemporary Developments in NMR

Preface to Second Edition

During the eleven years since publication of the first edition of this book, applications of NMR in chemistry and biochemistry have mushroomed. With the discovery of new NMR phenomena and the spectacular development of NMR instrumentation, the types of problems amenable to solution by NMR are steadily increasing. For example, two major areas that had only limited NMR study eleven years ago—solid state phenomena and biochemical processes—are now among the most exciting and rapidly growing fields of NMR research. The advent of routine carbon-13 NMR spectrometers has had a major impact on the use of NMR in organic structure elucidation, and the now routine high field multinuclear spectrometers promise to be of great value in inorganic and metallo-organic chemistry.

In revising this book, I have tried to retain the basic organization and presentation that proved successful in the first edition. Large sections on basic principles, chemical shifts, coupling constants, and analysis of complex spectra have been changed only slightly. On the other hand, many parts have been expanded substantially—for example, carbon-13, nuclear Overhauser effect, relaxation mechanisms, and use of superconducting magnets, each of which was treated only briefly in the first edition. Fourier transform methods, which were covered in one paragraph in the first edition, now take up an entire chapter. As in the first edition, problems are given at the ends of most chapters, with answers to selected problems provided in Appendix D. The original collection of proton spectra of "unknowns" (Appendix C) has been augmented with a number of carbon-13 spectra.

Preface to First Edition

Few techniques involving sophisticated instrumentation have made so rapid an impact on chemistry as has nuclear magnetic resonance. Within five years after the discovery that NMR frequencies depended upon the chemical environments of nuclei, commercial instruments capable of resolving resonance lines separated by less than 0.1 part per million (ppm) were available. Chemists immediately found NMR to be a valuable tool in structure elucidation, in investigations of kinetic phenomena, and in studies of chemical equilibria. Rapid developments in our understanding of NMR phenomena and their relation to properties of chemical interest continue today unabated, and dramatic instrumental developments have improved resolution and sensitivity by factors of ~50 from the first commercial instruments. Today more than 1500 NMR spectrometers are in use, and the scientific literature abounds in reference to NMR data.

In the course of teaching the background and applications of NMR both to graduate students and to established chemists who wanted to learn more of this technique, I have felt the need for a textbook at an "intermediate" level of complexity—one which would provide a systematic treatment of those portions of NMR theory most needed for the intelligent and efficient utilization of the technique in various branches of chemistry and yet one which would avoid the mathematical detail presented in the several excellent treatises on the subject.

In this book I have attempted to present an explanation of NMR theory and to provide sufficient practical examples of the use of NMR to permit the reader to develop a clear idea of the many uses—and the limitations—of this technique. Many practical points of experimental methods are discussed, and pitfalls pointed out. A large collection of problems and spectra of "unknown" compounds of graded difficulty permits the student to test his knowledge of NMR principles. Answers to selected problems are given. I have not attempted to include large compendia of data, but ample literature references and lists of data tabulations and reviews should permit the reader to locate the specialized data needed for specific applications. Many of the literature references are to recent reviews or to other books, rather than to original articles, since the references are intended to provide guides to further reading, not to give credit for original contributions. Under these circumstances an author index would be pointless and has not been included.

Acknowledgments

I wish to thank T. C. Farrar, who generously allowed the use of a number of excerpts from our joint book "Pulse and Fourier Transform NMR." I appreciate the help and advice of R. J. Highet in obtaining the ^{13}C spectra in Appendix C, of C. L. Fisk and H. Shindo for helping with several figures; and of R. Wasylishen for suggesting several problems.

I also wish to acknowledge permission from the following publishers to reproduce copyrighted material: American Chemical Society, Figs. 1.5, 1.6, 3.5, 8.4 and 11.4; American Institute of Physics, Figs. 4.9, 4.10, 4.13, 7.24, 8.2, 9.5, 9.8, 9.9, 10.10, and 11.3; the Faraday Society, Fig. 9.3; Pergamon Press, Fig. 13.1; Springer–Verlag, Fig. 7.3; and Varian Associates, Figs. 3.10 and 3.11.

Chapter 1

Introduction

1.1 Historical

Many atomic nuclei behave as though they are spinning, and as a result of this spin they possess angular momentum and magnetic moments. These two nuclear properties were first observed indirectly in the very small splittings of certain atomic spectral lines (hyperfine structure). In 1924 Pauli[1] suggested that this hyperfine structure resulted from the interaction of magnetic moments of nuclei with the already recognized magnetic moments of electrons in the atoms. Analysis of the hyperfine structure permitted the determination of the angular momentum and magnetic moments of many nuclei.

The concept of nuclear spin was strengthened by the discovery (through heat capacity measurements) of *ortho* and *para* hydrogen[2]— molecules that differ only in having the two constituent nuclei spinning in the same or opposite directions, respectively.

In the early 1920s Stern and Gerlach[3] had shown that a beam of atoms sent through an inhomogeneous magnetic field is deflected according to the orientation of the electron magnetic moments relative to the magnetic field. During the 1930s refinements of the Stern–Gerlach technique permitted the measurement of the much smaller values of *nuclear* magnetic moments.[4] A major improvement in this type of experiment was made by Rabi and his co-workers[5] in 1939. They sent a beam of hydrogen molecules through first an inhomogeneous magnetic field and then a homogeneous field, and they applied radio-frequency (rf) electromagnetic energy to the molecules in the homogeneous field. At a sharply defined frequency, energy was absorbed by the molecular beam and caused a small but measurable deflection of the beam. This actually was the first observation of nuclear magnetic resonance, but such studies were performed only in molecular beams under very high vacuum. It was not until 1946 that nuclear magnetic resonance was found in bulk materials (solids or liquids). In that year Purcell and his co-workers at Harvard reported nuclear resonance absorption in paraffin wax,[6] while Bloch and his colleagues at Stanford found nuclear resonance in liquid water.[7] (They received the 1952 Nobel Prize for their discovery.) When we speak of nuclear magnetic resonance, we are really thinking of the kind of NMR discovered by Bloch and Purcell; that is, nuclear magnetic resonance in bulk materials.

The early work in NMR was concentrated on the elucidation of the basic phenomena and on the accurate determination of nuclear magnetic moments. NMR attracted little attention from chemists until, in 1949 and 1950, it was discovered that the precise resonance frequency of a nucleus depends on the state of its chemical environment.[8] In 1951 separate resonance lines were found for chemically different protons in the same molecule.[9] The discovery of this so-called *chemical shift* set the stage for the use of NMR as a probe into the structure of molecules; this is the aspect of NMR that we shall explore in this book.

1.2 High Resolution NMR

It is found that chemical shifts are very small, and in order to observe such shifts one must study the material in the right state of aggregation. In solids, where intermolecular motion is highly restricted, internuclear interactions cause such a great broadening of resonance lines that chemical shift differences are masked. In solution, on the other hand, the rapid molecular tumbling causes these interactions to average to zero, and sharp lines are observed. Thus there is a distinction between *broad line NMR* and *high resolution,* or narrow line, NMR. We shall deal almost exclusively with the latter. (With sophisticated methods that we shall mention later, it is possible in some cases to mask the effects of internuclear interactions in solids and thus obtain relatively narrow lines.)

An NMR spectrum is obtained by placing a sample in a homogeneous magnetic field and applying electromagnetic energy at suitable frequencies. In Chapter 2 we shall examine in detail just how NMR spectra arise, and in Chapter 3 we shall delve into the procedures by which NMR is studied. Before we do so, however, it may be helpful to see by a few examples the type of information that can be obtained from an NMR spectrum.

Basically there are three quantities that can be measured in a high res-

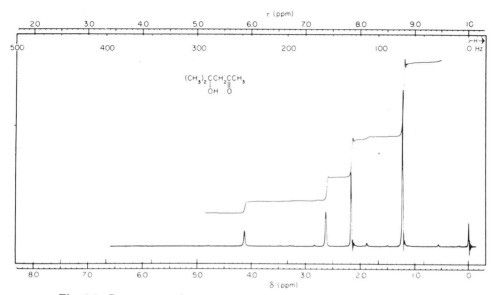

Fig. 1.1 Proton magnetic resonance spectrum of 4-hydroxy-4-methyl-2-pentanone (diacetone alcohol). Assignments of lines to functional groups as follows: $\delta = 1.23$, $(CH_3)_2$; 2.16, $CH_3C{=}O$; 2.62, CH_2; 4.12, OH. (For definition of δ scale, see Chapter 4.)

olution NMR spectrum: (1) frequencies, (2) areas, and (3) widths or shapes of the resonance lines. Figure 1.1 shows the spectrum of a simple compound, diacetone alcohol. This spectrum, as well as the others shown in this chapter, arises only from the resonance of the hydrogen nuclei in the molecule. (We shall see in Chapter 2 that we normally obtain a spectrum from only one kind of nucleus and discriminate against the others.) The line at zero on the scale below the spectrum is a reference line (see Chapters 3 and 4). Each of the other lines can be assigned to one of the functional groups in the sample, as indicated in the figure. The step function shown along with the spectrum is an integral, with the height of each step proportional to the area under the corresponding spectral line. There are several important features illustrated in this spectrum: First, the chemical shift is clearly demonstrated, for the resonance frequencies depend on the chemical environment, as we shall study in detail in Chapter 4. Second, the areas under the lines are different and, as we shall see when we examine the theory in Chapter 2, the area of each line is proportional to the number of nuclei contributing to it. Third, the widths of the lines are different; in particular, the line due to the OH is considerably broader than the others. We shall examine the reasons for different line widths in Chapters 2, 8, and 11.

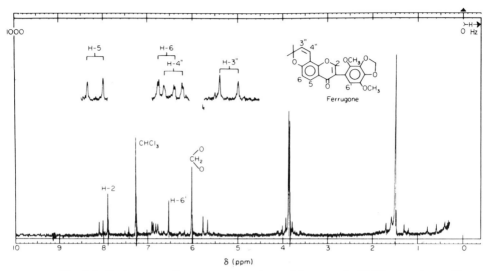

Fig. 1.2 Proton magnetic resonance spectrum of ferrugone in $CDCl_3$, showing multiplets due to spin–spin coupling between protons 5 and 6 and between protons 3″ and 4″. Assignments to functional groups: $\delta = 1.5$, CH_3; ~3.85, OCH_3; ~5.7, $H_{3''}$; 6.0, OCH_2O; 6.55, $H_{6'}$; 6.8, H_6; 6.9, $H_{4''}$; 7.27, $CHCl_3$; 7.9, H_2; ~8.1 H_6 (Highet[10]).

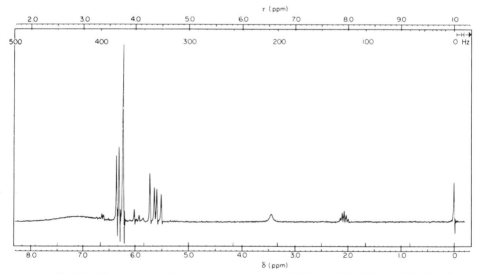

Fig. 1.3 Proton magnetic resonance spectrum of $CH_2{=}CH{-}C(O){-}NH_2$ (acrylamide) in acetone-d_6. Assignments: $\delta = 5.5$–6.7, $CH_2{=}CH$; 7.1 (very broad), NH_2; acetone-d_5, 2.07; impurities, 3.45, 6.0.

The spectrum in Fig. 1.1 is particularly simple. A more typical spectrum—that of a natural product, ferrugone—is given in Fig. 1.2. This spectrum consists of single lines well separated from each other, as were the lines in Fig. 1.1, and of simple multiplets. (The inset shows the multiplets on an expanded abscissa scale.) The splitting of single lines into multiplets arises from interactions between the nuclei called *spin–spin coupling*. This is an important type of information obtainable from an NMR spectrum. In Chapter 5 we shall inquire into the origin of spin coupling and what information of chemical value we can get from it.

Figure 1.3 shows the spectrum of a simple molecule, acrylamide. The three vinyl protons give rise to the 12-line spectrum at $\delta = 5.5–6.7$, which shows little regularity in spacing or intensity distribution. A spectrum of this sort must be analyzed by procedures that we shall discuss in detail in Chapter 7. However, the appearance of such complex spectra can be altered substantially by examining the sample at a different magnetic field

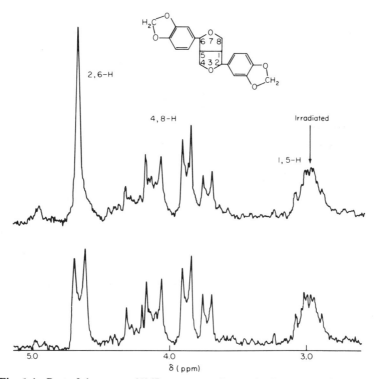

Fig. 1.4 Part of the proton NMR spectrum of sesamin. Bottom, ordinary spectrum; top, with additional radio-frequency irradiation in the vicinity of the complex multiplet at the right of the spectrum.

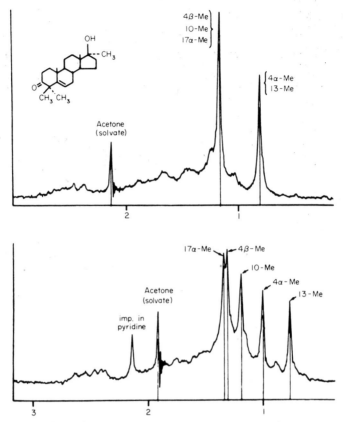

Fig. 1.5 Portions of the proton NMR spectra of 4,4,17α-trimethyl-17β-hydroxy-5-androsten-3-one in CDCl$_3$ (upper) and in pyridine (lower). Reprinted with permission from G. Slomp and F. MacKellar, *J. Am. Chem. Soc.* **82,** 999 (1960). Copyright by the American Chemical Society.

strength. Often by increasing the field (and the corresponding observation frequency) by a sufficient amount the apparently irregular spacings of lines give way to more readily discerned simple multiplets. In Chapters 4 and 5 we shall see why NMR spectra are dependent on magnetic field strength, while in Chapter 3 we shall look into many aspects of NMR instrumentation.

A powerful method for unraveling complex spectra is *double reso-nance,* in which two radio frequencies are applied to the sample simultaneously. Figure 1.4 shows the results of one type of double resonance experiment. Application of an intense rf field at the frequency of the complex multiplet near the right of the spectrum causes the doublet near the left end of the spectrum to collapse to a single line, while the remainder of

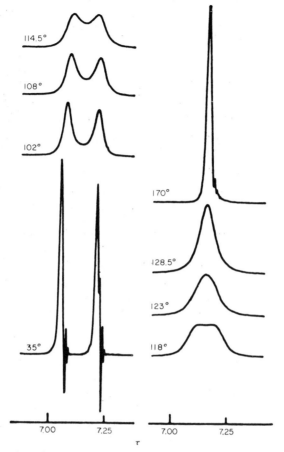

Fig. 1.6 Portion of the proton NMR spectrum (60 MHz) of N,N-dimethylformamide at various temperatures. Reprinted with permission from F. A. Bovey, *Chem. Eng. News* **43**, 98 (August 30, 1965). Copyright by the American Chemical Society.

the spectrum is unchanged. The theory and several different applications of double resonance will be covered in Chapter 9.

Often the appearance of an NMR spectrum is strongly dependent on intermolecular interactions (or "medium effects"), as well as on molecular structure. For example, the upper portion of Fig. 1.5 gives the spectrum of a steroid obtained in solution in deuterochloroform. There are five methyl groups in the molecule, but only two distinct methyl peaks are observed because of accidental coincidences in chemical shifts. On the other hand, the spectrum obtained in pyridine (lower portion of Fig. 1.5) shows all five expected lines. In Chapter 12 we shall explore the effects of solvent interactions and of hydrogen bonding.

The ways in which nuclei *relax,* or return to their equilibrium state after some perturbation, will be taken up briefly in Chapter 2 and discussed more fully in Chapter 8. A great deal of useful information on molecular structure and dynamics can be extracted from an analysis of nuclear magnetic relaxation processes.

Although most of our discussion of experimental techniques will be concentrated in Chapter 3, one aspect —*Fourier transform methods*—is so important in current NMR practice that we shall devote all of Chapter 10 to an exposition of the rationale of the technique and its advantages. With Fourier transform methods it is possible to obtain more than an order of magnitude increase in sensitivity over the conventional procedures.

NMR spectra are often influenced strongly by rate phenomena. Figure 1.6 shows a portion of the spectrum of N,N-dimethylformamide, in which there is hindered rotation of the $N(CH_3)_2$ group around the CN amide bond. At low temperature there are two distinct peaks for the two separate methyl groups, but with increasing temperature these peaks broaden and eventually coalesce, as the rate of internal rotation of the dimethylamino group increases. In Chapter 11 we shall inquire into the reasons for this type of behavior and the kind of kinetic information that can be obtained from the spectra.

NMR is increasingly being used as a method for quantitative analysis. Unlike many other types of spectroscopy, the area under an NMR absorption line (or multiplet) is directly proportional to the number of nuclei contributing to it, regardless of their chemical nature. So NMR is inherently a quantitative method, and recent instrumental improvements have made possible the realization of this potential. We shall comment briefly in Chapter 13 on some of the specific aspects of quantitative analysis by NMR.

Chapter 2

The Theory of NMR

2.1 Nuclear Spin and Magnetic Moment

Nuclei that act as though they are spinning possess angular momentum, which is quantized in units of $\hbar$, where $\hbar$ is Planck's constant divided by 2π. For the maximum observable component of angular momentum, we may write

$$p = I\hbar = \frac{Ih}{2\pi}. \qquad (2.1)$$

The constant of proportionality, which is either an integer or half-integer, is given the symbol I and is referred to as the nuclear spin quantum number or more commonly the *nuclear spin*. We can classify nuclei, then, according to their nuclear spins. There are a number of nuclei that have $I = 0$ and hence possess no angular momentum. This class of nuclei includes all those that have both an even atomic number and an even mass number; for example, ^{12}C, ^{16}O, and ^{32}S. These nuclei, as we shall see, cannot experience magnetic resonance under any circumstances. Appendix B lists the nuclear spins of all isotopes that have $I \neq 0$. A few of the more common nuclei are

$$I = \tfrac{1}{2}: \quad {}^1H, {}^3H, {}^{13}C, {}^{15}N, {}^{19}F, {}^{31}P;$$

$$I = 1: \quad {}^2H(D), {}^{14}N;$$

$$I > 1: \quad {}^{10}B, {}^{11}B, {}^{17}O, {}^{23}Na, {}^{27}Al, {}^{35}Cl, {}^{59}Co.$$

Those nuclei that have $I \geq 1$ have nonspherical nuclear charge distribution and hence an electric quadrupole moment Q. We shall consider the effect of the quadrupole moment later. Our present concern is with *all* nuclei that have $I \neq 0$, since each of these possesses a magnetic dipole moment, or a *magnetic moment,* μ. We can think of this moment qualita-

9

tively as arising from the motion (spinning) of a charged particle. This is an oversimplified picture, but it nevertheless gives qualitatively the correct results that (a) those nuclei that have a spin have a magnetic moment and (b) the magnetic moment is collinear with the angular momentum vector. We can express these facts by writing

$$\boldsymbol{\mu} = \gamma \mathbf{p}. \tag{2.2}$$

(We use the customary bold face type to denote vector quantities.) The constant of proportionality γ is called the *magnetogyric ratio* and is different for different nuclei, since it reflects nuclear properties not accounted for by the simple picture of a spinning charged particle. (Sometimes γ is termed the gyromagnetic ratio.) While p is a simple multiple of $\hbar$, μ and hence γ are not and must be determined experimentally for each nucleus (usually by an NMR method). Values of μ are given in Appendix B.

2.2 Classical Mechanical Description of NMR

In considering the interaction of a magnetic moment with an applied magnetic field, we can use either a classical mechanical or a quantum mechanical treatment. Each has some advantages: particularly for the understanding of transient effects and exchange processes, it is very convenient to use a classical approach; whereas in discussing chemical shifts and spin couplings, it is necessary to use energy levels resulting from a quantum treatment.

Consider the interaction of a magnetic moment $\boldsymbol{\mu}$ with a magnetic field $\mathbf{H_0}$ (typically 10 to 80 T, or 10,000 to 80,000 G).* As shown in Fig. 2.1, the moment lies at some angle θ with respect to the field. The magnetic interaction between $\mathbf{H_0}$ and $\boldsymbol{\mu}$ generates a torque *tending* to tip the moment toward $\mathbf{H_0}$. Because the nucleus is spinning, the resultant motion does not change θ but rather causes the magnetic moment to *precess* around the magnetic field, as indicated by the dashed path traced out by the end of $\boldsymbol{\mu}$. (The situation is entirely analogous to the precession of a spinning top in the earth's gravitational field.)

Mathematically the torque $\mathbf{L}$ is given by classical magnetic theory as

$$\mathbf{L} = \boldsymbol{\mu} \times \mathbf{H_0}. \tag{2.3}$$

---

* The proper symbol for magnetic induction in SI units is $\mathbf{B}$, but virtually all NMR literature, as well as commercially available chart paper, continues to use $\mathbf{H}$. Strictly speaking, $\mathbf{H}$ refers to the applied magnetic field, while $\mathbf{B}$ is the actual induced magnetization.

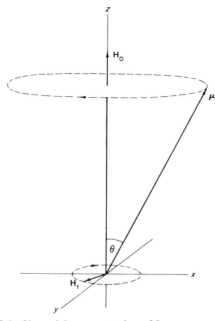

Fig. 2.1 Vectorial representation of Larmor precession.

From classical mechanics[13]

$$\frac{d\mathbf{p}}{dt} = \mathbf{L},\tag{2.4}$$

so

$$\frac{d\mathbf{p}}{dt} = \boldsymbol{\mu} \times \mathbf{H}_0.\tag{2.5}$$

Substituting Eq. (2.2), we have

$$\frac{d\boldsymbol{\mu}}{dt} = \gamma\frac{d\mathbf{p}}{dt} = \gamma\boldsymbol{\mu} \times \mathbf{H}_0.\tag{2.6}$$

Since the length of $\boldsymbol{\mu}$ is constant, Eq. (2.6) expresses the fact that the motion of $\boldsymbol{\mu}$ occurs with only a single degree of freedom, namely, a precession about $\mathbf{H}_0$.[14] Such a precession of $\boldsymbol{\mu}$ with angular velocity and direction given by $\boldsymbol{\omega}_0$ is expressed by

$$\frac{d\boldsymbol{\mu}}{dt} = \boldsymbol{\omega}_0 \times \boldsymbol{\mu}.\tag{2.7}$$

Comparing Eqs. (2.6) and (2.7), we see that

$$\boldsymbol{\omega}_0 = -\gamma \mathbf{H}_0. \tag{2.8}$$

Thus the nuclear moment precesses about $\mathbf{H}_0$ with a frequency

$$\nu_0 = \frac{|\boldsymbol{\omega}_0|}{2\pi} = \frac{\gamma}{2\pi} H_0. \tag{2.9}$$

Equation (2.8) (as well as (2.9)), often called the *Larmor equation,* is the basic mathematical expression for NMR. The precession (or Larmor) frequency is directly proportional to the applied magnetic field and is also proportional to γ (or μ), which varies from one nucleus to another (see Appendix B).*

One significant feature of the Larmor equation is that the angle θ does not appear. Hence the nucleus precesses at a frequency governed by its own characteristic properties and that of the magnetic field. On the other hand, the *energy* of this spin system does depend on θ, since

$$E = -\boldsymbol{\mu} \cdot \mathbf{H}_0 = -\mu H_0 \cos \theta. \tag{2.10}$$

If a small magnetic field $\mathbf{H}_1$ is placed at right angles to $\mathbf{H}_0$, and is made to rotate about $\mathbf{H}_0$ at a frequency ν_0 (see Fig. 2.1), then $\boldsymbol{\mu}$ experiences the resultant of $\mathbf{H}_0$ and $\mathbf{H}_1$ and θ changes by $d\theta$. Energy is thus absorbed from the field $\mathbf{H}_1$ into the nuclear spin system. If $\mathbf{H}_1$ rotates at any frequency $\nu \neq \nu_0$, then it is alternately in and out of phase with $\boldsymbol{\mu}$, and no net energy absorption occurs. Hence the absorption of energy is a resonance phenomenon, sharply tuned to the natural nuclear precession frequency.

In practice, the rotating field $\mathbf{H}_1$ is obtained from a linearly polarized electromagnetic field that results from the passage of electric current at frequency ν through a coil. If this field is polarized along the x axis, it may be thought of as resulting from two equal fields counterrotating in the xy plane, as indicated in Fig. 2.2. Mathematically

$$\begin{aligned} H_r &= (H_1)_x \cos 2\pi\nu t + (H_1)_y \sin 2\pi\nu t, \\ H_l &= (H_1)_x \cos 2\pi\nu t - (H_1)_y \sin 2\pi\nu t. \end{aligned} \tag{2.11}$$

Obviously the sum of H_r and H_l has only an x component. With respect to the precessing nuclei, the counterrotating field is at a frequency $2\nu_0$ away and may be ignored.

---

* By analogy to the electron magnetic moment, which is expressed in Bohr magnetons, a *nuclear Bohr magneton* μ_0 is defined as

$$\mu_0 \equiv e\hbar/2Mc = 5.0500 \times 10^{-24} \quad \text{erg/gauss},$$

where e and M are the charge and mass of the proton. In Appendix B values of μ are given in units of μ_0.

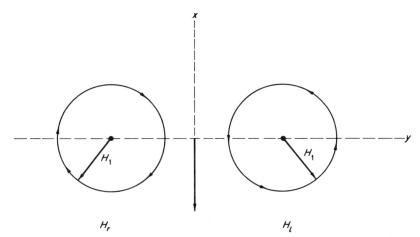

Fig. 2.2   Resolution of a plane (linearly) polarized wave into two counterrotating components.

We might say a few words about the orders of magnitude involved in NMR. Applied fields H_0 of 10–80 kilogauss (kG) are commonly employed; Larmor frequencies with such fields are in the radio-frequency (rf) range of several megahertz (MHz).* The values of ν_0 for a field H_0 of 14,000 G are given for a few important nuclei in Fig. 2.3, and other values of ν_0 (for a 10,000-G field) are listed in Appendix B.

In conventional spectroscopy the spectrum is scanned by varying the frequency of incident radiation. In NMR either the radio frequency or the magnetic field may be varied, since by the Larmor relation (Eq. (2.9)) ν_0 is proportional to H_0. In either case the range needed to encompass all the resonances from a given type of nucleus is small—usually less than a few kilohertz or a few gauss. Therefore, *in a given experiment we study the resonance of only one type of nucleus.*

The magnitude selected for H_1 depends on factors that we shall take up later in this chapter. For many NMR experiments H_1 is less than a milligauss; for others it may range as high as a few gauss. In any event, H_1 is much smaller than H_0.

2.3 Quantum Mechanical Description of NMR

As with other branches of spectroscopy, an explanation of many aspects of NMR requires the use of quantum mechanics. Fortunately, the particular equations needed are simple and can be solved exactly.

* The hertz (Hz) is now by international agreement preferred for the unit of frequency rather than the older cycle per second (cps); 1 Hz ≡ 1 cps.

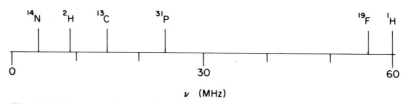

Fig. 2.3 Resonance frequencies for several nuclei in a magnetic field of approximately 14,100 G.

In quantum theory, the energy of interaction between a magnetic moment and an applied field, given by Eq. (2.10), appears in the Hamiltonian operator $\mathcal{H}$,

$$\mathcal{H} = -\boldsymbol{\mu} \cdot \mathbf{H}_0. \tag{2.12}$$

By substituting from Eqs. (2.1) and (2.2), we have

$$\mathcal{H} = -\gamma\hbar\mathbf{H}_0 \cdot \mathbf{I}. \tag{2.13}$$

Here $\mathbf{I}$ is interpreted as an operator. From the general properties of spin angular momentum in quantum mechanics,[15] it has been shown that the solution of this Hamiltonian gives energy levels in which

$$E_m = -\gamma\hbar m H_0. \tag{2.14}$$

The quantum number m may assume the values

$$-I, -I + 1, \ldots, I - 1, I.$$

There are thus $2I + 1$ energy levels, each of which may be thought of as arising from an orientation of $\boldsymbol{\mu}$ with respect to $\mathbf{H}_0$ such that its projection on $\mathbf{H}_0$ is quantized (see Fig. 2.4). For the particularly important case of $I = \frac{1}{2}$ there are just two energy levels; in this case we often speak of a nuclear spin as having "flipped" from an orientation with the field to one opposed to the field. The energy separation between the states is linearly dependent on the magnetic field.

In these terms NMR arises from transitions between energy levels, just as in other branches of spectroscopy. Transitions are induced by the absorption of energy from an applied electromagnetic field. The presence of this field is treated by adding to the Hamiltonian a term expressing interaction between the spin system and the applied radio-frequency field H_1:

$$\mathcal{H}' = 2\mu_x H_1 \cos 2\pi\nu t. \tag{2.15}$$

From the well-established results of time-dependent perturbation theory[16]

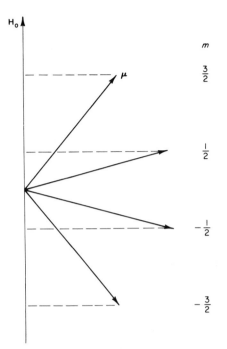

Fig. 2.4 The $2I + 1$ orientations of $\boldsymbol{\mu}$ with respect to $\mathbf{H_0}$ and the quantization of the projection of $\boldsymbol{\mu}$ on $\mathbf{H_0}$. The case illustrated is $I = \frac{3}{2}$.

we find that the probability of transition per unit time between levels m and m' is

$$P_{mm'} = \gamma^2 H_1^2 |(m|I_x|m')|^2 \delta(\nu_{mm'} - \nu). \qquad (2.16)$$

$2H_1$ is the magnitude of the radio-frequency field applied in the x direction, with H_0 in the z direction; $(m|I_x|m')$ is the quantum mechanical matrix element of the x component of the nuclear spin operator and is zero unless $m = m' \pm 1$; and $\delta(\nu_{mm'} - \nu)$ is the Dirac delta function, which is zero unless $\nu_{mm'} = \nu$. The frequency corresponding to the energy difference between states m and m' $\nu_{mm'}$, as given by the Bohr relation,

$$\nu_{mm'} = \frac{\Delta E_{mm'}}{h} = \frac{\gamma H_0 |m' - m|}{2\pi}. \qquad (2.17)$$

Several important points are contained in Eq. (2.16). First, the transition probability increases with γ and with the applied field H_1.* (This

---

* While the power absorbed varies as H_1^2 according to Eq. (2.16), the observed NMR signal, which is proportional to an induced *voltage* in a coil (see Chapter 3), varies linearly with H_1.

latter relation will be modified in Section 2.5.) Second, the matrix element furnishes the selection rule $\Delta m = \pm 1$, so that transitions are permitted only between adjacent energy levels and thus give only a single line at a frequency

$$\nu = \frac{\gamma}{2\pi} H_0. \tag{2.18}$$

Third, the resonance condition is expressed in the delta function. Actually the delta function would predict an infinitely sharp line, which is unrealistic; therefore, it is replaced by a line shape function $g(\nu)$, which has the property that

$$\int_0^\infty g(\nu) \, d\nu = 1. \tag{2.19}$$

(In practice $g(\nu)$ often turns out to be Lorentzian or approximately Lorentzian in shape.) Equation (2.16) becomes then

$$P_{mm'} = \gamma^2 \, H_1^2 |(m|I_x|m')|^2 g(\nu). \tag{2.20}$$

For nuclei with $I = \frac{1}{2}$ there is only one transition, so Eq. (2.20) becomes[16]

$$P = \frac{1}{4}\gamma^2 \, H_1^2 g(\nu). \tag{2.21}$$

2.4 Effect of the Boltzmann Distribution

The tendency of nuclei to align with the magnetic field and thus to drop into the lowest energy level is opposed by thermal motions, which tend to equalize the populations in the $2I + 1$ levels. The resultant equilibrium distribution is the usual compromise predicted by the Boltzmann equation. For simplicity we shall consider only nuclei with $I = \frac{1}{2}$, so that we need include only two energy levels, the lower corresponding to $m = -\frac{1}{2}$ and the upper to $m = +\frac{1}{2}$.* We shall designate the levels by the subscripts $-$ and $+$, respectively. For the $I = \frac{1}{2}$ system the Boltzmann equation is

$$\frac{n_+}{n_-} = \exp\left(-\frac{\Delta E}{kT}\right). \tag{2.22}$$

By substitution of the values of E from Eq. (2.14), and by introducing

---

* The level with $m = -\frac{1}{2}$ is of lower energy if the field H_0 is taken along the negative z axis, and m refers to the projection of I on the z axis. This convention is consistent with that employed in Chapter 7.

Eqs. (2.1) and (2.2), we find that this becomes

$$\frac{n_+}{n_-} = \exp\left(-\frac{\gamma\hbar H_0}{kT}\right) = \exp\left(-\frac{2\mu H_0}{kT}\right). \tag{2.23}$$

For small values of the argument in the exponential the approximation $e^{-x} = 1 - x$ may be employed to show that the fractional excess population in the lower level is

$$\frac{n_- - n_+}{n_-} = \frac{2\mu H_0}{kT}. \tag{2.24}$$

For ^{1}H, which has a large magnetic moment, in a field of 14,000 G this fractional excess is only about 1×10^{-5} at room temperature.*

One consequence of this slight excess population in the lower level is the appearance of a very small macroscopic magnetic moment directed along $\mathbf{H}_0$. The mean value $\bar{\mu}$ is given by the weighted average of the oppositely directed moments from the two states:

$$\bar{\mu} = \frac{n_-}{n_+ + n_-} \mu + \frac{n_+}{n_+ + n_-}(-\mu). \tag{2.25}$$

From the Boltzmann distribution, Eq. (2.23), together with the fact that $2\mu H_0/kT \ll 1$, we see that

$$\frac{n_-}{n_+ + n_-} \approx \frac{1}{2}\left(1 + \frac{\mu H_0}{kT}\right), \qquad \frac{n_+}{n_+ + n_-} \approx \frac{1}{2}\left(1 - \frac{\mu H_0}{kT}\right). \tag{2.26}$$

Then

$$\bar{\mu} = \frac{1}{2}\left(1 + \frac{\mu H_0}{kT}\right)\mu - \frac{1}{2}\left(1 - \frac{\mu H_0}{kT}\right)\mu = \frac{\mu^2 H_0}{kT}. \tag{2.27}$$

For N nuclei per unit volume, the total magnetization is N times as large, and the *volume magnetic susceptibility* is

$$\kappa = \frac{N\bar{\mu}}{H_0} = \frac{N\mu^2}{kT}. \tag{2.28a}$$

For N nuclei of spin I this result can be generalized[17] to

$$\kappa = \frac{N(I + 1)\mu^2}{3IkT}. \tag{2.28b}$$

For protons in water at room temperature this is about 3×10^{-10}. This nuclear paramagnetic susceptibility is ordinarily completely masked by the diamagnetic susceptibility due to the electrons, which is about 10^{-6}, but has been measured at very low temperature.

---

* This very small difference in population occurs because the energy levels are only slightly separated from each other. In this case ΔE is only $\sim$6 *millicalories*.

The near equality of population in the two levels is an important factor in determining the intensity of the NMR signal. According to the Einstein formulation, the radiative transition probability between two levels is given by[3]

$$P_+ \propto B_+\rho(\nu)n_-,$$
$$P_- \propto B_-\rho(\nu)n_+ + A_-n_+. \tag{2.29}$$

P_+ and P_- are the probabilities for absorption and emission, respectively; B_+ and B_- are the coefficients of absorption and of *induced* emission, respectively; A_- is the coefficient of *spontaneous* emission; and $\rho(\nu)$ is the density of radiation at the frequency that induces the transition. Einstein showed that $B_+ = B_-$, while $A_- \propto \nu^3 B_-$. As a result of this strong frequency dependence, spontaneous emission (fluorescence), which usually dominates in the visible region of the spectrum, is an extremely improbable process in the rf region and may be disregarded. Thus the *net* probability of absorption of rf energy, which is proportional to the strength of the NMR signal, is

$$P \propto B\rho(\nu)(n_- - n_+). \tag{2.30}$$

The small value of $(n_- - n_+)$ accounts in large part for the insensitivity of NMR relative to other spectroscopic methods (see Chapter 3).

2.5 Spin–Lattice Relaxation

It is important now to consider the manner in which the Boltzmann distribution is established. Again we shall for simplicity treat only the case $I = \frac{1}{2}$. If initially the sample containing the nuclear spin system is outside a polarizing magnetic field, the difference in energy between the two levels is zero, and the populations n_+ and n_- must be equal. When the sample is placed in the field, the Boltzmann distribution is not established instantaneously. Since spontaneous emission is negligible, the redistribution of population must come about from an interaction of the nuclei with their surroundings (the "lattice"). As we shall see, this is a nonradiative first-order rate process characterized by a "lifetime," T_1, called the *spin–lattice relaxation time*. The origin of this process may be seen in the following: Let $n = (n_- - n_+)$ be the difference in population; let $n_0 = (n_- + n_+)$; let $W_+ =$ the probability for a nucleus to undergo a transition from the lower to the upper level as a result of an interaction with the environment, and let W_- be the analogous probability for the downward transition. Unlike the radiative transition probabilities, W_+ and W_- are not equal; in fact, at equilibrium, where the number of upward and downward transitions are equal,

$$W_+ n_- = W_- n_+.$$ (2.31)

From Eqs. (2.23) and (2.31),

$$\frac{W_+}{W_-} = \left(\frac{n_+}{n_-}\right)_{eq} = \exp\left(-\frac{2\mu H_0}{kT}\right).$$ (2.32)

Using the same approximations as in Section 2.4 for $2\mu H_0/kT \ll 1$, and defining W as the mean of W_+ and W_-, we may write

$$\frac{W_+}{W} = \frac{(n_+)_{eq}}{n_0/2} = 1 - \frac{\mu H_0}{kT},$$
$$\frac{W_-}{W} = \frac{(n_-)_{eq}}{n_0/2} = 1 + \frac{\mu H_0}{kT}.$$ (2.33)

The total rate of change of n is

$$\frac{dn}{dt} = \frac{dn_-}{dt} - \frac{dn_+}{dt} = 2\frac{dn_-}{dt}.$$ (2.34)

But by definition of W_+ and W_-,

$$\frac{dn_-}{dt} = n_+ W_- - n_- W_+.$$ (2.35)

So, from Eqs. (2.33) to (2.35),

$$\frac{dn}{dt} = -2W\left(n - n_0\frac{\mu H_0}{kT}\right).$$ (2.36)

By introducing Eq. (2.26), we obtain

$$\frac{dn}{dt} = -2W(n - n_{eq}).$$ (2.37)

This rate equation describes a first-order decay process, characterized by a rate constant $2W$. If we define a time T_1,

$$T_1 = \frac{1}{2W},$$ (2.38)

and integrate, we obtain

$$n - n_{eq} = (n - n_{eq})_{t=0} \exp\left(-\frac{t}{T_1}\right).$$ (2.39)

Thus T_1 serves as a measure of the rate with which the spin system comes

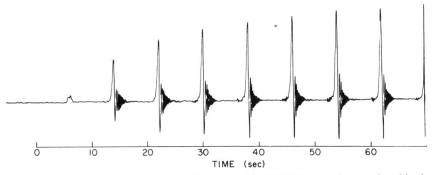

Fig. 2.5 Spin–lattice relaxation in benzene (60 MHz). The sample was placed in the magnetic field H_0 at $t = 0$ and the spectrum scanned repetitively using a small value of H_1.

into equilibrium with its environment and hence is called the spin–lattice relaxation time.

The magnitude of T_1 is highly dependent on the type of nucleus and on factors such as the physical state of the sample and the temperature. For liquids T_1 is usually between 10^{-2} and 100 sec, but in some cases may be in the microsecond range. In solids T_1 may be much longer—sometimes days. The mechanisms of spin–lattice relaxation and some chemical applications will be taken up in Chapter 8.

If T_1 is long enough (> 10 sec), the effect of spin–lattice relaxation may be seen readily in a series of scans made immediately after the sample is placed in the magnetic field. An example is given in Fig. 2.5. Shorter T_1's must be measured by more elaborate techniques, to be discussed in Chapter 8.

2.6 Line Widths

We pointed out in Section 2.3 that an NMR line is not infinitely sharp, and we assumed some function $g(\nu)$ as the line shape. The existence of spin–lattice relaxation implies that the line must have a width at least as great as can be estimated from the uncertainty principle:

$$\Delta E \cdot \Delta t \sim h. \tag{2.40}$$

Since the average lifetime of the upper state cannot exceed T_1, this energy level must be broadened to the extent of h/T_1, and thus the half-width of the NMR line resulting from this transition must be *at least* of the order of $1/T_1$.

There are, however, other processes that can increase line widths substantially over the value expected from spin–lattice relaxation. The most important of these involves the interaction between nearby nuclear magnetic moments. The field experienced by one nuclear moment and caused by another of magnitude μ at a distance r is proportional to μ/r^3. A given nucleus experiences a field of this magnitude from each of its neighbors, and each such contribution may either augment the field applied to the sample or detract from it, depending on the orientation of each magnetic moment with respect to the applied field. It is apparent that not all nuclei in a macroscopic sample will experience the same internuclear field; thus the resonance frequencies of the nuclei will differ, and the observed line will be broadened. The theory will be taken up in more detail in Chapter 8, where it will be shown that this interaction averages almost completely to zero when the molecules in the sample are in rapid random motion with respect to the applied field. Thus this interaction, which is the major source of line width in solids, is negligible in most liquids or solutions containing small molecules. For polymers the situation is intermediate. Typical line widths are: liquids, <0.1–10 Hz; polymers, 10–10^4 Hz; solids, 10^4–10^5 Hz.

When the line width is greater than that predicted from T_1, it is convenient to define another time T_2 (shorter than T_1), which is called *the spin–spin relaxation time*. T_2 is defined so as to be consistent with the uncertainty principle order of magnitude relation

$$\nu_{1/2} \, T_2 \sim 1, \tag{2.41}$$

where $\nu_{1/2}$ is the width of the line at half maximum intensity. For consistency with T_2 introduced in an alternative manner (see Section 2.9), we can formulate the precise definition of T_2 in terms of the line shape function $g(\nu)$ as follows:

$$T_2 \equiv \tfrac{1}{2}[g(\nu)]_{\text{max}}. \tag{2.42}$$

We can easily show that this apparently arbitrary definition of T_2 gives the desired relationship expressed in Eq. (2.41). If the line has a Lorentzian shape, as is predicted theoretically and usually verified experimentally, then $g(\nu)$ obeys the equation

$$g(\nu) = \frac{a}{b^2 + (\nu - \nu_0)^2}, \tag{2.43}$$

where ν_0 is the frequency of the peak, and a and b are constants. By introducing Eqs. (2.42) and (2.43) into Eq. (2.19), we find that

$$T_2 = \frac{1}{\pi \nu_{1/2}}. \tag{2.44}$$

For a line shape other than Lorentzian the definition of T_2 given in Eq. (2.42) leads to a relation between T_2 and $\nu_{1/2}$ differing in the constant factor but otherwise consistent with the uncertainty principle relation (cf. Problem 7).

The "natural" widths of very sharp NMR lines cannot be measured directly because even the best existing apparatus has a magnetic field inhomogeneity equivalent to $\sim 0.05-0.1$ Hz.

2.7 Saturation

In the presence of an rf field the fundamental rate equation for spin–lattice relaxation (2.37) must be modified by including a term like that in Eq. (2.30), which expresses the fact that the rf field causes net upward transitions proportional to the difference in population n. The resultant differential equation is

$$\frac{dn}{dt} = -\frac{n - n_{eq}}{T_1} - 2nP. \tag{2.45}$$

Substituting for P from Eq. (2.21) and solving the differential equation, we find that the steady-state value n_{ss} is

$$n_{ss} = \frac{n_{eq}}{1 + \frac{1}{2}\gamma^2 H_1^2 T_1 g(\nu)}. \tag{2.46}$$

The decrease in n below the Boltzmann equilibrium value naturally decreases the magnitude of the NMR signal below what would be expected. This process is called *saturation* and in extreme cases can lead to a virtual disappearance of an NMR signal. The denominator of Eq. (2.46) is largest when $g(\nu)$ is at its maximum value. Substituting for $g(\nu)_{max}$ from Eq. (2.42), we obtain

$$n_{ss} = \frac{n_{eq}}{1 + \gamma^2 H_1^2 T_1 T_2} = n_{eq}Z. \tag{2.47}$$

The quantity Z, which is defined by Eq. (2.47), is called the *saturation factor*. We shall refer to this factor in Section 3.7, when we consider the practical effects of saturation on signal strength.

2.8 Macroscopic Magnetization

We have now explained many fundamental aspects in terms of the behavior of nuclear magnetic moments subject to quantum restrictions whereby nuclear spin orientations seem to be quantized along $\mathbf{H_0}$. On the other hand, in Section 2.2 we developed a simple classical picture of the precessing nucleus free to assume *any* orientation relative to $\mathbf{H_0}$. How do we rationalize these two approaches? As Slichter[18] points out in his lucid description of NMR, quantum theory does not require that a magnetic moment actually be oriented along a discrete direction, as indicated in Fig. 2.4; rather its orientation is described by a linear combination of the "allowed" orientations. In quantum mechanical terms the expectation value of any observable property, such as the orientation of $\boldsymbol{\mu}$, is described in terms of a combination of the basic allowed states. Thus, for a single nucleus with $I = \frac{1}{2}$, it is not strictly correct to say that it necessarily flips from one spin orientation to the other.

In practice, however, we never deal with a single nucleus, but with an ensemble of identical nuclei. Even in the smallest sample we can imagine using, there will be a significant fraction of Avogadro's number of nuclei. It is helpful, then, to define a *macroscopic* magnetization $\mathbf{M}$ as the vector sum of the individual magnetic moments. As shown in Fig. 2.6a, an ensemble of identical nuclei precessing about H_0 (taken, as usual, along the z axis) have random phase in the x and y directions when they are at equilibrium. The resultant macroscopic magnetization is then oriented along the z direction and has a value M_0, which is responsible for the small nuclear magnetic susceptibility calculated in Eq. (2.28).

An imposed rf field $\mathbf{H_1}$ at the Larmor frequency can be treated as acting on $\mathbf{M}$, rather than on each individual magnetic moment. The result, as discussed in more detail in the following section, is that $\mathbf{M}$ is tipped

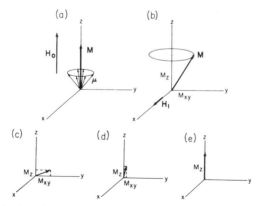

Fig. 2.6 Behavior of the macroscopic magnetization $\mathbf{M}$. (a) Formation of $\mathbf{M}$ from the individual magnetic moments of the nuclei. (b) Tipping of $\mathbf{M}$ by action of the rf field $\mathbf{H_1}$ causes reduction in M_z and creation of M_{xy}. (c)–(e) M_{xy} returns to zero, and M_z increases to its equilibrium value of M_0 by relaxation processes.

away from the z axis, M_z is thus reduced, and a component M_{xy} is generated in the xy plane (see Fig. 2.6b). The existence of this component implies some phase coherence in the xy plane, which was lacking at equilibrium. As the nuclei exchange energy with each other (spin–spin relaxation), they gradually lose phase coherence, and M_{xy} decays back to its equilibrium value of zero (Fig. 2.6c,d). At the same time the nuclei lose energy to their surroundings (spin–lattice relaxation), and M_z relaxes back to its original value of M_0 (Fig. 2.6e).

2.9 The Bloch Equations: Nuclear Induction

In Bloch's original treatment of NMR[7] he postulated a set of phenomenological equations that accounted successfully for the behavior of the macroscopic magnetization **M** in the presence of an rf field. We shall merely outline the approach used. In Eq. (2.6) we expressed the effect of an applied magnetic field on a nuclear magnetic moment. By summing over all moments we can write a similar equation for **M**:

$$dM/dt = \gamma M \times H. \tag{2.48}$$

Here **H** is any magnetic field—fixed (H_0) or oscillating (H_1). By expanding the vector cross product we can write a separate equation for the time derivative of each component of **M**:

$$
\begin{aligned}
dM_x/dt &= \gamma(M_y H_z - M_z H_y), \\
dM_y/dt &= \gamma(-M_x H_z + M_z H_x), \\
dM_z/dt &= \gamma(M_x H_y - M_y H_x).
\end{aligned}
\tag{2.49}
$$

By the usual convention, the component $H_z = H_0$, a fixed field, while H_x and H_y represent the rotating rf field, as expressed in Eq. (2.11). To account for relaxation Bloch assumed that M_z would decay to its equilibrium value of M_0 by a first-order process characterized by a time T_1 (the *longitudinal* relaxation time), while M_x and M_y would decay to zero with a first-order time constant T_2 (the *transverse* relaxation time). Overall, then the Bloch equations become

$$
\begin{aligned}
dM_x/dt &= \gamma(M_y H_0 + M_z H_1 \sin \omega t) - M_x/T_2, \\
dM_y/dt &= \gamma(M_z H_1 \cos \omega t - M_x H_0) - M_y/T_2, \\
dM_z/dt &= -\gamma(M_x H_1 \sin \omega t + M_y H_1 \cos \omega t) - (M_z - M_0)/T_1.
\end{aligned}
\tag{2.50}
$$

The Bloch equations can be solved analytically only under certain limiting conditions. The one in which we shall be most interested at

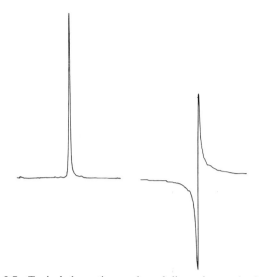

Fig. 2.7 Typical absorption mode and dispersion mode signals.

present is the "slow passage" case, in which H_1 is small, and the rate of sweeping through the spectrum is slow enough so that a steady state is established in which the time derivatives of Eq. (2.50) are zero. Under these conditions it can be shown[16] that the rf field H_1 produces a magnetization in the *xy* plane, which leads to two measurable signals: (1) the *absorption* signal, a component 90° out of phase with H_1, which has a Lorentzian line shape; and (2) the *dispersion* signal, which is in phase with H_1. The shapes of these signals are shown in Fig. 2.7. By appropriate electronic means (see Section 3.3) we can select one of these two signals for study. Usually the absorption mode is used, but the dispersion mode has some advantages (Section 3.4).

Bloch showed that the rotating magnetization M_{xy} can induce a small rf signal in a coil placed along the *y* axis, while the field H_1 is supplied by a coil on the *x* axis. The induced signal can be amplified and detected to give the NMR spectrum. This procedure, *nuclear induction,* is used in many NMR spectrometers (see Section 3.1).

2.10 The Rotating Frame of Reference

Both the mathematical treatments and pictorial representations of many physical phenomena can often be simplified by referring to a coordinate system that moves in some way, rather than one fixed in the labora-

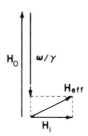

Fig. 2.8 Formation of H_{eff} in a frame rotating at ω (radians/sec) from the static field H_0 and the rf field H_1.

tory. We shall find it convenient to define a coordinate system, or frame of reference, that rotates about H_0 at a rate ν_{rf} revolutions/sec, or ω_{rf} radians/sec, where the frequencies are those of the rf field H_1. In this rotating frame of reference H_1 is then fixed, usually along the x' axis (the x axis of the rotating frame). An ensemble of nuclear magnetic moments precesses at the angular frequency $\omega = \gamma H_0$, from the Larmor equation. If $\omega = \omega_{rf}$, then as viewed in the rotating frame the nuclear moments are not moving about H_0. Thus in this particular rotating frame there *appears* to be no torque on M; that is, the analog of Eq. (2.48) for this rotating frame becomes

$$\left(\frac{dM}{dt}\right)_{rot} = \gamma M \times H_{rot} = 0. \tag{2.51}$$

Since the value of M is unchanged, it appears to an observer in the rotating frame that the magnetic field H_{rot} acting on the magnetization is zero. We can now generalize to nuclear moments precessing at $\omega \neq \omega_{rf}$. In the rotating frame they precess at $\omega_0 - \omega_{rf}$ radians/sec, and they thus conform to the relation

$$\left(\frac{dM}{dt}\right)_{rot} = \gamma M \times H_{rot} = \omega_0 - \omega_{rf}. \tag{2.52}$$

In this general case it appears that

$$H_{rot} = H_0 - \omega/\gamma, \tag{2.53}$$

with H_{rot} along the z' axis of the rotating frame.

We have not yet taken into account the field H_1, which is fixed along the x' axis. Actually M experiences the vector sum of H_{rot} and H_1. The *effective field* H_{eff}, acting on M in the rotating frame is then

$$H_{eff} = (H_0 - \omega/\gamma) + H_1. \tag{2.54}$$

The vector addition is depicted in Fig. 2.8. We can now state the general rule: **M** *always responds to* $\mathbf{H}_{\text{eff}}$. Far from resonance $|\mathbf{H} - \omega/\gamma| \gg |\mathbf{H}_1|$, so $\mathbf{H}_{\text{eff}} \approx \mathbf{H}_0$; at resonance $\mathbf{H}_{\text{eff}} = \mathbf{H}_1$.

2.11 Adiabatic Passage; Ringing

To understand many NMR phenomena we must recognize that the rate at which resonance is approached can be quite important. A very slow change of magnetic field or frequency is called *adiabatic,* and the adiabatic theorem tells us that if the rate of change is slow enough that

$$\frac{dH_0}{dt} \ll \gamma H_1^2, \tag{2.55}$$

M remains aligned with $\mathbf{H}_{\text{eff}}$.[19] Far from resonance, where $\mathbf{H}_{\text{eff}} \approx \mathbf{H}_0$ we know that **M** is, at equilibrium, aligned with $\mathbf{H}_0$; under conditions of adiabatic passage, then, **M** slowly tips with $\mathbf{H}_{\text{eff}}$ until at resonance it is aligned along $\mathbf{H}_1$.*

High resolution spectra are normally obtained with scan rates that ideally conform to slow passage conditions and hence to adiabatic conditions. In practice, however, scan rates are such that the magnetization cannot quite follow $\mathbf{H}_{\text{eff}}$, and after resonance some magnetization is left in the xy plane. The result is that the resonance line, instead of having a symmetric Lorentzian shape, shows *ringing* or "wiggles" after the line, as indicated in Fig. 2.9. This effect is easily understood from Fig. 2.6. Suppose that the rf frequency is held constant at ν_0, and the spectrum is scanned by increasing the magnetic field so that the Larmor frequency of the nuclei increases through the value ν_0. Immediately after the resonance condition is passed, the magnetization is depicted by Fig. 2.6b, with a component rotating at the Larmor frequency in the xy plane. H_1 is still rotating in the xy plane at ν_0, and as the Larmor frequency increases due to the steadily increasing field H_0, there is interference between H_1 and M_{xy}, leading to the beat pattern typical of two close-lying frequencies. As M_{xy} decays with a time constant T_2, the envelope of the ringing pattern should in principle furnish a measure of T_2. In practice, the decay is often due principally to inhomogeneities in the magnetic field. Because of field inhomogeneity, nuclei in different portions of the sample experience slightly different values of H_0, hence have different Larmor frequencies. Thus

---

* When the scan rate is slow enough to be adiabatic, but fast enough so that no significant relaxation occurs during passage through the line, the condition of *adiabatic rapid passage* is fulfilled. This is an important NMR excitation method for broad line studies but is rarely used in high resolution NMR.

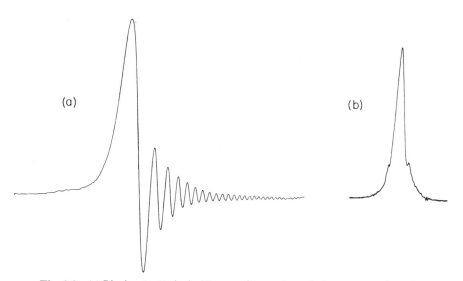

(a)   (b)

Fig. 2.9 (a) Ringing (or "wiggles") seen after moderately fast passage through resonance. Direction of scan is from left to right. (b) Same spectral scan at a slower rate, approaching that of "slow passage." Discontinuities near the base of the line are due to spectrometer instabilities at the very slow scanning speed.

they not only undergo resonance at a slightly different H_0, leading to a
somewhat broadened line, but after resonance they get out of phase more
quickly (i.e., in a time less than T_2). The decay rate then is characteristic
of field homogeneity, rather than the molecular T_2. The appearance of
ringing is often a useful practical criterion of homogeneity, with more intense and longer-lived ringing indicating a more homogeneous field.

Problems

1. Using the values for μ and I given in Appendix B, verify that the resonance frequencies in Appendix B for ^1H, ^2H, ^{14}N, and ^{31}P are correct.

2. Find values for $(n_- - n_+)/n_-$ for ^{19}F, ^{31}P, and ^{15}N at (a) 14,100 G and
   300°K; (b) 14,100 G and 5°K; and (c) 50,000 G and 300°K.

3. Find the value of the volume nuclear paramagnetic susceptibility κ for
   PF_5 at $-90°C$ and at 2°K. Assume a density of 1.0 at both temperatures.

4. Derive an equation for signal strength as a function of time that can be used to calculate T_1.

5. Use the equation derived in Problem 4 and the data in Fig. 2.5 to calculate T_1 for benzene.

6. Fill in the details of the derivation of Eq. (2.37) from Eq. (2.34).

7. Derive an expression in terms of T_2 for the width at half-height of a Gaussian-shaped line.

8. Use Eq. (2.47) to find the maximum value of H_1 that can be used for protons with $T_1 = T_2 = 4$ sec if Z is to be maintained as large as 0.95. (Note that Eq. (2.47) is derived for slow passage and that in practice larger values of H_1 are used with correspondingly faster sweep rates.)

9. Find H_{eff}/H_1 and the direction of $\mathbf{H}_{eff}$ relative to $\mathbf{H}_1$ for protons resonating at exactly 100 MHz, where $H_1 = 1$ mG and is 1 kHz off resonance.

Chapter 3

Instrumentation and Techniques

3.1 Basic NMR Apparatus

The basic instrumentation needed for NMR spectroscopy is shown schematically in Fig. 3.1. The essential components are as follows:

(1) A *magnet,* usually capable of producing a field of at least 14,000 G. In general, a higher field is desirable for increased sensitivity (Section 2.4) and for reasons to be discussed in Chapter 4. The magnet may be one of three types: a permanent magnet; an electromagnet; or a superconducting solenoid. Permanent magnets are generally simpler and cheaper, but are limited to about 1.4 T (14,000 G) for a moderate gap and 2.1 T for a narrow gap. They do not possess the flexibility of the more commonly used electromagnets. Electromagnets can go to very high field strengths (~ 10 T), but for high resolution NMR they are generally limited to about 2.4 T. An electromagnet requires a power supply of high stability and must be water cooled. Superconducting magnets now provide high resolution NMR at 8.5 T, and higher fields are rapidly becoming available. They are quite stable and require no power once energized. However, they are expensive to purchase and to operate, since liquid helium is required as a refrigerant. However, with efficient dewars, helium consumption has been reduced to the point where operating expenses are often less than for a large electromagnet (2.4 T), which requires substantial electrical energy and cooling water.

(2) A source of rf power, the *transmitter*. This may be designed for use over a range of frequencies (a *frequency synthesizer*) if a number of different nuclei are to be studied, or may be crystal controlled at a single frequency.

(3) A *probe,* which fits into the magnet and holds the sample, as well

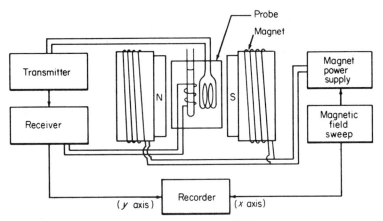

Fig. 3.1 The essential components of an NMR spectrometer system.

as electrical coils to carry rf power to and from the sample. The probe may be of the single coil or double coil (crossed coil) type. In the latter, one coil is attached to the transmitter, while the other, which is placed at right angles to the first, picks up the induced signal from the resonant change in nuclear magnetization. (The crossed coil type is illustrated in Fig. 3.1.)

(4) A *receiver*, which amplifies, detects, and filters the NMR signal.

(5) *Sweep circuitry* to vary either the frequency or the magnetic field in a controlled manner. (The latter is shown in Fig. 3.1.) For NMR pulse methods, to be described in detail in Chapter 10, this sort of sweep circuit is not needed.

(6) A *recorder* for displaying the spectrum. In some cases the spectral data may alternatively be put into a digital computer for suitable processing.

3.2 Requirements for High Resolution NMR

In practice the components listed in the preceding section actually involve apparatus that is mechanically and electronically quite sophisticated. Particularly for the study of the narrow lines that make up the NMR spectrum of a liquid or solution (high resolution NMR), there are very stringent requirements on the magnet and the electronic systems.

Homogeneity. Adequate magnetic field homogeneity across the sample is the *sine qua non* for NMR. As we saw in Section 2.6, NMR lines from small molecules in the liquid phase are usually <0.6 Hz in width, and as we shall see in Chapter 5, lines separated from each other by <1 Hz can frequently provide information of chemical value. For proton resonance, rf frequencies of 60–500 MHz are usually used; hence a resolution of at least 5×10^{-9} is required. This phenomenal homogeneity requirement is realized by several means.

First, the sample volume is restricted. In most instruments the "effective volume" of the sample (i.e., that within the rf receiver coil) is restricted to about 0.04 cm³. The sample is usually placed in a cylindrical tube, and considerably more sample may actually be required in practice; this point will be taken up in Section 3.5.

Second, electromagnets and permanent magnets must have pole pieces that are large in diameter relative to the width of the air gap. All parts must be machined and aligned carefully, and the pole faces must be polished almost to optical flatness. Superconducting magnets are fabricated in solenoidal form, so they have no pole pieces, but the miles of superconducting wire that go into such a magnet must be of extremely high quality and must be wound precisely.

Third, electromagnets are often *cycled,* that is, increased for a few minutes to a field higher than that desired for the NMR studies. Because of hysteresis effects this procedure substantially reduces the field gradients across a sample. A magnet that has not been adequately cycled has a field gradient of a "domed" shape, as indicated in Fig. 3.2a. The center of the sample, containing the bulk of the nuclei, comes into resonance then at the lowest applied field as the spectrum is scanned, while the remainder of the sample resonates at a higher applied field. The result is a line shape of the form shown in Fig. 3.2a. A field that has been overcycled has a "dished" gradient and gives a line shape of the form in Fig. 3.2b. Magnets operating at high field strength (near 2.4 T) are generally not cycled, but satisfactory field homogeneity can be obtained as described below.

Fourth, the temperature of the cooling water for an electromagnet and the air temperature for any magnet (especially a permanent magnet) must be carefully controlled. Usually the magnet yoke is insulated to decrease thermal fluctuations.

Fifth, electrical coils of carefully chosen geometry are placed on the pole faces of the magnets, and small dc currents are passed through them. The small magnetic fields thus generated are adjusted to compensate any inhomogeneity in the main magnetic field. With these so-called *shim coils,* linear field gradients in the x, y, and z directions can be reduced by about an order of magnitude, and a second-order, or "curvature," control

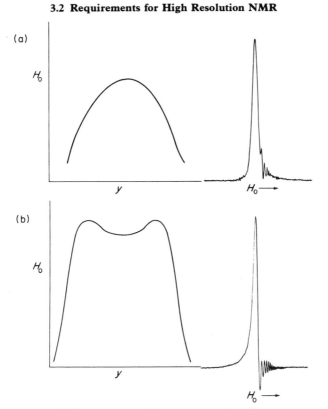

Fig. 3.2 Typical field gradients and line shapes resulting from improper cycling of an electromagnet: (a) undercycled; (b) overcycled.

permits in effect a fine control over the cycling. For magnets that cannot be cycled, additional coils provide correction of higher-order field gradients. Some spectrometers are equipped with a device that automatically adjusts the current in one of the shim coils to maintain optimum homogeneity.

Finally, a considerable improvement in *effective* homogeneity is achieved by spinning the sample tube about its axis. If the field gradient across the sample in a direction transverse to the spinning axis is ΔH, then it has been shown[20] that a spinning rate

$$R > \frac{\gamma \Delta H}{2\pi} \quad \text{rev/sec} \tag{3.1}$$

averages out much of the inhomogeneity by causing each portion of the sample to move periodically through the entire gradient. The sample thus behaves as though it experiences only the average field rather than the en-

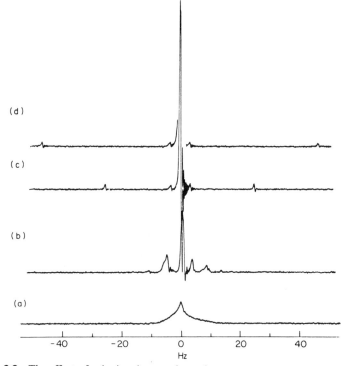

Fig. 3.3 The effect of spinning the sample on the proton resonance spectrum of tetra-methylsilane (TMS). (a) without spinning, showing a line about 3.5 Hz wide; (b) spinning at the slow rate of about 4 rev/sec; (c) spinning rate 25 rev/sec; (d) spinning rate 46.5 rev/sec. (The small peaks seen in (c) and (d) near the base of the principal line are not spinning side-bands, but arise from molecules of TMS containing ²⁹Si. See Section 7.25 for details.)

tire range of field. This procedure is so effective that it is used almost uni-versally in high resolution NMR work. Sample spinning is usually accom-plished by means of a small air turbine mounted on the probe. Rotation of the order of 30 rev/sec is typical.

There is one practical drawback to the spinning technique: the peri-odic spinning modulates the magnetic field and leads to the appearance of *sidebands* (i.e., "images" of the spectral peaks) symmetrically placed and separated by the spinning frequency and integral multiples of it. Spinning sidebands can usually be reduced to less than 1% of the ordinary peak in-tensity by proper adjustment of the electrical shim coils and by use of high-precision sample tubes and spinning apparatus. Spinning sidebands can easily be recognized by their change in position when the spinning speed is altered. A higher speed not only causes them to move farther from the parent peak but also reduces their intensity. Examples of spinning effects and spinning sidebands are shown in Fig. 3.3.

Stability. A highly homogeneous field is of little practical value in measuring sharp line NMR spectra if there are significant fluctuations or drift of field or frequency during the period of observation. The first real efforts to achieve adequate stability for high resolution NMR were aimed at stabilizing the radio frequency with a crystal controlled oscillator. For an electromagnet the magnetic field was stabilized independently by a flux stabilizer, which senses changes in the field and applies correction currents through the magnet power supply. Permanent magnets require good thermostatting to insure stability, while superconducting magnets are inherently stable when they are in the persistent mode.

The greatest stability is achieved, not by controlling the rf and magnetic field separately, but rather by controlling their ratio. Because of the Larmor relation, an NMR signal itself can provide excellent field/frequency stability when it is applied to an appropriate feedback loop. The means for accomplishing such control will be taken up in Section 3.4.

3.3 Modulation and Phase Sensitive Detection

Many features of NMR instrumentation depend on the periodic modulation of either the radio frequency or the magnetic field. Because of the Larmor relation, the two types of modulation often give essentially the same results. Field modulation is usually employed because it is simpler, but often we express the results as though the rf were modulated.

It has been shown[21] that the modulation of the field at an audio frequency that is large compared with the widths of NMR lines (a frequency typically in the range 1–20 kHz) results in the appearance of sidebands at multiples of the audio frequency. By altering the *modulation index,* which is a measure of the amount of audio power applied, the intensity distribution in the sidebands may be varied. Fig. 3.4 shows a schematic example of the appearance of modulation sidebands in a multiline spectrum.

Almost all high resolution NMR spectrometers now employ *phase sensitive detection* of the NMR signal, in both the rf and the audio range. This is accomplished by using a signal from the transmitter as a reference and electronically detecting the amplified output signal from the NMR probe that is either in phase or 90° out of phase with the transmitter signal. The rf phase detection thus selects the absorption or dispersion mode of the NMR signal (see Section 2.9), while the audio phase detection discriminates against spurious signals at other frequencies (e.g., erratic

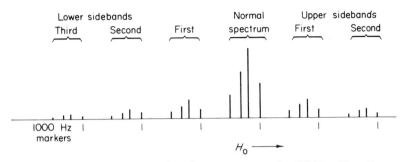

Fig. 3.4 Schematic representation of a spectrum spanning 500 Hz with audio modulation of 1000 Hz.

changes in amplifier gain or transmitter level). This procedure thereby eliminates much low-frequency noise and leads to a more stable spectral base line.

Baseline stabilization can be achieved more elegantly by other means, as well. One method is to gate the transmitter and receiver on alternately at 10,000–50,000 times/sec. This technique eliminates leakage of transmitter power to the receiver, and generates sidebands at the gating frequency. By phase detecting at that frequency the same benefits accrue as with audio modulation.

3.4 Field/Frequency Control

Several methods are used for "locking" the magnetic field to the radio frequency, but the basic principle is always the same: the signal from the NMR line to be used as a lock is detected in the dispersion mode, and this signal is used to adjust either the magnetic field or the rf. As shown in Fig. 2.7, the dispersion mode signal changes sign precisely at the resonance condition, so that the feedback control loop obtains the necessary information on direction as well as magnitude of drift.

The method of locking may be either external or internal and either homonuclear or heteronuclear. An *external lock* refers to one in which the NMR line that provides the lock signal comes from a substance contained in a spatially separate rf coil from the sample being studied. Usually a material giving a single strong resonance is used (e.g., water). If the nuclear species providing the lock (e.g., ^1H) is the same as that being studied, the lock is said to be *homonuclear;* otherwise it is *heteronuclear.* An external lock has the advantage of being completely independent of the electronic circuitry for the sample and is normally not interrupted by change of sample. It has the disadvantage of providing a lock dependent on the magnetic field at some spatial separation from the sample itself (often 1 cm or so distant). The stability of an external lock system may be of the order of 2×10^{-8}/hr.

An *internal lock* uses a signal from some material within the rf coil that surrounds the sample. Usually the locking substance is dissolved in the sample solution, or it may be the solvent itself. In some cases it is physically separated from the sample solution by being placed into a capillary or annulus in the sample tube. An internal lock, either homonuclear or heteronuclear, usually provides much better stability than an external lock—1×10^{-9}/hr is typical. Most instruments now employ an internal lock, with an external lock also available in some instances. For study of nuclei with low sensitivity and/or low natural abundance a heteronuclear internal lock is usually used; for example, almost all ^{13}C spectrometers use the ^{2}H resonance of a deuterated solvent for the lock.

Once a lock is established some method is needed for sweeping either frequency or magnetic field for conventional NMR studies. Usually two different rf signals excite sample and lock resonances, the sample signal being detected in absorption mode, the lock signal in dispersion mode. The exciting rf may come from two different oscillators or frequency synthesizers, or (for a homonuclear lock) they may represent two different audio modulated sidebands. Usually the lock frequency is held constant and the sample frequency varied (*frequency sweep*), but the sample frequency is sometimes fixed and the lock frequency varied to cause the magnetic field to change in a controlled manner (*field sweep*).

3.5 Signal/Noise and Size of Sample

Several factors determine the signal/noise ratio in NMR and hence limit the minimum size of sample that may be studied. We have already seen in Section 2.8 that an NMR signal for a given nucleus should in principle increase quadratically with field strength, so that large values of H_0 are preferred. The inherent sensitivity varies substantially from one nucleus to another, as indicated in Appendix B. The electronic circuits employed and the care used in manufacture of the probe are, of course, important factors. The "filling factor," that is, the fraction of the volume of the receiver coil that is actually filled by sample, not glass of the sample tube or air space, is especially important. The use of thin pieces of glass on which the coil is wound, as well as the use of thin-walled sample tubes, is mandatory to obtain a high filling factor.

From the practical standpoint, the commercially available spectrometers operating at 60 MHz display a signal somewhat above the noise level from a single sharp line due to protons present at about $0.005M–0.02M$. When possible concentrations of $>0.2M$ are usually preferred. Normally, for proton resonance a sample of about 0.4 ml is contained in a precision, thin-walled glass tube of about 5-mm outer diameter. In some instruments with large magnets larger-diameter sample tubes may be used to gain some sensitivity at the expense of a larger total amount of solution required. Microcells, in which the sample is contained in a spherically shaped cavity of $25–50$ μl can be used to reduce the total *amount* of sample required, but their greater wall thickness usually requires higher concentrations than the minimum values quoted above. Recently small capillary sample tubes containing only about 10 μl have come into use with specially wound receiver coils of very small diameter.

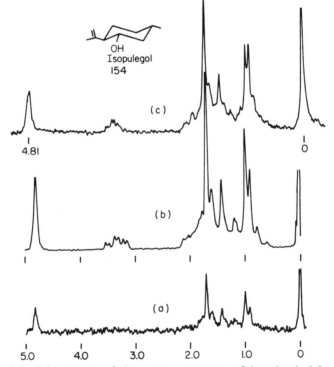

Fig. 3.5 Enhancement of the proton resonance of isopulegol (0.8 mg) by time averaging. (a) Single scan on Varian A-60 spectrometer with sweep rate of 1 Hz/sec; (b) average of 210 scans with sweep rate of 2 Hz/sec; (c) single scan on the more sensitive Varian HA-100 spectrometer with sweep rate of 1 Hz/sec. Reprinted with permission from R. E. Lundin *et al. Anal. Chem.* **38**, 291 (1966).

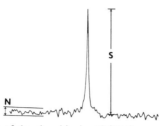

Fig. 3.6 Measurement of the signal/noise (S/N) from a spectral trace. Peak-to-peak noise is illustrated; root mean square (rms) noise is often approximated by dividing peak-to-peak noise by 2.5.

The greater the long-term stability of the spectrometer, the longer the time that can be spent in scanning a spectrum. A longer scan time permits additional electronic filtering to reduce some of the noise and thereby improve the signal/noise ratio. Particularly with internal lock spectrometers a scan duration of hours is possible. With very slow scans, however, the rf power must be kept low to avoid saturation (see Sections 2.7 and 3.7).

An alternative method of improving signal/noise by the expenditure of additional scanning time is the use of *time averaging*. Instead of displaying a spectrum on a chart, the spectral information is placed in digital form in the memory of computer. The information from repetitive scans is then added to that in the computer memory, so that after N scans, the signal is N times as great as would be obtained with one scan. Since noise is random, it can be shown[22] that the noise after N scans has only increased by $\sqrt{N}$, so that there is a signal/noise improvement of $\sqrt{N}$. An example is shown in Fig. 3.5.

A precise definition of signal/noise is needed for quantitative comparison of the performance of different instruments. Usually the "noise" refers to root mean square (rms) noise, rather than the maximum noise excursions (peak-to-peak noise), illustrated in Fig. 3.6. The rms noise should be calculated by evaluating the quantity $[(1/n)\Sigma_{i=1}^{n} q_i^2]^{1/2}$, where the sum is over a large number of points. Under conditions usually met in practice, it can be shown[24] that the peak-to-peak noise is a multiple of the rms noise. The exact multiplicative factor is subject to some question, but a value of 2.5 is reasonable and has been generally adopted by NMR spectrometer manufacturers.

The sensitivity of NMR spectrometers has improved gradually but quite substantially over the years, as higher magnetic fields have come into use and improved electronic circuits and components have become available. For example, Table 3.1 gives the sensitivity specifications quoted for various commercial instruments. Each refers to a standard sample in a 5-mm diameter sample tube and gives the signal/noise for a

Table 3.1

Signal Noise for Various Spectrometers[a]

Year	Spectrometer model	S/N
1961	A–60	6
1965	HA–100	30
1969	HR–220	80
1978	XL–200	300
1978	WH–360	800

[a] Proton specification: 1% ethylbenzene, largest peak in methylene quartet, single scan/ single pulse.

single scan. Improved methods of spectral acquisition and time averaging, which we discuss in the following section, provide still further gains in the sensitivity per unit time.

3.6 Fourier Transform Methods

The last 10 years have seen a revolution in NMR instrumentation with the introduction of Fourier transform (FT) methods to enhance sensitivity. The name stems from the mathematical methods used to analyze the data. They will be taken up in detail in Chapter 10. Three distinctly different FT methods have been introduced: (1) pulse excitation; (2) stochastic excitation; and (3) rapid scan correlation NMR. Of the three, pulse techniques[25] are by far the most commonly used; they are very versatile and are discussed in detail in Chapter 10. In this method no scan of magnetic field or frequency is employed; instead nuclei throughout the spectrum are excited simultaneously by a short rf pulse. As a result the time needed to acquire the spectral data is substantially reduced, and a number of repetitive pulses can be used in a short time with time averaging to improve signal/noise.

In stochastic excitation,[26] also, nuclei over the entire spectral range are excited simultaneously, this time by use of random or pseudo-random noise, which contains a wide range of frequency components. The saving in time is the same as that of the pulse method. The third FT method, rapid scan correlation NMR,[27] is closely related to the conventional slow scan techniques that we have discussed. However, as implied by the name, the scan is fast—often 1 kHz/sec—and the spectrum is highly distorted by excessive ringing (Section 2.11). However, the true slow passage spectrum can be obtained by applying mathematical operations, that are equivalent to cross correlation of the observed response with that for an infinitely narrow single line scanned under the same conditions. The sensitivity of the method is almost equivalent to the pulse and stochastic FT techniques. In practice the necessary calculations are usually carried out in digital computers using two Fourier transformations, and much of the programming for these computations is similar to that for pulse FT studies.

3.7 Intensity Measurements

In Section 2.4 we saw that the strength of an NMR signal, which is measured by the area under the NMR line, is proportional to the number of nuclei contributing to the line. Accurate measurements of these areas greatly facilitate the interpretation of spectra and also provide a means of conducting quantitative analyses. Most NMR spectrometers are equipped with electronic integrators, which record relative areas of spectral lines as a step function such as the one in Fig. 1.1. In principle such integrations can be accurate to within 1–2% provided signal/noise ratio is sufficiently high. To achieve this precision in practice it is necessary to observe several precautions. First, the phase detector must be carefully adjusted to

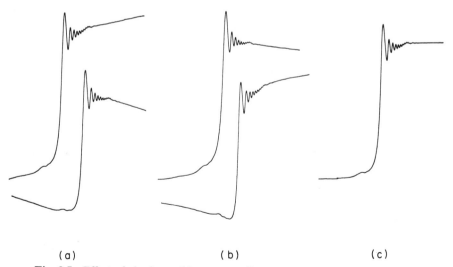

(a)   (b)   (c)

Fig. 3.7   Effect of phasing and integrator adjustments on a spectrum and on an integral: (a) amplifier balance incorrectly set; (b) phase incorrectly set; (c) correct adjustments.

insure that the signal is a pure absorption mode with no dispersion characteristics. Second, there must be no drift in the amplifying or integrating system. The proper adjustments are illustrated in Fig. 3.7. Third, the value of H_1 must be low enough to insure that appreciable saturation does not occur. Since the saturation parameter (Eq. (2.4)) depends on the values of T_1 and T_2, which can differ appreciably between chemically different nuclei of the same species (see Chapter 8), even the measurement of relative areas can be appreciably in error if saturation occurs. In practice, saturation is often somewhat less serious than the Bloch equations would indicate since they are derived for slow passage, and most NMR spectra are obtained with more rapid sweeps. In fact, if the scanning speed is sufficiently rapid and the rf power H_1 is kept low, the values of integrals can be made essentially independent of relaxation times. The sensitivity lost by using a more rapid scan can be regained by time averaging. In pulse FT experiments there is no scan rate to consider, but other factors are equally critical, as discussed in Chapter 10. When the spectral data are in digital form in a computer, as in an FT experiment, it is usually preferable to compute the integral numerically, rather than to rely on an electronic integrator.

3.8 References

As we shall see in detail in Chapter 4, the range of resonance frequencies encompassed by the chemical shifts for a given nucleus is very small relative to the resonance frequencies of the nuclei. The latter are in the range of $1 - 500$ MHz at the magnetic fields usually used, while the former seldom span more than several kilohertz. Within this range we wish to make measurements accurate to a small fraction of the line width, that is, in many cases to about 0.1 Hz or less. While it is possible with sufficiently elaborate instruments to measure frequencies in the megahertz range to < 0.1 Hz, there is no means known by which the magnetic field could be independently measured to the necessary accuracy of 1 part in about 10^8 or 10^9. Thus an *absolute* measure of the resonance condition is impossible for high resolution NMR, and all measurements reported are merely made relative to some agreed-upon standard or reference for each nucleus. Unfortunately, an "ideal" reference is not always available, and for experimental reasons it is not always convenient to use a given reference. Hence, NMR data have often been expressed in different ways, and it is important to know how a reference compound can be used and how to convert from one to another.

Two types of reference are used in NMR: *internal* and *external.** An internal reference is a compound, usually giving a sharp NMR line, that is dissolved directly in the sample solution under study. The reference is then dispersed uniformly at a molecular level through the sample. The magnetic field acts equally on the sample and reference molecules, so that a determination of the *difference* in resonance frequency between the two, together with an *approximate* knowledge of the magnetic field or the imposed rf frequency, is adequate for a meaningful measurement. Internal references are used most commonly and are generally advantageous. Provided the reference compound does not react chemically with the sample, the only serious drawback to an internal reference is the possibility that intermolecular interactions will influence the resonance frequency of the reference. Usually, by careful choice of relatively inert compounds this effect can be made small enough to be disregarded. The internal reference of choice for proton resonance is tetramethylsilane. Other references, as well as scales for expressing data, will be given in Chapter 4.

An external reference is a compound placed in a separate container from the sample. Usually an external reference is placed either in a small sealed capillary tube inside the sample tube or in the thin annulus formed by two precision coaxial tubes. In either case, the usual rapid sample rotation (Section 3.2) makes the reference signal appear as a sharp line superimposed on the spectrum of the sample. An external reference is advantageous in eliminating the possibility of intermolecular interactions or chemical reaction with the sample. Also, there are no problems with solubility of the reference in the sample solution. There is, however, a serious difficulty raised by the difference in bulk magnetic susceptibility between sample and reference. In all substances with completely paired electrons, the motion of the electrons in a magnetic field is such as to make the substance diamagnetic, that is, repellant to a magnetic field. The magnetization per unit volume induced in the sample is given by

$$\mathbf{M} = \kappa\mathbf{H}_0, \tag{3.2}$$

where κ, the *volume magnetic susceptibility,* is negative for all diamagnetic materials. For the sample, κ_S is approximately the weighted average of the values for solute and solvent; for sufficiently dilute solutions this is nearly the value for the solvent. But the external reference in general will have a value $\kappa_R \neq \kappa_S$. As a result, the magnetic field experienced by a molecule in the sample will be slightly different from that experienced by

---

* Internal and external references should not be confused with internal and external locks, as discussed in Section 3.4. The functions of locks and references are different, but in some cases a single compound can serve as both an internal lock substance and a reference.

a molecule in the reference. Standard electromagnetic theory[28] shows that these fields $(H_0)_S$ and $(H_0)_R$, are

$$(H_0)_S = H_0[1 + (\tfrac{4}{3}\pi - \alpha)\kappa_S],$$
$$(H_0)_R = H_0[1 + (\tfrac{4}{3}\pi - \alpha)\kappa_R].$$

$$(3.3)$$

The quantity α is called the *shape factor* and depends upon the shape of the interface between sample and reference phases. If this interface is a sphere, $\alpha = \tfrac{4}{3}\pi$, so that the susceptibility correction reduces to zero. Spherical cells are sometimes used, but they are generally inconvenient, and imperfections in the glass wall can introduce spurious effects.[29] For the usual cylindrical sample tubes that are long relative to their diameter, and for the usual geometry used in permanent and electromagnets with H_0 perpendicular to the axis of the cylindrical sample tube (see Fig. 3.8), $\alpha = 2\pi$, and

$$(H_0)_S = H_0(1 - \tfrac{2}{3}\pi\kappa_S),$$
$$(H_0)_R = H_0(1 - \tfrac{2}{3}\pi\kappa_R).$$

$$(3.4)$$

Since all commonly used reference substances, as well as most samples, are diamagnetic, κ_S and κ_R are usually negative.

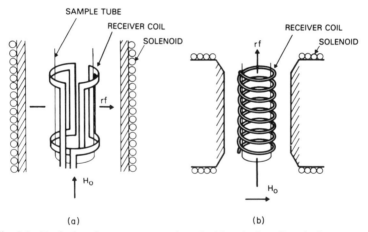

Fig. 3.8 Typical probe geometry employed with a single coil probe in (a) a superconducting magnet and (b) an iron core magnet. In each case the rf coil is fabricated in such a way that $\mathbf{H}_1$ is perpendicular to $\mathbf{H}_0$.

With a superconducting magnet, which is normally in the shape of a long solenoid, the most convenient axis for a cylindrical sample tube is along that of the solenoid, which is the direction of H_0. In this case[30] the analogs of the expressions in Eq. (3.4) are

$$(H_0)_S = H_0(1 + \tfrac{4}{3}\pi\kappa_S),$$
$$(H_0)_R = H_0(1 + \tfrac{4}{3}\pi\kappa_R). \tag{3.5}$$

The use of Eqs. (3.4) and (3.5) in correcting observed data will be taken up in Chapter 4 in connection with scales for expressing chemical shifts.

3.9 Magnetic Susceptibility Measurements

In order to use expressions of the sort given in the previous section, data for volume magnetic susceptibilities must be available. Magnetic susceptibilities have long been measured by "classical" means, such as with the use of a Gouy balance, and tabulations are given in chemistry handbooks. Often the accuracy of these data is insufficient for use in determination of NMR chemical shifts. Fortunately, the NMR method itself can provide more accurate data, in many cases with smaller amounts of material than needed for the classical methods.

If both a superconducting solenoid and an iron core magnet based NMR systems are available, then the magnetic susceptibility of a sample (liquid or solution) can be obtained from two measurements of the resonance frequency of a single line of the sample relative to an external reference (usually contained in a spinning, coaxial tube). From Eqs. (3.4) and (3.5), together with the Larmor relation, Eq. (2.9), we can easily derive the relation

$$\kappa_S - \kappa_R = \frac{1}{2\pi}\left[\frac{\nu_S^{sc}}{\nu_0^{sc}} - \frac{\nu_S^{ic}}{\nu_0^{ic}}\right], \tag{3.6}$$

where ν_S^{sc} and ν_S^{ic} are the frequencies (in Hz) of the sample resonance relative to the external reference, and ν_0^{sc} and ν_0^{ic} are the operating frequencies (in Hz) of the spectrometers using the superconducting and iron core magnets, respectively. Only one well established susceptibility value, κ_R, must be known from non-NMR determinations. (Care should be taken that the two spectrometers needed for this method are operated at the same temperature.)

Other NMR procedures for determining magnetic susceptibilities without the necessity of using two spectrometers have been proposed and their application discussed in some detail.[31] One of the most frequently used methods employs a precision coaxial cell, of the sort mentioned in Section 3.8. Analysis of the magnetic field distribution in such a cell[32] shows that when the cell is spun rapidly, as in normal NMR spectral measurements, the material in the outer annulus behaves as though it were in a simple tube, but when the cell is not spun, the substance in the annulus gives rise to a broad, two-peaked resonance. With a given liquid in the annulus, the separation of the two maxima is proportional to the magnetic susceptibility of the sample in the main (inner) tube. Thus by using several substances of known susceptibility in the inner tube, a linear calibration curve may be constructed for the particular cell being used, and subsequently the susceptibilities of other samples can easily be determined.[31a] Some of the precautions that must be taken in using the method have been discussed in detail.[31b,d]

3.10 Frequency Calibration

With adequate field/frequency control, spectra are usually recorded on precalibrated charts, and it is normally unnecessary to record a calibration marker other than the one internal or external reference peak with each spectrum. However, the built-in calibration must be checked and adjusted regularly, and for more accurate measurements, calibration markers should be recorded directly with the spectrum of the sample. The most convenient procedure involves the modulation of the field at a measured audio frequency so as to produce sidebands, as in Fig. 3.4. The spacing on the chart paper may thus be accurately determined. Often two sidebands of different frequency are used to bracket a peak of interest. With an internal lock spectrometer, the greatest accuracy is obtained by stopping the scan at the peak of interest and determining with a frequency counter the difference between the locking and observing frequencies.

For the accurate determination of the separation of two lines that are very close together (< 5 Hz) the method of "wiggle-beats" is sometimes used. The exponentially decaying ringing pattern shows a modulation envelope due to interference between the ringing patterns for the two or more evenly spaced peaks. The separation in seconds between successive modulation maxima can be shown to be the reciprocal of the separation (in Hz) between the peaks themselves.[33]

With pulse Fourier transform methods, frequencies are generally determined directly and accurately from the time base of the computer, and no independent calibration of the frequency scale is usually required.

3.11 Control of Sample Temperature

Most NMR spectrometers are equipped to vary the temperature of the sample by passing preheated or precooled nitrogen past the spinning sample tube. The commercially available instruments are usually limited, by the materials used in the probe, to temperatures between $-100°$ and $200°C$, but some high resolution probes have been constructed to operate at temperatures as low as $-190°$ and as high as $300°C$. At extreme temperatures some deterioration of magnetic field homogeneity is often found. It is important that adequate time be allowed for temperature equilibration of the system, and that the homogeneity adjustments be optimized at each temperature. Instrument calibration is also subject to variation with temperature. The uniformity of temperature through the sample and the stability of temperature control have not been thoroughly checked with different instruments, but constancy of the order of $\pm 2°C$ is usually claimed for commercial instruments. Often it is convenient to measure sample temperature by substituting a sample tube containing a liquid with resonances that are known to be temperature dependent. For proton resonance the spectra of methanol and of ethylene glycol are usually used for low and high temperature regions, respectively. Van Geet has discussed temperature measurement and provided calibration curves for these liquids.[34]

Changes in temperature of the sample will alter the Boltzmann distribution of spins in the various energy levels. For the case of $I = \frac{1}{2}$, Eqs. (2.24) and (2.30) show that a reduction in temperature from $27°C$ ($300°K$) to $-60°C$ causes an increase in signal intensity of 40%; a corresponding decrease in intensity is obtained on increasing sample temperature.

3.12 Useful Solvents

In addition to the desired solvation property itself, the major considerations in selecting a solvent for NMR spectroscopy are the minimization of interaction or reaction between solvent and sample and the avoidance of strong NMR signals from the solvent. For proton NMR a list of commonly employed solvents is given in Table 3.2. Solvents that contain no hydrogen atoms are, of course, desirable, but all the hydrogen-containing solvents listed in Table 3.2 are available commercially in deuterated form of $98-99.7\%$ isotopic purity. (Even so a small signal from the residual protons must be expected in the region indicated.) Chloroform-d is relatively inexpensive, and is the most commonly employed solvent for proton NMR. Samples with readily exchangeable hydrogen atoms (such as OH or NH groups) will, of course, lose these hydrogens to a solvent containing exchangeable deuterium atoms.

Many substances that are vapors at room temperature and atmospheric pressure may be used as NMR solvents in sealed tubes or at re-

Table 3.2

SOME USEFUL SOLVENTS FOR PROTON NMR

Solvent	δ (Residual proton signal)[a]
$CDCl_3$	7.27
CD_3OD	3.35, 4.8[b]
Acetone-d_6	2.05
D_2O	4.7[b]
p-Dioxane-d_8	3.55
Dimethylsulfoxide-d_6	2.50
Pyridine-d_5	6.98, 7.35, 8.50
Benzene-d_6	7.20
Acetic acid-d_4	2.05, 8.5[b]
CCl_4	—
CS_2	—
SO_2	—
Hexafluoroacetone	—

[a] δ is given in parts per million relative to tetramethylsilane (internal reference). See Chapter 4 for definition of δ.

[b] Highly variable, depending on solute and temperature. (See discussion of hydrogen bonding, Chapter 12.)

duced temperature. For example, SO_2 has a vapor pressure of about 3 atm at room temperature and can be easily contained in sealed thin-walled, 5-mm diameter NMR sample tubes. For use of SO_2 above room temperature, sample tubes with 1-mm wall thickness are used.

For NMR studies of nuclei other than hydrogen, suitable solvents that do not contain the nucleus being studied are usually readily available. Frequently, the use of two or more solvents can provide valuable information on molecular structure. A discussion of solvent effects will be given in Chapter 12.

3.13 Sampling Techniques

The selection of solvents and choice of concentration usually requires a compromise among several competing factors. For example, a relatively inert solvent (saturated hydrocarbon or CCl_4) is often desirable to minimize molecular interactions, but solubilities of many samples often require strongly interacting polar solvents. Likewise, a low concentration both minimizes solute–solute interactions and usually reduces viscosity, thus leading to sharper lines (see Chapter 8). However, improved signal/noise naturally results from higher concentrations. In practice most proton NMR spectra are obtained where possible with samples at a concentration of $0.1M$ or greater. The lower sensitivity or lower natural abundance, or both, of most other nuclei usually requires considerably higher concentrations.

Spectra of gases can be obtained, but pressures of several atmospheres are usually necessary in the absence of time averaging to obtain an adequate amount of sample. Solids must, of course, be dissolved to permit rapid molecular tumbling to average out the interactions that lead to signal broadening (see Sections 2.6 and 10.7).

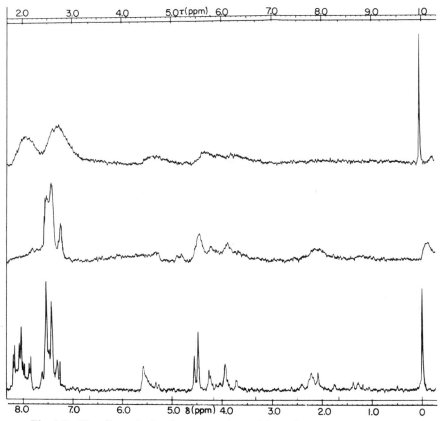

Fig. 3.9 The effect of a tiny ferromagnetic particle on the proton resonance spectrum of a benzoylated sugar. The top and middle curves are repeat runs with the particle present; the bottom curve is the spectrum with the particle removed.

For proton resonance, solvents containing exchangeable protons are limited in use. If a proton in the sample exchanges rapidly with a solvent proton, separate signals for the two types of protons are not observed; instead, only a single peak appears at the weighted average of the two frequencies (see Chapter 11). Because of the preponderance of solvent molecules this is essentially the solvent frequency. Even small amounts of hydrogen or hydroxyl ion or of water often catalyze exchange. At intermediate rates of exchange, lines are often broadened, sometimes beyond recognition. An illustration of exchange broadening was given in the OH peak in Fig. 1.1, and further examples are shown in Chapter 11.

The presence of paramagnetic materials (those containing unpaired electrons) in NMR samples often causes line broadening due to the decrease in spin–lattice or spin–spin relaxation time. (The effects will be discussed in Chapter 8.) Ions such as Fe^{3+}, Cu^{2+}, and Mn^{2+} can cause very substantial broadening of some NMR lines if the ions are complexed to the sample being studied. Even molecular oxygen from the atmosphere causes some broadening; hence for optimum resolution of narrow lines, the sample should be deoxygenated, either by evacuation or by bubbling nitrogen through the sample.

Small solid particles floating in the sample cause local magnetic field inhomogeneities and often lead to line broadening. The effect is especially pronounced if the particle is ferromagnetic. For example, a small piece of steel thinner than a fine hair can cause the deterioration of resolution and appearance of an erratic base line depicted in Fig. 3.9. The presence of a ferromagnetic particle is best confirmed by examining the sample under a bright light while moving a small permanent magnet ($\sim$ 100-G-field) beside the sample tube. The magnet can also be used to retain the particle in the tube while the solution is withdrawn into a long disposable pipet for transfer to another tube.* Low-retention microfilters are available for filtering NMR samples.[35]

As an aid to spectral interpretation it is often desirable to exchange all active hydrogen atoms in a sample by deuterium. This can be accomplished most readily by adding a drop of D_2O to the sample tube and shaking for a few minutes.[36] If the water is immiscible with the sample solution, the layers usually separate readily when the sample is spun, but care must be taken that small water droplets do not adhere to the wall of the sample tube in the region of the receiver coil, for a deterioration of

---

* The ubiquitousness of ferromagnetic particles in NMR samples is often not appreciated. In our laboratory we have found a substantial percentage (10–15%) of "routine" but fairly pure organic samples to contain tiny ferromagnetic particles, in some cases probably coming from steel spatulas.

resolution may result. A small peak due to water dissolved in the solvent may appear, in any event, since water is not completely insoluble in most organic solvents.

3.14 Micro Techniques

When the total quantity of sample available is limited, and when it is sufficiently soluble in a suitable NMR solvent, the best signal/noise is obtained by using a minimum volume of solvent and then confining the solution to the region of the receiver coil. Two different approaches have been used in designing sample cells for such small samples:

1. In order to use the ordinary receiver coil intended for 5-mm diameter sample tubes, it is desirable to employ a cell with a diameter as close to 5 mm as possible, while the height of the sample is restricted approximately to the height of the receiver coil, usually a few mm. In some cases it is possible to use only a small volume in an ordinary sample tube, but the discontinuities in magnetic susceptibility at the interfaces between sample and air, or sample and glass, cause severe loss in resolution. This difficulty can be overcome in principle by using a spherical cell (see Section 3.8), but in practice imperfections are often present that render such cells very difficult to use. The most successful of the spherical type cells consists of a thin-walled sphere holding about 25–30 μl, which is positioned carefully inside an ordinary cylindrical NMR tube. The outer NMR tube contains solvent to minimize the susceptibility effects. With luck and patience good spectra can be obtained with these cells.

2. The alternative approach, which has now been implemented with a number of commercial spectrometers, is to design a receiver coil to accommodate a cylindrical sample tube only 1.5–2 mm in diameter. With such a tube (which can be a disposable capillary), 15 μl of sample provides a long enough column to minimize susceptibility end effects. The use of capillary cells permits rather straightforward observation of very small amounts of material. An excellent discussion of many of the practical aspects of handling capillaries and small samples has recently been published.[37]

With pulse Fourier transform methods (Section 3.6) it is possible to make several thousand pulse repetitions per hour to improve signal/noise by coherent time averaging. An overnight run of a sample of molecular weight about 200 can produce an acceptable ^{1}H spectrum with about 1 μg in a solution of 15 μl. For example, Fig. 3.10 shows the ^{1}H spectrum of 1.0 μg of cortisone acetate. At this level, handling of the sample and care

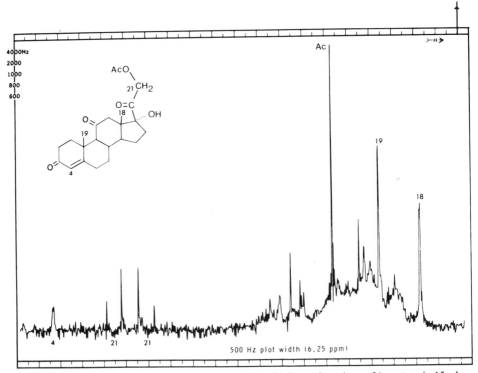

Fig. 3.10 Proton NMR spectrum (80 MHz) of 1.0 μg of cortisone-21-acetate in 15 μl CDCl$_3$. The spectrum was obtained with a 1.7-mm diameter sample tube, 30,000 pulses (45° flip angle) in 14 hr, and the residual HDO peak was subtracted (Shoolery[38]).

to avoid impurities (including traces of water) in the solvent become critical.

Even for the inherently less sensitive nucleus ^{13}C, which is only 1.1% abundant as well, samples at or below 1 mg are now being studied with capillary cells. Fig. 3.11 gives an illustration of the capability of present instruments in producing useful ^{13}C spectra from small samples.

With the improved sensitivity of modern FT-NMR instruments, it is possible to separate sufficient amounts of sample by chromatographic methods. For example, the effluent from a gas chromatograph can be passed through an appropriate capillary tube, which is cooled to condense the sample and sealed at one end. Solvent can then be added and a spectrum obtained.

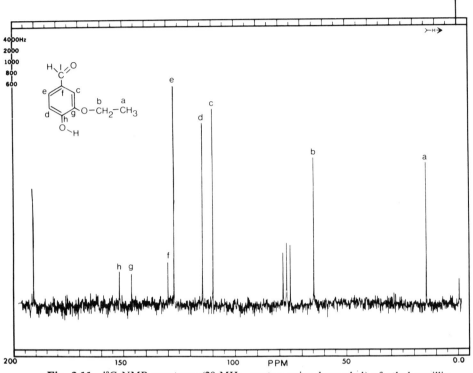

Fig. 3.11 ^{13}C NMR spectrum (20 MHz, proton noise-decoupled) of ethyl vanillin, 500 μg in 7.5 μl CDCl$_3$, 1.7-mm diameter sample tube, 64,000 pulses (40° flip angle), 16 hr accumulation (Shoolery[38]).

Problems

1. Most NMR spectrometers with conventional iron core magnets provide for spinning the sample about an axis designated as the y axis, while the field H_0 is placed along the z axis. Which electrical shim coils would you expect to require the most careful adjustment—those correcting for inhomogeneity along the x, y, or z axis? Why?

2. Modulation of radio frequencies can be accomplished by altering the *frequency* or the *amplitude* of the rf wave. Show by trigonometric relations that a modulation of the *amplitude* of a wave described by the relation $A \cos(2\pi\nu_0 t)$ by an audio wave described by $B \cos(2\pi\nu_m t)$ results

in the appearance of two sidebands at frequencies $\nu_0 + \nu_m$ and $\nu_0 - \nu_m$.

3. External lock proton NMR spectrometers have a control that must usually be adjusted for each sample to place the internal reference TMS at the zero position on the chart paper, while internal lock spectrometers do not need such adjustment. Why is there a difference between these two instruments in this respect?

4. What sensitivity enhancement can theoretically be obtained by time averaging over a weekend (5 P.M. Friday to 9 A.M. Monday) relative to a single 500-sec scan?

Chapter 4

Chemical Shifts

4.1 The Origin of Chemical Shifts

The chemical shift, which is really the cornerstone of chemical applications of NMR, has its origin in the magnetic screening produced by electrons. Thus a nucleus experiences not the magnetic field that is applied to the sample (H_0), but rather the field after it has been altered by the screening or shielding of the electrons surrounding the nucleus. Since electrons are magnetic particles also, their motion is influenced by the imposition of an external field; in general, the motion induced by an applied field is in a direction so as to oppose that field (Lenz's law). Thus at the nucleus the magnetic field is

$$H(\text{nucleus}) = H_0 - \sigma H_0 = H_0(1 - \sigma). \tag{4.1}$$

The screening factor, or *shielding factor,* σ is found to be small (roughly 10^{-5} for protons and $< 10^{-3}$ for most other nuclei); σ is actually a second-rank tensor, so that the magnitude of the shielding depends on the orientation of the molecule relative to the applied field. For molecules in rapid motion, however, the directional (anisotropic) part of σ averages out, so that σ may be treated as a simple number. (Effects of the anisotropy will be mentioned in Sections 7.25 and 8.3.) Actually σ cannot generally be determined experimentally, for that would require measurements of a "bare" nucleus stripped of all electrons. For most applications it is sufficient to make all measurements relative to some agreed-upon reference compound.

4.2 Reference Compounds

As we saw in Section 3.8, both internal and external references are used, and for various reasons different chemical substances may be se-

Table 4.1

FREQUENTLY USED REFERENCE COMPOUNDS

Nucleus	Reference compounds
^{1}H	$Si(CH_3)_4$ TMS; aqueous solutions: $(CH_3)_3Si(CH_2)_3SO_3^-Na^+$ DSS, $(CH_3)_3Si(CD_2)_2COO^-Na^+$ TSP, acetone, dioxane, acetonitrile
^{13}C	$Si(CH_3)_4$ TMS, CS_2; aqueous solutions: dioxane, acetonitrile
^{19}F	CCl_3F, C_6F_6, C_4F_8
^{31}P	P_4O_6, H_3PO_4 (85% in water)
^{14}N, ^{15}N	$(CH_3)_4N^+Cl^-$, NH_3, NH_4^+, NO_3^-, CH_3NO_2
^{17}O	H_2O
^{11}B	$BF_3{:}O(C_2H_5)_2$, BF_3, BCl_3

lected for use as a reference. There is at present general agreement that for ^{1}H resonance, data should be reported with respect to an internal reference of tetramethylsilane (TMS) for nonaqueous solvents in which TMS is soluble. For aqueous solutions no single reference is predominant. Almost all ^{13}C data are reported relative to the carbon resonance of TMS as an internal reference. For many other nuclei there is as yet no widespread agreement on satisfactory reference compounds, so it is essential that the reference be clearly stated whenever data are presented in publications. Table 4.1 lists a few of the more commonly used reference compounds for several frequently studied nuclei. By means of double resonance techniques (Chapter 9) it is possible to refer chemical shifts of nuclei other than hydrogen to the ^{1}H chemical shift of TMS.

4.3 Chemical Shift Scales

NMR data are usually *measured* in frequency units (hertz) from the chosen reference, as pointed out in Sections 3.3 and 3.10. In cases where complex spectra occur or where spin–spin coupling constants are to be given, the results should be *reported* in frequency units as well. For chemical shifts, however, the use of frequency units has the disadvantage that the reported chemical shift is dependent on the value of the magnetic field (and through the Larmor relation, Eq. (2.9), on the rf frequency), since the chemical shift is induced by the field (see Eq. (4.1)). Hence it is customary and highly desirable to *report* chemical shifts in the dimensionless unit of parts per million (ppm), which is independent of the rf frequency or magnetic field strength. We can see the relation between such a dimensionless scale and the shielding factor. With the inclusion of the shielding factor, the Larmor Eq. (2.9) relating the resonance frequency and the *applied* magnetic field (H_0) is

$$\nu_0 = \frac{\gamma}{2\pi} H_0(1 - \sigma). \tag{4.2}$$

We can now distinguish two cases: (a) Suppose the field is held fixed at H_0 and the frequency is varied to scan the spectrum. The resonance frequencies of sample and reference are

$$\nu_s = \frac{\gamma}{2\pi} H_0(1 - \sigma_S), \qquad \nu_R = \frac{\gamma}{2\pi} H_0(1 - \sigma_R). \tag{4.3}$$

The chemical shift on the dimensionless scale is usually given the symbol δ. If we define

$$\delta \equiv \frac{\nu_S - \nu_R}{\nu_R} \times 10^6, \tag{4.4}$$

then by substitution of Eq. (4.3), we have

$$\delta = \frac{\sigma_R - \sigma_S}{1 - \sigma_R} \times 10^6 \approx (\sigma_R - \sigma_S) \times 10^6, \tag{4.5}$$

since $\sigma \ll 1$. (b) Suppose, alternatively, that the radio frequency is held constant at ν_0 and the magnetic field varied. The resonances of sample and reference will then be found at the values of the applied field given by

$$\nu_0 = \frac{\gamma}{2\pi} H_S(1 - \sigma_S), \qquad \nu_0 = \frac{\gamma}{2\pi} H_R(1 - \sigma_R). \tag{4.6}$$

To obtain consistency with the foregoing definition of δ (Eq. (4.4)) we must define

$$\delta = \frac{H_R - H_S}{H_R} \times 10^6. \tag{4.7}$$

Then by substitution of Eq. (4.6) into (4.7), we obtain

$$\delta = \frac{\sigma_R - \sigma_S}{1 - \sigma_S} \times 10^6 \approx (\sigma_R - \sigma_S) \times 10^6. \tag{4.8}$$

Two important points should be noted from the results of the last paragraph. First, a more *highly shielded* nucleus (larger σ) will show its resonance at a *higher applied field* when the field is scanned, but at a *lower frequency* when frequency is varied. Second, a sample that is *less* *shielded* than the reference will be assigned a *larger* value of the *chemical shift* δ.*

It is apparent that the definitions of δ in Eqs. (4.4) and (4.7) could both have had sample and reference reversed, so that the value of δ would increase with increasing shielding. Unfortunately, both conventions have been used. For ¹H resonance the protons in TMS prove to be more shielded than almost all other protons; hence the definition embodied in Eqs. (4.4) and (4.7) is generally used, since it permits virtually all chemical shifts to be expressed as positive numbers. For some other nuclei, however, the reference compounds often employed have relatively unshielded nuclei so that the alternative convention has often been chosen, again to permit mostly positive numbers to be used for the chemical shifts. There is a growing trend among NMR spectroscopists to use the convention in Eqs. (4.4) and (4.7), which assigns a higher value of δ to nuclei at higher frequency. Recently the American Society for Testing and Materials and the International Union of Pure and Applied Chemistry have endorsed this convention for all nuclei.[40,41] We shall adhere to this internationally adopted convention in this book. However, since both conventions are used in the literature, it is essential that the user of NMR data ascertain in each case the convention that has been employed.

With the accepted convention applied to ¹H NMR, a larger chemical shift δ (measured with respect to TMS) implies a nucleus that is less shielded and hence resonates at lower field when the magnetic field is varied. During the early period of NMR, when almost all spectrometers operated with a magnetic field sweep, it seemed more convenient to think of chemical shifts on a scale that increases with increasing magnetic field. To accomplish this without using largely negative numbers, an alternative scale, the τ scale, was introduced and gained wide popularity. This is a dimensionless scale that assigns to TMS (internal reference) the value of exactly 10 ppm. Most ¹H chemical shifts, as we shall see in Section 4.5, lie within 10 ppm to the low field side of TMS; thus most τ values are between 0 and 10. The τ scale is obviously related to the δ scale, as defined in Eqs. (4.4) and (4.7), by

$$\tau = 10 - \delta. \tag{4.9}$$

Since many tabulations of proton chemical shifts have been presented in τ values, we include a τ scale on many of the spectra in this book, even though it is not an officially recommended scale.

---

* Although the definitions in Eqs. (4.4) and (4.7) are satisfactory for most purposes, a small error is introduced with the assumption that $\sigma \ll 1$. Since $\sigma_S \neq \sigma_R$, these two definitions are not in principle consistent.[39] Recently Eq. (4.4) has been adopted as the formal definition of δ.[40]

4.4 Magnetic Susceptibility Correction

In Section 3.8 we saw that the magnetic field experienced by a molecule depends on the bulk magnetic susceptibility of the medium in which the molecule resides. If molecules of sample and reference compounds are in a single solution (internal reference), the same magnetic susceptibility effect applies to both; hence the difference in magnetic field between the two is independent of the susceptibility of the solution. However, if an interface exists between sample and reference (external reference), we found in Eq. (3.4) that there is a difference in field between sample and reference given by

$$(H_0)_S - (H_0)_R = H_0[\tfrac{2}{3}\pi(\kappa_R - \kappa_S)]. \tag{4.10}$$

(It should be recalled that Eqs. (3.4) and (4.10) apply to the usual cylindrical sample tube oriented with its axis perpendicular to the direction of H_0, and that κ_S and κ_R are negative for ordinary diamagnetic materials.) By substituting the expressions for $(H_0)_S$ and $(H_0)_R$ from Eq. (3.4) in place of H_0 in Eq. (4.3), and by using the recommended definition of δ in Eq. (4.4), we find that

$$\delta = \frac{H_0(1 - \tfrac{2}{3}\pi\kappa_S)(1 - \sigma_S) - H_0(1 - \tfrac{2}{3}\pi\kappa_R)(1 - \sigma_R)}{H_0(1 - \tfrac{2}{3}\pi\kappa_R)(1 - \sigma_R)} \times 10^6. \tag{4.11}$$

Since κ's are of the order of 10^{-6} and σ's are of the order of 10^{-4} to 10^{-6}, terms in the numerator of Eq. (4.11) involving products of κ and σ can be dropped, and in the denominator both κ_R and σ_R are negligible relative to unity. Thus we obtain

$$\delta = [(\sigma_R - \sigma_S) + \tfrac{2}{3}\pi(\kappa_R - \kappa_S)] \times 10^6. \tag{4.12}$$

Hence, the true chemical shift is

$$\delta(\text{true}) \equiv (\sigma_R - \sigma_S) \times 10^6 = \delta(\text{measured}) - \tfrac{2}{3}\pi \times 10^6(\kappa_R - \kappa_S). \tag{4.13}$$

(Obviously, if the alternative definition of $\delta = (\sigma_S - \sigma_R) \times 10^6$ is used, κ_R and κ_S in Eq. (4.13) must be interchanged.)

For the geometry normally used with a superconducting solenoid, Eq. (3.5) must be used in place of Eq. (3.4) in the above treatment, leading to the relation

$$\delta(\text{true}) = (\sigma_R - \sigma_S) \times 10^6 = \delta(\text{measured}) + \tfrac{4}{3}\pi \times 10^6(\kappa_R - \kappa_S). \tag{4.14}$$

For pairs of diamagnetic materials, $\kappa_R - \kappa_S$ is typically 0–0.4×10^{-6}, thus introducing a susceptibility correction of ~ 1 ppm. For nuclei with large chemical shift ranges (e.g., ^{13}C, ^{19}F, ^{31}P) the accuracy of much of the data often has not warranted making susceptibility corrections, but mod-

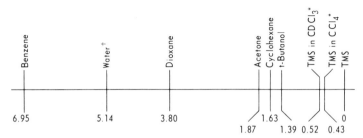

Fig. 4.1 Relative resonance frequencies for various ^{1}H reference compounds. Except where solutions are indicated, all data refer to the undiluted liquid. *For dilute solutions (<1% by volume); † temperature-dependent. These data were obtained experimentally with cylindrical sample tubes in the usual geometric arrangement for iron core magnets.

ern spectrometers provide sufficiently accurate data that susceptibility corrections may be desirable when an external reference is used.

For proton resonance the small range of chemical shifts makes it mandatory to correct for susceptibility if an external reference is used. If the sample solution is sufficiently dilute, conversions to internal TMS reference from various external references may be made readily by means of measured frequency differences for TMS in different solvents. Some useful figures for such conversions are given in Fig. 4.1. For example, from Fig. 4.1 we see that a proton with $\delta \approx 2.13$ ppm downfield from external acetone would have $\delta = 2.37$ ppm with respect to external cyclohexane, and if the proton in question were in a dilute solution in CCl_4, it would have $\delta = 3.57$ ppm relative to *internal* TMS. Further examples of interconversion of scales are given in the problems at the end of the chapter.

Becconsall *et al.*[30] have pointed out that by using both a superconducting magnet and an iron core magnet the magnetic susceptibility correction can be made implicitly. From Eqs. (4.13) and (4.14) we can obtain

$$\delta_{true} = \tfrac{1}{3}[\delta_{meas}^{sc} + 2\delta_{meas}^{ic}]. \qquad (4.15)$$

The superscripts refer to superconducting and iron core magnets. The two measurements must, of course, be made with the same external reference at the same temperature. This method (as well as others that require magnetic susceptibility corrections) is desirable in principle, but the extra effort of making two measurements with different spectrometers renders it impractical for most studies. Where small shifts are significant, as in the study of weak intermolecular interactions (see Chapter 12), the technique is valuable.

Most ^{1}H chemical shifts are measured with respect to TMS as an internal reference, and the bulk of the ^{13}C chemical shifts are now measured this way also. For other nuclei the magnitude of the effect of intermolecular interactions on reference compounds has not been established, and external references are often preferred.

4.5 Empirical Correlations of Chemical Shifts

As we shall see in the next section, the bases for chemical shifts can be accounted for theoretically in principle, but a priori calculations in general cannot at present provide exact values for these quantities. For predictions of chemical shifts for nuclei in particular chemical environments, it is therefore necessary to rely very largely on empirical correlations between observed shifts and chemical structure. Figures 4.2–4.7 provide a general orientation of the orders of magnitude involved and the effects of various functional groups for the chemical shifts of six of the most widely studied nuclei. These figures are meant to be illustrative of the usual values observed. There may be individual compounds in some cases that fall outside the ranges given; likewise, it is often possible to restrict the range more narrowly if specific classes of compounds are considered.

The most striking feature of Figs. 4.2–4.7 is the very small range of chemical shifts for hydrogen nuclei (~ 12 ppm) relative to the much larger range (but still small on an absolute basis) for the heavier nuclei. As we shall see in Section 4.6, this difference is associated with the presence of 2p electrons in the heavier nuclei and their absence in hydrogen. We shall discuss other features of these figures in the following sections.

4.6 Theory of Chemical Shifts

For a single free *atom* in a spherically symmetric S electronic state, Lamb[56] showed that the effect of an imposed magnetic field is to induce an electron current that leads to a shielding factor,

$$\sigma_D = \frac{4\pi e^2}{3mc^2} \int_0^\infty r\rho(r) \, dr. \qquad (4.16)$$

Here $\rho(r)$ is the density of electrons as a function of radial distance from the nucleus, and e, m, and c are the usual fundamental constants.

For molecules the Lamb theory is inadequate because it assumes that the electrons are free to move in any direction, whereas in a molecule electronic motion is severely restricted. Ramsey[57] used second-order perturbation theory to develop a formula that in principle accounts for the

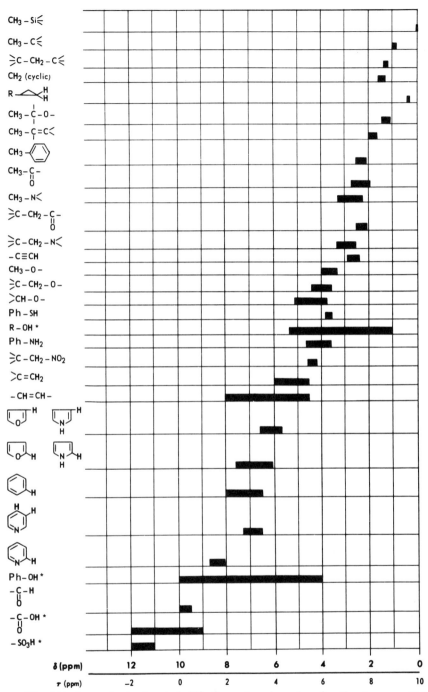

Fig. 4.2 Approximate chemical shifts for protons in various functional groups. Data refer to proton given in boldface type and are taken from various sources.[42-44] * Chemical shift highly dependent on hydrogen bonding (see Chapter 12); also influenced by exchange effects (see Chapter 11). Reference: TMS (internal).

62

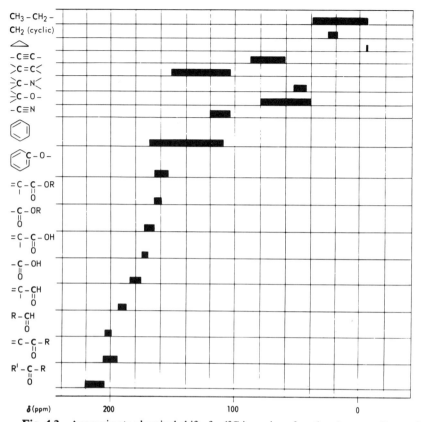

Fig. 4.3 Approximate chemical shifts for ^{13}C in various functional groups. Data refer to ^{13}C given in boldface type and are taken from the books by Stothers[45] and Levy and Nelson.[46] Reference: TMS.

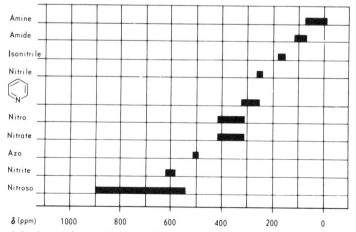

Fig. 4.4 Approximate chemical shifts for ^{14}N and ^{15}N in various functional groups. Data from Randall and Gillies[47] and Webb and Witanowski.[48] Reference: NH_3.

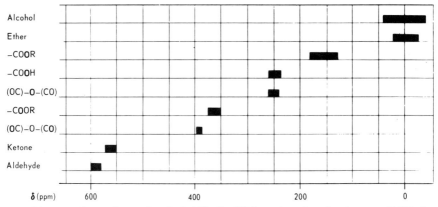

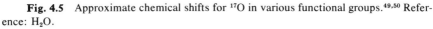

Fig. 4.5 Approximate chemical shifts for ^{17}O in various functional groups.[49,50] Reference: H_2O.

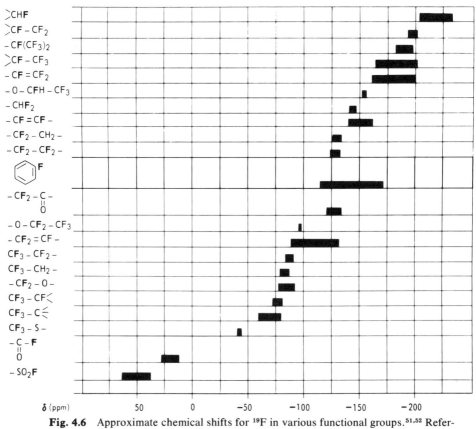

Fig. 4.6 Approximate chemical shifts for ^{19}F in various functional groups.[51,52] Reference: CCl_3F.

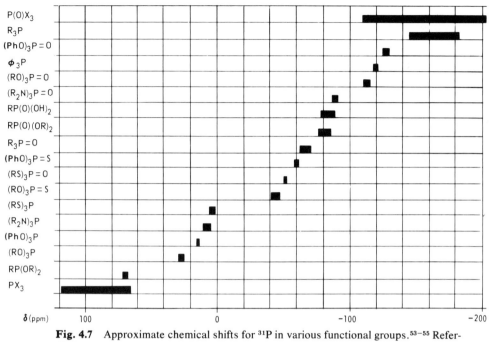

Fig. 4.7 Approximate chemical shifts for ^{31}P in various functional groups.[53-55] Reference: P_4O_6.

shielding factor in molecules. Ramsey's expression for σ, which is not reproduced here, is the sum of two terms.[58] The first, σ_D, is essentially the Lamb expression and is often called the diamagnetic term since $\sigma_D > 0$. It leads to a shielding of the nucleus. The second term, σ_P, is negative and is sometimes referred to as the temperature-independent paramagnetic term. (This distinguishes it from the temperature-dependent paramagnetism that results from unpaired electrons.) This term corrects for the fact that the electrons in a molecule are not disposed with spherical symmetry about the nucleus in question. Thus the presence of p electrons near the nucleus is an important factor in determining σ_P. The mathematical expression for σ_P includes in the denominator energies of excitation, that is, differences in energy between ground and excited electronic states. For most molecules Ramsey's expression cannot be evaluated to any degree of accuracy because insufficient data are available for the energies of all the excited states and the electron distribution. For the hydrogen molecule it has been possible to make a reliable semiempirical evaluation for both σ_D and σ_P. The result is

$$\sigma = \sigma_D + \sigma_P = 32.1 \times 10^{-6} + (-5.5 \times 10^{-6})$$
$$= 26.6 \times 10^{-6}.$$

The Ramsey formulation is probably most useful in furnishing a framework for discussing chemical shifts, and the factors that are important. In general, we might list five terms that could contribute to σ: (a) the diamagnetic contribution from the atom in question, σ_D; (b) the paramagnetic contribution from the atom in question, σ_P; (c) the effect of neighboring atoms; (d) interatomic currents; and (e) the effect of external electric fields.

Contribution (e) arises in practice primarily from neighboring molecules and will be discussed in connection with solvent effects in Chapter 12. For protons, the paramagnetic contribution (b) is likely to be small because a hydrogen atom has large electronic excitation energies and no low-lying p orbitals. Consequently, it is generally considered that term (b) is negligible for protons. For other nuclei, however, the paramagnetic term is often the dominant term. We shall place emphasis first on chemical shifts of protons in the following treatment of effects (a), (c), and (d), with a discussion of heavier nuclei in Section 4.10.

4.7 Effect of Electron Density

Term (a) in the preceding section can be evaluated approximately for protons by the Lamb formula to give

$$\sigma(\text{local diamagnetic}) \approx 20 \times 10^{-6} \, \lambda, \qquad (4.17)$$

where λ is the effective number of electrons in the 1s orbital of the hydrogen. For a completely screened hydrogen atom, λ would approach 1; for a hydrogen ion, it would be 0. Thus the local diamagnetic effect is in the range of a few parts per million, which is just the range of chemical shifts observed for protons.

Equation (4.17) suggests that there ought to be some sort of correlation between the shielding factor and the electron density around the hydrogen. For example, a more acidic proton, such as the OH proton in phenol, should be less shielded than the corresponding less acidic proton in an alcohol. This is indeed found to be the case, the chemical shift for the OH proton of phenol occurring about 4 ppm to lower field than that of ethanol. (Hydrogen bonding can alter these chemical shifts substantially, as we shall see in Chapter 12.)

The presence of a formal charge near a magnetic nucleus leads to a substantial shielding or deshielding. For example, Fig. 4.8 shows the spectra of 1-methylcytosine and its hydrochloride. In the hydrochloride the positive charge is shared primarily between positions 3 and 7, but the

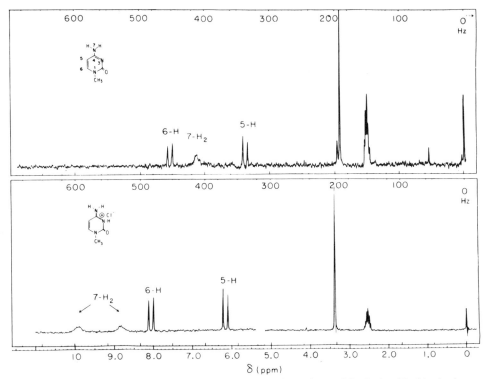

Fig. 4.8 Proton resonance spectra of 1-methylcytosine and its hydrochloride, both obtained at 60 MHz at room temperature in solution in dimethylsulfoxide-d_6. Note downfield shifts of all peaks on formation of the cation. The separate lines for the two NH_2 protons in the cation arise from hindered rotation of the amino group. (See Chapter 11 for further details.)

effect can be seen on all proton resonances, which are shifted downfield. Similar, but larger, effects are found with charged aromatic systems. Comparison of the chemical shifts in the symmetric molecules $C_5H_5^-$, C_6H_6, and $C_7H_7^+$ shows that a charge localized on one of the carbon atoms of benzene would result in a shielding or deshielding of the attached proton by about 9 ppm/electron.[59] This value may be used to calculate from NMR data the approximate change in charge density resulting from the introduction of an electron-donating or electron-withdrawing substituent. In general, electron densities obtained this way have been in rather good agreement with those calculated from molecular orbital treatments.

A strongly electronegative atom or group attached to or near a magnetic nucleus has the expected effect of deshielding the nucleus. Thus rough correlations are found between chemical shift and electronegativity

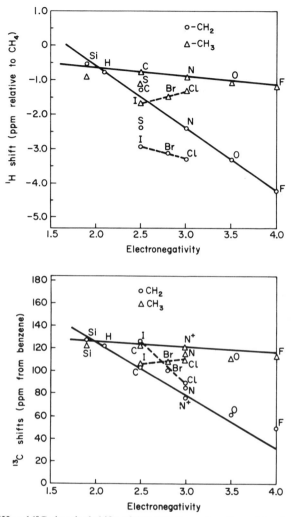

Fig. 4.9 ^{1}H and ^{13}C chemical shifts of CH_3CH_2X as a function of the electronegativity of X (Spiesecke and Schneider[60]).

of substituents. Figure 4.9 shows typical results for CH_3CH_2X. Note the parallelism between the correlations for ^{1}H and ^{13}C chemical shifts, but the vastly different range of δ encompassed.

Substitution on an aromatic ring causes changes in shielding of protons resulting from addition or withdrawal of charge. The generalizations used by chemists in predicting electron density at ortho, meta, and para positions apply in large measure to NMR spectra, as indicated for some typical substituents in Fig. 4.10. To a large extent, substituent effects are

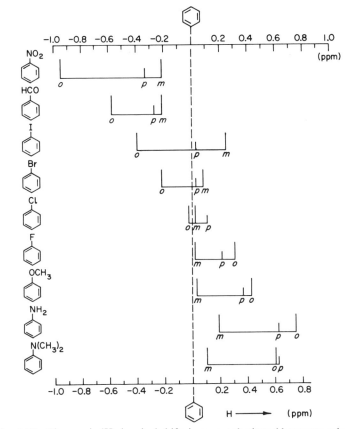

Fig. 4.10 Changes in ^1H chemical shifts in monosubstituted benzenes relative to benzene. Values are given as $(\delta_{benzene} - \delta_{subs})$ (Spiesecke and Schneider[61]).

approximately additive for aromatic systems. Extensive studies have provided tables of substituent contributions.[62] Such empirical methods of predicting chemical shifts are quite useful, but their limitations should be recognized. In aromatic systems, for example, additivity provides surprisingly good results for many meta and para disubstituted benzenes, but gives only fair agreement with experiment for substituted benzenes containing appreciable dipole moments.

4.8 Magnetic Anisotropy and Chemical Shifts

While variation in electron density around a proton is probably the most important factor influencing its chemical shift, many exceptions are

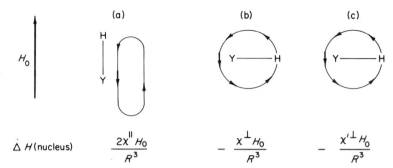

Fig. 4.11 Secondary magnetic field generated at the hydrogen nucleus due to the diamagnetism of the neighboring group Y. Arrows indicate closed lines of flux due to a point magnetic dipole located at the center of electric charge of Y but shown slightly displaced for clarity. (a) H—Y bond parallel to H_0; (b) (c) H—Y bond perpendicular to H_0. Parts (b) and (c) are identical if Y has axial symmetry.

found to a correlation between δ and electron density. To account for these cases, we must consider the induced magnetic fields that have their origins in atoms or functional groups near the atom in question.

Suppose we consider a simple system, H—Y, where Y is an atom or a more complex part of a molecule. If a magnetic field H_0 is imposed on this molecule, the electrons around Y are forced to move in some fashion, and as a result there is a magnetic dipole moment μ_Y generated at Y. The magnitude of μ_Y is

$$\mu_Y = \chi_Y \mathbf{H}_0. \tag{4.18}$$

where χ_Y is the magnetic susceptibility of Y. Since $\chi_Y < 0$ for diamagnetic materials, μ_Y points in a direction opposite $\mathbf{H}_0$. We can obtain better insight into the effect of μ_Y on the local field at the proton if we look separately at the H—Y bond oriented along three mutually perpendicular directions, one of which coincides with that of $\mathbf{H}_0$. The situation is depicted in Fig. 4.11. Since field strength arising from a point magnetic dipole varies inversely as the cube of the distance from the dipole, the increment of magnetic field experienced by the proton H due to the induced moment at Y is the average of the three contributions shown in Fig. 4.11.

$$\Delta H(\text{nucleus}) = \frac{H_0}{3R^3}(2\chi^{\|} - \chi^{\perp} - \chi'^{\perp}), \tag{4.19}$$

where R is the distance between the proton H and the center of electric charge in Y. The notations $^{\|}$ and $^{\perp}$ refer to the direction of the H—Y bond relative to H_0. The factor of 2 in the parallel component arises from the spatial degeneracy present in this orientation; that is, the lines of flux illustrated in the plane of the page are duplicated in a plane perpendicular to the page. The orientations depicted in Fig. 4.11b, c are, in general, different, since one represents a view of the "edge" of group Y and the other the "face" of Y.

Equation (4.19) shows that the magnitude of the field increment at the proton depends on a *magnetic anisotropy* in Y; that is, a lack of equality of the three susceptibility components. (This is sometimes referred to as the *neighbor anisotropy effect*.) The field increment from Eq. (4.19) thus results in a change in shielding

$$\Delta\sigma = -\frac{1}{3R^3}(2\chi^{\parallel} - \chi^{\perp} - \chi'^{\perp}). \tag{4.20}$$

One particularly simple example of this effect occurs in $HC{\equiv}CH$, where the anisotropy arises from the freedom of the electrons in the triple bond to circulate at will around the axis of the molecule. If Y in Fig. 4.11 represents the $C{\equiv}CH$ fragment, then $\chi^{\parallel}$ is large in magnitude because the flow of electrons around the bond generates a moment along the H—Y axis. On the other hand, the electrons are less likely to circulate perpendicular to the H—Y axis because they then would cut through chemical bonds. Consequently $\chi^{\perp}$ and $\chi'^{\perp}$ are small (and equal because of the axial symmetry). Keeping in mind that both $\chi^{\parallel}$ and $\chi^{\perp}$ are negative, we expect from Eq. (4.20) a large *positive* $\Delta\sigma$. Thus the resonance is predicted to be at a higher field than it would be in the absence of this large neighbor anisotropy effect. A comparison of the proton chemical shifts of the series C_2H_6, $CH_2{=}CH_2$, and $HC{\equiv}CH$ shows that the $HC{\equiv}CH$ line is only 0.6 ppm lower in field than the line of C_2H_6, while C_2H_4 lies 4.4 ppm lower in field than C_2H_6. On electronegativity grounds alone C_2H_2 should be lower in field than C_2H_4.

Nothing in our discussion has required H to be chemically bonded to Y. In general, where the proton in question does not lie on one of the principal axes of magnetic susceptibility, the neighbor anisotropy effect may be calculated from the relation[63]

$$\Delta\sigma = \frac{1}{3R^3}[(2 - 3\cos^2\theta)(\chi^{\parallel} - \chi^{\perp}) - (\chi^{\parallel} - \chi'^{\perp})], \tag{4.21}$$

where the proton is at a distance R from the center of the anisotropic group or bond in a direction inclined at an angle θ to the direction of $\chi^{\parallel}$. If $\chi^{\perp} = \chi'^{\perp}$, Eq. (4.21) becomes simpler:

$$\Delta\sigma = \frac{1}{3R^3}(1 - 3\cos^2\theta)(\chi^{\parallel} - \chi^{\perp}). \tag{4.22}$$

Susceptibility differences of various bonds may be estimated empirically by comparisons of data from different molecules or may be calculated with quantum mechanical models. Since the theoretical calculations require severe approximations, most attention has been devoted to the empirical approach. Here, however, two major difficulties occur: first, the isolation of neighbor anisotropy effects from other effects; and second, the approximations of a point magnetic dipole centered at some often arbitrary position in group Y. As a result, rather different values of magnetic anisotropies are sometimes reported for such important groups as C—H, C—C, C=C, and C=O. For the C=O group, for example, there is general agreement that a proton directly above the plane of the   C=O group is shielded and a proton at certain positions in the plane (such as an aldehyde proton) is deshielded; but the magnitudes of the anisotropies are disputed.[64] A generalized version of Eq. (4.21) which is based on a magnetic dipole of finite length, rather than a point dipole, has been derived and applied to a number of molecules of known geometry.[65] This treatment should provide more reliable values of χ's since it is valid at shorter distances than Eq. (4.21) (which should not be used for $R < 3$ A).

A commonly recognized factor in determining chemical shifts is called the "C—C bond effect," which refers to the decrease in proton shielding with addition of C—C bonds, for example,

$$CH_3X, \qquad C—CH_2X, \qquad \begin{matrix} C \\ \diagdown \\ C \diagup \end{matrix} CHX.$$

The origin of this effect is not entirely clear; it may be due to the neighbor

Table 4.2

CHEMICAL SHIFTS[a] OF PROTONS α TO
OXYGEN ATOMS[b]

X	CH_3X	$C—CH_2X$	$\begin{matrix} C \\ \diagdown \\ C \diagup \end{matrix}CHX$
OR	3.29	3.40	
OH	3.38	3.56	3.85
OC_6H_5	3.73	3.90	
OCR ‖ O	3.65	4.10	5.01
$OC—C_6H_5$ ‖ O	3.90	4.23	5.12
$OCCF_3$ ‖ O	3.96	4.34	

[a] In ppm relative to TMS (internal).
[b] Jackman.[66]

anisotropy effect or it may be an intrusion of the paramagnetic effect, to which we referred earlier. Examples of the C—C bond effect may be seen in Fig. 4.2 and Table 4.2.

4.9 Ring Currents

A special, and quite important, type of anisotropy effect occurs when it is possible to have *inter*atomic circulation of electrons within a molecule. A circulation of π electrons around the periphery of an aromatic ring, for example, gives rise to a "ring current" and resultant induced shielding effects.*

If an aromatic ring is oriented perpendicular to the applied field $\mathbf{H_0}$, as in Fig. 4.12, the π electrons are relatively free to circulate around the ring and will move in a direction such that the magnetic moment resulting from their motion opposes the applied field. This moment may be pictured as a point magnetic dipole at the center of the ring, with the dipole field falling off as the cube of the distance. If we construct closed lines of magnetic flux, as in Fig. 4.12, we find that the sign of the ring current effect is highly dependent upon geometry. An aromatic proton in the plane of the ring experiences a ring current field that enhances the applied field; hence its

---

* In principle, one could consider the sum of magnetic anisotropies for the atoms of the ring, but in practice the inclusion of ring currents as a separate contribution to the chemical shift is useful.

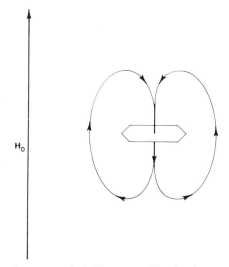

H_0

Fig. 4.12 Secondary magnetic field generated by the ring current in benzene.

resonance occurs at a lower applied field than might otherwise be anticipated. In fact, the chemical shift of benzene is about 1.7 ppm lower in field than that of ethylene, which is very similar from the standpoint of hybridization and electronegativity. A proton held over the aromatic ring, on the other hand, would be expected to experience an upfield shift.

The magnitude of the shielding due to a ring current may be estimated from a point dipole calculation[67] or more accurately from a model that treats the π electrons as rings above and below the atomic plane.[68] (In the calculations, averages are taken over all orientations of the aromatic ring in the magnetic field, not just the one shown in Fig. 4.12.) The ring model may be used to calculate semiquantitatively the ring current effect at various positions relative to a benzene ring. The results of such a calculation have been tabulated[69] and are illustrated in the contour diagram of Fig. 4.13.

The ring model may also be used to interpret data on such molecules as porphyrins, whose large ring currents lead to substantial downfield shifts for protons outside the ring and a very large upfield shift for the NH protons inside the electron ring.

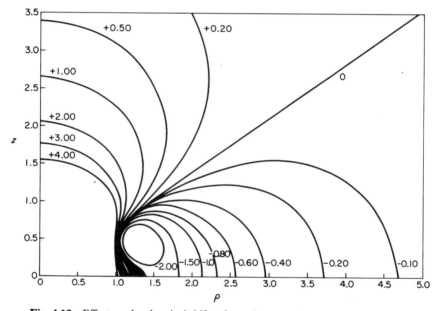

Fig. 4.13 Effect on the chemical shifts of a nucleus at various positions due to the ring current in benzene. The plot represents one quadrant of a plane passing normally through the center of the ring, which lies horizontally. A positive sign denotes an upfield contribution to the chemical shift; ρ and z are in units of the benzene C—C distance, 1.39 Å (Johnson and Bovey[68]).

4.10 Nuclei Other Than Hydrogen

Our discussion of the factors affecting chemical shifts has focused almost entirely on proton resonance, but the various effects that we listed are also applicable to other nuclei with similar magnitudes. However, the paramagnetic effect σ_P is much larger for other nuclei than it is for the proton, as pointed out in Section 4.6. In fact, the paramagnetic effect is the dominant feature for most other nuclei and accounts qualitatively for the much larger ranges of chemical shifts, as illustrated in Figs. 4.2–4.7. For example, the difference in chemical shift between F_2 and F^- of approximately 500 ppm is attributed to σ_P. For a spherically symmetric ion, σ_P can be ignored, but in a molecule such as F_2 with considerable p bond character it is very significant. Similar conclusions can be drawn for ^{13}C, ^{31}P, and many other nuclei.

The very much larger range of chemical shifts for nuclei other than hydrogen suggests that their chemical shifts should reflect often subtle differences in chemical structure, and this is indeed found to be true. For example ^{19}F, which has been studied extensively, provides a sensitive probe of changes in the electron density in many molecules, as illustrated in Fig. 4.6. Since ^{19}F has a range of chemical shifts greater than 250 ppm and a large magnetogyric ratio, the chemical shift range in Hz is larger than that for any other commonly studied nucleus.

The study of ^{31}P provides much useful information on the structures of both organic and inorganic compounds. As shown in Fig. 4.7, trivalent ^{31}P is generally much less shielded than the pentavalent nucleus, but even within the class of common organic phosphates chemical shifts vary over many ppm and correlate with structural parameters.

Several elements have more than one isotope amenable to NMR study. For example, ^1H, ^2H (deuterium, D), and ^3H (tritium) can all be investigated. The radioactivity of ^3H is an obvious disadvantage, and only a handful of studies have been carried out, but there is promise of an upsurge of interest in this nucleus. Deuterium has a sensitivity less than one per cent of that of ^1H, and its natural abundance is very low. Nevertheless, for some studies, especially in biochemical systems, ^2H NMR of isotopically enriched samples is proving of great value. Most studies of deuterium NMR take advantage of the relaxation behavior of this nucleus, which has a small quadrupole moment (see Chapter 8).

Nitrogen NMR can be studied with ^{14}N or ^{15}N. Both have very low inherent sensitivity (about 10^{-3} as great as ^1H). Nitrogen-14 is over 99% abundant, but it has a rather large quadrupole moment, which, as we shall see in Chapter 8, usually leads to rapid relaxation and very broad lines.

Nitrogen-15 has a spin of $\frac{1}{2}$, hence no quadrupole moment, but its natural abundance of less than 0.4% makes its study very difficult. Together with a few other nuclei, ^{15}N has a *negative* magnetogyric ratio; i.e., its magnetic moment points in the direction opposite its angular momentum vector.* A negative magnetogyric ratio has some important consequences in double resonance experiments (see Section 9.4), and as a result it can in some circumstances further increase the difficulty in observing ^{15}N NMR.

4.11 Carbon-13

The low natural abundance (1.1%) and low inherent sensitivity of ^{13}C (only 1.6% as sensitive as ^{1}H) have made ^{13}C NMR difficult. However, advances in techniques, especially Fourier transform and double resonance methods (which we shall cover in detail in later chapters) now permit routine study of ^{13}C at natural abundance.

Figure 4.3 shows that the chemical shift range in ppm for ^{13}C is about 20 times as great as that for ^{1}H, so that ^{13}C NMR clearly offers great promise for the investigation of organic molecules. Although the paramagnetic shielding term is dominant for ^{13}C, the chemical shifts in Fig. 4.3 show a remarkable qualitative similarity in sequence to those of ^{1}H in Fig. 4.2. Alkanes are highly shielded, alkenes and aromatics much less so. Carbonyl carbons are quite deshielded, with chemical shifts about 160–220 ppm from TMS. Polar substituents generally have a large deshielding effect on adjacent carbons. Magnetic anisotropy and ring current effects, as we have seen, seldom exceed 1 ppm, and hence are of much less relative importance in determining ^{13}C chemical shifts than for proton shifts. However, other factors, such as steric and stereochemical effects, have been identified in many instances.

4.12 Tabulations of Chemical Shifts and Spectra

In this chapter we have considered some of the major effects that determine chemical shifts. It is worthwhile reiterating the point made in Section 4.5 that theory alone is insufficient for predicting accurate chemical shifts and that recourse must be had to empirical data. It is beyond the scope of this book to include extensive tabulations of data beyond those

---

* Si and the commonly studied isotopes of Ag, Cd, and Sn are among other nuclei with negative magnetogyric ratios. See Appendix B for a complete list.

few examples in the preceding sections. However, as an aid in finding chemical shift data, we present the following summary of a number of the more extensive compilations of chemical shift data or complete spectra:

1. "Varian High Resolution NMR Spectra Catalog," Vols. 1 and 2. Varian Associates, Palo Alto, California. An excellent collection of 700 proton resonance spectra of many types of compounds, together with indexes by name, functional groups, and chemical shifts.

2. API-TRC compilation of NMR spectra. Thermodynamics Research Center, Texas A and M Univ., College Station, Texas. A continuing compilation of proton resonance spectra of hydrocarbons and other compounds.

3. F. A. Bovey, "NMR Data Tables for Organic Compounds." Wiley (Interscience), New York, 1967. A compilation of ^1H chemical shifts and spin–spin coupling constants for more than 4200 organic compounds.

4. G. Slomp and J. G. Lindberg, Chemical shifts of protons in nitrogen-containing organic compounds, *Anal. Chem.* **39**, 60 (1967). A correlation of chemical shift with functional groups, based on data for 2300 protons.

5. Sadtler Standard NMR Spectra." Sadtler Research Laboratories, Philadelphia, Pennsylvania. A continuing compilation of more than 10,000 ^1H and 4000 ^{13}C spectra.

6. N. F. Chamberlain, "The Practice of NMR Spectroscopy." Plenum Press, New York, 1974. Extensive correlation charts for ^1H NMR, based on 24,000 chemical shifts and 10,000 coupling constants.

7. J. G. Grasselli and W. M. Ritchey (eds.), "Atlas of Spectral Data and Physical Constants for Organic Compounds." CRC Press, Cleveland, Ohio, 1975. Guide to ^1H and ^{13}C NMR spectra among 21,000 organic compounds listed.

8. J. B. Stothers, "^{13}C NMR Spectroscopy." Academic Press, New York, 1972. A virtually complete summary of ^{13}C data prior to 1970.

9. L. F. Johnson and W. C. Jankowski, "^{13}C NMR Spectra: A Collection of Assigned, Coded and Indexed Spectra." Wiley (Interscience), New York, 1972. Uses same indexing system as compilation No. 1.

10. E. Breitmeier, G. Haas, and W. Voelter, "Atlas of Carbon-13 NMR Data." Heyden, London. A continuing series, currently with data for 2000 compounds.

11. R. F. Zürcher, *Helv. Chim. Acta* **46**, 2054 (1963). An extensive tabulation of chemical shifts of the angular methyl protons in steroids.

12. H. Suhr, "Anwendungen der Kernmagnetischen Resonanz in der Organischen Chemie." Springer, New York, 1965. Many useful tabulations of chemical shifts.

13. W. Brügel, "NMR Spectra and Chemical Structure." Academic Press, New York, 1967. A compilation of proton chemical shifts and spin–spin coupling constants for 88 general classes of organic compounds; also some data on ^{19}F, ^{31}P, and ^{11}B.

14. G. A. Webb and M. Witanowski (eds.), "Nitrogen NMR." Plenum Press, New York, 1973. Six chapters covering ^{14}N and ^{15}N chemical shifts, coupling constants and relaxation effects.

15. L. E. Mohrmann and B. L. Shapiro, *J. Chem. Phys. Res. Data,* **6**, 919 (1977). A critically evaluated compilation of ^1H chemical shifts in substituted benzenes.

16. J. W. Emsley and L. Phillips, Fluorine chemical shifts, *Prog. NMR Spectrosc.* **7**, 1 (1971). A comprehensive tabulation (526 pages).

17. ^{13}C NMR Spectral Search System. NIH–EPA Chemical Information System. An on-line computer retrieval system, containing 5000 ^{13}C spectra.

18. Preston NMR Abstracts. Preston Publ., Niles, Illinois. A continuing series of abstracts of NMR literature with a computer retrieval system.

19. W. Bremser, L. Ernst, and B. Franke, "Carbon–13 NMR Spectral Data." Verlag Chemie, New York, 1978. A microfiche collection of 10,000 ^{13}C spectra.

20. W. Brügel, "Handbook of NMR Spectral Parameters." Heyden, London, 1979. A three-volume compendium of proton chemical shifts and coupling constants for more than 7500 compounds.

21. M. Shamma and D. M. Hindenlang, "Carbon–13 NMR Shift Assignments in Alkaloids and Amines." Plenum Press, New York, 1979.

4.13 Empirical Estimation of Chemical Shifts

When the spectra of suitable model compounds are not available, an approximate calculation of the expected chemical shift of a proton in a given environment may be helpful. To the extent that substituent effects are additive, a table of constants may be prepared for each substituent on the carbon bearing the proton in question. Most attempts at such tabulations[70-72] for ^{1}H have met with only limited success, since many complexities of chemical shifts and the presence of through-space interactions are not accounted for by such a simple treatment.

For ^{13}C chemical shifts empirical estimations have been extremely successful. Because of the much larger range of chemical shifts for this nucleus, an estimate to 1 ppm or so can often be quite adequate for assignments of lines or predictions of unknown spectra, whereas a useful estimate for proton shifts would often need to be within 0.1 ppm. Hence, magnetic anisotropy effects and other factors of the order of a few tenths of a ppm become relatively unimportant for ^{13}C. A theoretical rationale for ^{13}C shifts has been developed, which shows that electronic effects that attenuate rapidly through chemical bounds are, in general, responsible for alterations of ^{13}C shifts.

With the large amount of reliable data on ^{13}C chemical shifts that has become available, it has been possible to derive empirical equations with additive substituent constants. Several slightly different approaches have been used.[73-75] For example, the equation of Lindemann and Adams[74] for paraffins is based on 22 parameters that reflect the effects of carbon atoms as far as five bonds away from the carbon whose chemical shift is to be predicted. The effect of polar substituents (e.g., OH, Cl, COOR) can also be predicted; as anticipated, the predominant effect is on the carbon to which the substituent is attached, with a very rapid attenuation at adjacent carbons.

For paraffins the ^{13}C chemical shift of the kth carbon can be represented by

$$\delta(C_k) = B_s + N_3 C_s + N_4 D_s + M_2 A_{s2} + M_3 A_{s3} + M_4 A_{s4}, \quad (4.23)$$

where s is the number of carbon atoms bonded to the kth carbon; N_3 and N_4 are the numbers of carbon atoms 3 bonds and 4 bonds away from the

Table 4.3

^{13}C CHEMICAL SHIFT PARAMETERS FOR PARAFFINS

B_1	6.80	B_2	15.34	B_3	23.46	B_4	27.77
C_1	−2.99	C_2	−2.69	C_3	−2.07	C_4	0.68
D_1	0.49	D_2	0.25	D_3	0	D_4	0
A_{12}	9.56	A_{22}	9.75	A_{32}	6.60	A_{42}	2.26
A_{13}	17.83	A_{23}	16.70	A_{33}	11.14	A_{43}	3.96
A_{14}	25.48	A_{24}	21.43	A_{34}	14.70	A_{44}	7.35

kth carbon, respectively; M_2, M_3, and M_4 are the numbers of carbon atoms bonded to the kth carbon and having 2, 3, and 4 attached carbons, respectively. The other symbols represent empirical constants, the values of which are given in Table 4.3.

The effect of substituting a polar group for a CH_3 in an alkane can be estimated from the substituent constants in Table 4.4. Examples of the use of these empirical relations are given in Problems 9, 11, and 12, along with the solution to Spectrum 3 in Appendix D.

These empirical relations are quite useful in making assignments of observed spectral lines to individual carbon atoms in acyclic molecules.

Table 4.4

^{13}C CHEMICAL SHIFT PARAMETERS FOR
POLAR SUBSTITUENTS[a]

Substituent	Chemical shift change (ppm)[b]		
	C-1	C-2	C-3
OR[c]	45	−3	−1
OH	40	1	−1
OCOR	43	−2	−1
NH_2	20	2	−1
Cl	23	2	−1
F	61	−1	−2
COX[d]	15	−5	0
COOR	10	−1	−1
COOH	12	−3	−1
CN	−2	−1	−1

[a] Levy and Nelson.[46]

[b] Chemical shift change on replacing CH_3 by the substituent on C-1.

[c] R = alkyl

[d] X = Cl or NR_2

For alicyclic and aromatic molecules other empirical relations have been developed. While generally providing somewhat less quantitative agreement with experiment, these relations are nevertheless very useful in the interpretation of ^{13}C spectra. Stothers[45] and Levy and Nelson[46] provide many examples.

4.14 Isotope Effects on Chemical Shifts

Small but significant changes in chemical shifts are often found on isotopic substitution. The most commonly investigated situation is one in which the isotopic substitution is made in an atom near (1 to 3 chemical bonds removed) the atom being studied. For example, $\delta(^{13}C)$ decreases by about 0.3 ppm on going from $^{13}CHCl_3$ to $^{13}CDCl_3$ (a "one-bond" effect),[46] while $\delta(^{19}F)$ decreases by 0.47 ppm in going from $CF_2{=}CH_2$ to $CF_2{=}CD_2$.[76] Isotope effects are smaller in ^1H resonance, in keeping with the smaller total range of chemical shifts, but even here significant effects can be seen. For example, $\delta(^1H)$ decreases by 0.019 between CH_4 and CH_3D (a two-bond effect), while the three-bond effect in going from $CHF{=}CHF$ to $CHF{=}CDF$ is only 0.005 ppm.[76]

The cause of isotope effects on chemical shifts is not entirely established. Changes in vibrational amplitude are believed to play a major role, but the precise manner in which the effects arise is not yet settled. The possibility of a "zero-bond" isotope effect, that is, a change in chemical shift for the nucleus being studied, is of particular interest, but is difficult to investigate. One study of ^{14}N and ^{15}N chemical shifts showed that there is no measurable isotope effect.[77]

4.15 Paramagnetic Species

Metallo-organic compounds in which the metal is diamagnetic display chemical shifts for proton resonance that cover a range only slightly larger than that found for other organic molecules. If the metal is paramagnetic, however, chemical shifts for protons often cover a range of 200 ppm, and for other nuclei the range can be much greater. These large chemical shifts arise from either a *contact interaction* or a *pseudocontact interaction*. The former involves the transfer of some unpaired electron density from the metal to the ligand. This unpaired spin density can cause positive or negative chemical shifts, depending on the electron distribution and electron spin correlation effects.

The pseudocontact interaction (perhaps more appropriately called a dipolar interaction) arises from the magnetic dipolar fields experienced by a nucleus near a paramagnetic ion. The effect is entirely analogous to the magnetic anisotropy discussed in Section 4.8. It arises only when the g tensor of the electron is anisotropic; *i.e.*, for an axially symmetric case, $g_\parallel \neq g_\perp$. The g value for an electron is defined as

$$g = \nu_0/\beta_0 H_0, \tag{4.24}$$

where β_0 is the Bohr magneton and ν_0 and H_0 the electron resonance frequency and magnetic field, respectively. This anisotropy in g leads to anisotropic magnetic susceptibility, $\chi_\parallel \neq \chi_\perp$, and by Eq. (4.22) a nucleus experiences a shift inversely proportional to R^3.

Both contact and dipolar shifts from unpaired electrons are temperature dependent, normally varying approximately as $1/T$. The presence of unpaired electrons usually causes rapid nuclear relaxation and leads to line broadening (see Sections 2.6 and 8.5). High resolution NMR in paramagnetic complexes can be observed only in cases where the relaxation time is favorable. The theory and applications of such studies have been covered in considerable detail.[78]

4.16 Lanthanide Shift Reagents

The large chemical shifts caused by paramagnetic species have been exploited in *shift reagents,* which contain a paramagnetic ion attached to a ligand that can in turn complex with the molecule being studied.[79] The object is to induce large alterations in the chemical shifts of the latter molecule, while minimizing paramagnetic line broadening. The most successful ions in this regard are certain lanthanides, which have such a short relaxation time for the unpaired electron that little line broadening occurs (see Sections 8.7 and 8.8). The mechanism of action of the lanthanides is principally by the pseudo-contact mechanism, which falls off in a predictable manner with distance $(1/R^3)$.

The most commonly used shift reagents employ Eu^{3+}, Pr^{3+}, or Yb^{3+} as the paramagnetic ion in a chelate of the form

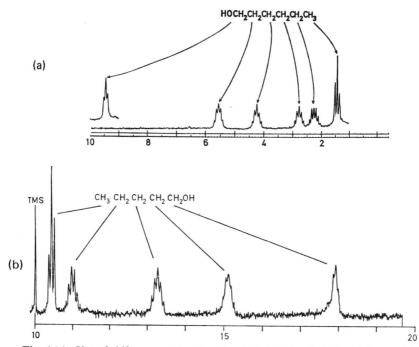

Fig. 4.14 Use of shift reagents to alter chemical shifts in alcohols. (a) Proton NMR spectrum (100 MHz) of *n*-hexanol in CCl_4 in the presence of Eu(dpm)$_3$. (b) Proton NMR spectrum (100 MHz) of *n*-pentanol in CCl_4 in the presence of Pr(dpm)$_3$ (LaMar *et al.*[78]).

with R = $C(CH_3)_3$ (dpm) or R = $CF_3CF_2CF_2$ (fod). The important properties of the ligand are adequate solubility in organic solvents and significant complexing ability with nucleophilic functional groups. (Ideally the complex should be 1 : 1, but often two substrate molecules complex with one molecule of the shift reagent.) In some cases other factors may be overriding in the choice of a reagent; e.g., ligand chirality when it is desired to form a complex with only one of a pair of optical isomers.

As indicated in Eq. (4.22), the direction of shift depends on the anisotropy in the susceptibility and on the angle between the principal axis of susceptibility and the vector **R** to the nucleus. For Eu^{3+} the induced shifts are normally downfield (to higher δ) and for Pr^{3+} they are upfield. On occasion some nuclei may lie at an angle $\theta > 55°$, so that the factor $(3 \cos^2 \theta - 1)$ changes sign, and shifts occur in the opposite directions.

Figure 4.14 shows typical changes in chemical shifts that are found with shift reagents. As anticipated, the magnitude of the shift is largest for nuclei (in this case protons) that are closest to the site of binding of the reagent (here, the oxygen). Since coupling constants are generally unaffected by shift reagents, spectra that are complex often become simplified and are amenable to first-order analysis (see Section 5.3).

Spectral simplification as an aid to analysis is probably the most common purpose for using shift reagents, but with appropriate quantitative consideration of the relative shifts of different nuclei it is often possible to obtain valuable information on molecular conformation. Both distance and angular factors must be taken into account. Several computer programs, based on Eq. (4.22), have been written to evaluate the consistency of various conformations with observed lanthanide induced shifts (LIS). Quantitative use of lanthanide shift reagents depends on the presence of only pseudo-contact interactions. For ^1H NMR, contact interactions with most lanthanides have generally been found to be negligible but for carbons near the site of complexation, appreciable contact shifts have been found in some cases so care must be used in interpreting LIS results.

Problems

1. The methylene protons of ethanol in CCl_4 have a chemical shift δ measured as 215 Hz from TMS (internal reference) at 60 MHz rf field. Express δ in parts per million and give τ (ppm) for these protons. How many hertz from TMS would you expect the chemical shift of these protons to be at 100 MHz? At $H_0 = 10,000$ G?

2. The difference in chemical shift between the α and β protons of naphthalene in dioxane solution has been reported as 14.34 Hz at 40 MHz. Express the difference in parts per million. Do you expect the α or β protons to resonate at lower field? Why?

3. The chemical shift of dioxane in CCl_4 is $\tau = 6.43$ ppm. What would be the chemical shift of dioxane in CCl_4 when measured with respect to external dioxane as a reference in an iron core magnet?

4. The methyl resonance lines of three chemically related substances (A, B, C), each dissolved in CCl_4, are reported as follows: A, $\tau = 9.06$ ppm; B, $\delta = 61$ Hz at 60 MHz, with respect to internal TMS; C, $\delta = -0.20$ ppm, with respect to external cyclohexane. Which of the three actually resonates at highest field? (Assume normal geometry of an iron core magnet.)

5. Benzene in CCl_4 (very dilute solution) has its proton resonance at $\tau =$

2.734 ppm. Express this chemical shift relative to external benzene as measured with the usual geometry of an iron core magnet. What portion of the difference is attributable to magnetic susceptibility effects? (Use susceptibility data from the "Handbook of Chemistry and Physics".)

6. The proton chemical shift of $CHCl_3$, measured with respect to external benzene, is 49.5 Hz at 60 MHz ($CHCl_3$ at lower field). The ^{13}C chemical shift of $CHCl_3$ is 52 ppm upfield from external benzene. What percentages of these reported chemical shifts are due to magnetic susceptibility effects?

7. From Fig. 4.13, what is the effect of the ring current on a proton at a distance 3.1 Å above the plane of a benzene ring and with a projected distance in the plane of 1.8 Å from the center of the ring?

8. With the aid of the proton chemical shift correlation chart, Fig. 4.2, deduce the structures of the molecules giving Spectra 1, 2, and 5, Appendix C.

9. With the aid of the ^{13}C chemical shift correlation chart, Fig. 4.3, and the empirical relations of Section 4.13, deduce the structures of the molecules giving Spectra 3–4, Appendix C.

10. Use the ^1H and ^{13}C data in Spectra 6a,b, Appendix C, to deduce the molecular structure.

11. Predict the ^{13}C spectra of (a) $(CH_3)_2CHCH_2CH_3$ and (b) $(CH_3)_3CCH_2CH_3$.

12. Predict the ^{13}C spectrum of $(CH_3)_3CCH_2CN$.

Chapter 5

Electron-Coupled Spin–Spin Interactions

5.1 Origin of Spin–Spin Coupling

From the discussion in Chapters 2 and 4, one might anticipate that an NMR spectrum would be made up of a number of single lines of different areas and widths, each arising from one or more chemically discrete nuclei. Actually, most NMR spectra consist not only of individual lines, but also of groups of lines termed *multiplets*. The multiplet structure arises from interactions between nuclei which cause splitting of energy levels and hence several transitions in place of the single transition expected otherwise. This type of interaction is commonly called *spin–spin coupling*.

There is another kind of spin–spin coupling that we described in Section 2.6: the magnetic dipole–dipole interaction between two different nuclear moments. We found that the magnitude of this dipole–dipole interaction is proportional to $1/R^3$, where R is the distance between the nuclei, but that it depends also on the angle between $\mathbf{R}$ and $\mathbf{H_0}$. When the nuclei are in molecules that are in rapid, random motion, as are most small molecules in solution, this interaction averages almost completely to zero. The coupling interaction in which we are now interested is normally manifested in solution; hence it must arise from a mechanism that is independent of the rotation of the molecule.

Ramsey and Purcell[80] suggested a mechanism for the coupling interaction that involves the electrons that form chemical bonds. Consider, for example, two nuclei, A and B, each with $I = \frac{1}{2}$. Suppose nucleus A has its spin oriented parallel to $\mathbf{H_0}$. An electron near nucleus A will tend to orient its spin antiparallel to that of A because of the tendency of magnetic mo-

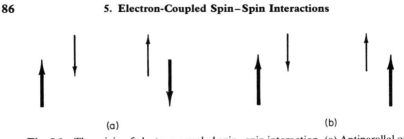

(a)   (b)

Fig. 5.1   The origin of electron-coupled spin–spin interaction. (a) Antiparallel orientation of nuclear spins; (b) parallel orientation of nuclear spins. Nuclear spins are denoted by boldface arrows, electron spins by lightface arrows.

ments to pair in antiparallel fashion. If this electron is in an orbital with another electron, then by the Pauli exclusion principle the spin of the second electron must be antiparallel to the first or parallel to that of nucleus A. Now if the second electron is near nucleus B, it will tend to orient the spin of B. Thus, information about the spin orientation of A is transmitted to B via the bonding electrons. In this situation, which is illustrated in Fig. 5.1a, the most favorable (i.e., lowest energy) situation occurs when nuclei A and B are antiparallel to each other. This does not mean that the opposite situation, where the spins of A and B are parallel, does not occur. These magnetic interactions are small, so that the parallel orientation of spins A and B (see Fig. 5.1b) is a state of only slightly higher energy and occurs with almost equal probability (i.e., in about half the molecules). Thus when nucleus A undergoes resonance and "flips" its spin orientation with respect to H_0, the energy of its transition depends on the initial orientation of B relative to A, and two spectral lines result, the difference in their frequency being proportional to the energy of interaction (coupling) between A and B.

The foregoing explanation of the origin of spin–spin coupling does not depend on the molecule being in an external field. Unlike the chemical shift, which is induced by and hence proportional to the applied field, spin–spin coupling is characteristic of the molecule. The magnitude of the interaction between nuclei A and B is given by the spin–spin coupling constant J_{AB}, which is always expressed in hertz (a unit of convenient magnitude, which is directly proportional to energy).

Spin coupling can occur where two nuclei are bonded together, such as $^{13}C—H$ or $^{31}P—H$, or where several bonds intervene, such as $H_A—^{12}C—^{12}C—H_B$. In general, then, the spin coupling information is carried by electrons through chemical bonds, not through space. (Possible exceptions, especially with nuclei other than hydrogen, have been suggested.)

From this electron spin polarization mechanism the magnitude of the coupling is *generally* expected to decrease as the number of intervening bonds increases. Magnitudes of couplings will be discussed in later sections.

For nuclei with $I > \frac{1}{2}$, more lines result from spin interactions, since there are $2I + 1$ possible orientations of a nuclear spin I relative to the applied field. Thus for the molecule HD (deuterium has $I = 1$), the proton resonance consists of three lines, while the deuterium resonance consists of two lines. It will be recalled from Section 2.1, however, that nuclei with $I > \frac{1}{2}$ possess a nuclear electric quadrupole moment. In an asymmetric electrical environment such nuclei usually relax rapidly and, as shown in Chapters 8 and 11, rapid relaxation can "decouple" two spin-coupled nuclei and lead to loss of multiplet structure. The halogens Cl, Br, and I almost always relax rapidly, as do most of the heavier nuclei with $I > \frac{1}{2}$. Nitrogen-14 and ^2H sometimes relax fast enough to be partially or completely decoupled.

5.2 Coupling between Groups of Equivalent Nuclei

Let us consider the case in which there is coupling involving a set of equivalent nuclei—for example, CH_3CHO, where coupling occurs between the CH_3 protons and the aldehyde proton. As a result, any one of the CH_3 protons "senses" the *two* possible orientations of the aldehyde proton spin, and the CH_3 line is split into a doublet, as shown in Fig. 5.2. The aldehyde proton, however, experiences *four* possible orientations of the CH_3 proton spins—all three protons oriented parallel to the field ("up" in Fig. 5.2), two protons up and one down, two down and one up, or all three down. These two intermediate situations can occur in three ways, as depicted in Fig. 5.2, so that relative probability of the four states is $1:3:3:1$. The aldehyde proton resonance, then, consists of four lines corresponding to these four different states and two of them will be three times as intense because these states are three times as probable.

One might well inquire at this point about the failure to observe additional splitting of lines due to spin coupling between pairs of CH_3 protons. While such *coupling* between equivalent protons does indeed exist, it does not lead to any observable *splitting* of lines. We shall say more about this point in the next section and in Chapter 7.

Our discussion thus far of the origin of spin–spin coupling has been only a qualitative exposition of the general mechanism for this important interaction. The analysis of spin–spin multiplets of the sort given in Fig. 5.2 is applicable only under certain conditions. This type of treatment, which is termed *first-order analysis,* is so widely used (and fre-

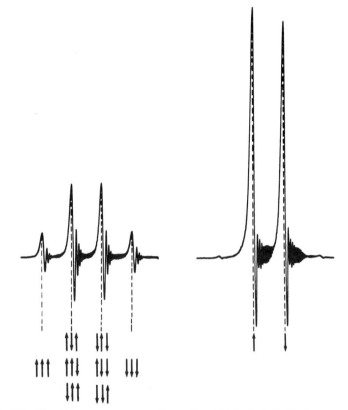

Fig. 5.2 The proton resonance spectrum of acetaldehyde: left, CHO resonance; right, CH₃ resonance. Under the CH₃ resonance is a schematic representation of the two spin orientations of the CHO proton, while the four orientations of the CH₃ protons are indicated beneath the CHO resonance.

quently misused) that we shall devote the next section to a detailed explanation.

5.3 First-Order Analysis

Spectra arising from coupled nuclei—both individual nuclei and groups of equivalent nuclei—may be treated by first-order analysis *only* when two conditions are satisfied:

1. *The chemical shift difference between nuclei (or groups of nuclei) must be much larger than the spin coupling between them.* For this com-

parison the chemical shift difference, $\nu_A - \nu_B$, and the spin coupling constant J_{AB} must clearly be expressed in the same units. The unit of frequency hertz (Hz; or cycles per second, cps), is universally employed, Since chemical shifts expressed in frequency units increase linearly with applied field (and radio frequency), spectra obtained at higher field strength are more likely to adhere to this relation than those obtained at lower field strength.

2. *Coupling must involve groups of nuclei that are magnetically equivalent, not just chemically equivalent.* Nuclei are said to be *chemically* equivalent when they have the same chemical shift, usually as a result of molecular symmetry (e.g., the 2 and 6 protons, or the 3 and 5 protons in phenol) but occasionally as a result of an accidental coincidence of shielding effects. Nuclei in a set are *magnetically* equivalent when they all possess the same chemical shift *and* all nuclei in the set are coupled equally with *any* other single nucleus in the molecule. Thus, in the tetrahedral molecule difluoromethane (I) H_a and H_b *are* magnetically equivalent since they are by symmetry equally coupled to F_a and they are equally coupled to F_b. On the other hand, in 1,1-difluoroethylene (II) H_a

(I)   (II)

and H_b *are not* magnetically equivalent, since H_a and F_a are coupled by $J(\text{cis})$, while H_b and F_a are coupled by $J(\text{trans})$, and in general $J(\text{cis}) \neq J(\text{trans})$.*

When there is *rapid* internal motion in a molecule, such as internal rotation or inversion, the equivalence of nuclei should be determined on an overall average basis, rather than in one of the individual conformations. For example, in CH_3CH_2Br, the three CH_3 protons are magnetically equivalent because they couple equally on the average with each of the methylene protons, even though in any one of the three stable conformations they would be magnetically nonequivalent. Further discussion of rate phenomena and conformational isomers will be presented in Chapters 7 and 11.

---

* The term "equivalent nuclei" has been widely used, in some cases to denote chemical equivalence and in others magnetic equivalence. Other terms have been suggested to attempt to avoid confusion: for example, "isochronous" for nuclei with the same chemical shift. We shall always specify "chemically equivalent" or "magnetically equivalent" when there is any ambiguity.

When first-order analysis is applicable, the number of components in a multiplet, their spacing, and their relative intensities can be determined easily from the following rules:

1. A nucleus or group of nuclei coupled to a set of n nuclei with spin I will have its resonance split into $2nI + 1$ lines. For the common case of $I = \frac{1}{2}$ there are then $n + 1$ lines.

2. The relative intensities of the $2nI + 1$ lines can be determined from the number of ways each spin state may be formed. For the case of $I = \frac{1}{2}$ the intensities of the $n + 1$ lines correspond to the coefficients of the binomial theorem, as indicated in Table 5.1.

3. The $2nI + 1$ lines are equally spaced, with the frequency separation between adjacent lines being equal to J, the coupling constant.

4. Coupling between nuclei within a magnetically equivalent set does not affect the spectrum.

Two examples of first-order spectra are given in Figs. 5.3 and 5.4. The necessity for magnetic equivalence is graphically demonstrated in Fig. 5.3, while the requirement that $(\nu_A - \nu_B) \gg J_{AB}$ is portrayed in Fig. 5.4. First-order analysis is usually considered applicable (for sets of magnetically equivalent nuclei, of course) when $(\nu_A - \nu_B)/J_{AB} > 7$; however,

Table 5.1

RELATIVE INTENSITIES OF FIRST-ORDER MULTIPLETS
FROM COUPLING WITH n NUCLEI OF SPIN $\frac{1}{2}$

n	Relative intensity								
0					1				
1					1	1			
2				1	2	1			
3				1	3	3	1		
4			1	4	6	4	1		
5		1	5	10	10	5	1		
6		1	6	15	20	15	6	1	
7	1	7	21	35	35	21	7	1	
8	1	8	28	56	70	56	28	8	1

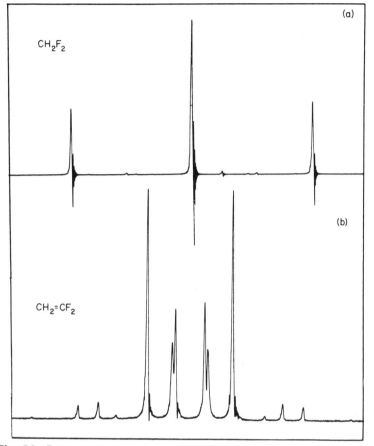

Fig. 5.3 Proton resonance spectra of (a) CH_2F_2 and (b) $CH_2\!\!=\!\!CF_2$ at 60 MHz. First-order analysis is applicable in (a) but not in (b). (The very weak lines in (a) and (b) are due to spinning sidebands and to an impurity.)

when $7 < (\nu_A - \nu_B)/J_{AB} < 20$, there is some distortion of intensities from the pattern given in Table 5.1, but the multiplet is still recognizable. The deviation in intensities always occurs in the direction of making the lines near the center of the overall spectrum more intense and those toward the edges less intense.

When two or more couplings are present that may be treated by the first-order rules, a repetitive procedure can be used. For example, Fig. 5.5 gives an illustration of the repetitive application of first-order analysis. Usually it is convenient to consider the largest coupling first, but it is immaterial to the final result. Since this procedure is widely used,

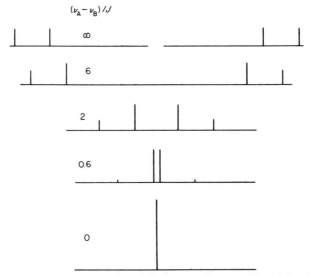

Fig. 5.4 Schematic representation of the spectra of two nuclei with $I = \frac{1}{2}$ as a function of the ratio of chemical shift to spin coupling constant.

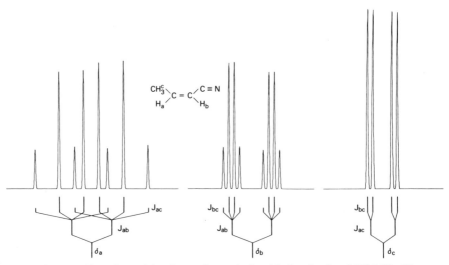

Fig. 5.5 Use of repetitive first-order analysis with the simulated 220 MHz ^{1}H spectrum of cis-CH$_3$CH=CHCN. The olefinic coupling constant is 11 Hz, while the 3-bond and 4-bond couplings to the methyl protons are 6.7 and 1.5 Hz, respectively. Note that the relative intensities of the first-order analysis are carried through each step. (The slight intensity deviations in the simulated spectrum from strict first-order predictions are due to the finite chemical shift differences.)

Fig. 5.5 should be studied carefully, along with further examples in the problems at the end of this chapter.

5.4 Signs of Coupling Constants

In our discussion of the simple mechanism of electron-coupled spin–spin interactions in Section 5.1, we showed that the state in which two coupled nuclei have antiparallel spin orientations has a lower energy than the one in which the spins are parallel. Chemical bonding and the interactions of nuclear spins are not always so simple, however, and in some cases the lower energy state is the one in which the spins of the coupled nuclei are parallel. We distinguish between these two situations by referring to the first system as possessing a *positive* coupling constant ($J > 0$) and to the second as having a *negative* coupling constant ($J < 0$).

When the first-order conditions are obeyed, the signs of the coupling constants can never be obtained directly from the spectrum. In more complex cases, however, the *relative* signs of the various coupling constants within the molecule sometimes influence the appearance of the spectrum and hence can be determined from the observed spectrum. We shall discuss this topic in more detail in Chapter 7. Double resonance techniques (Chapter 9) can also be used very effectively to obtain relative signs of coupling constants. The *absolute* signs of J's cannot be found from ordinary high resolution NMR spectra, but there are overwhelming reasons (based on both theory and more sophisticated NMR experiments) for believing that all one-bond $^{13}C–H$ coupling constants are positive. Absolute values of other coupling constants are usually based on the $^{13}C–H$ coupling being >0.

The signs of coupling constants are of considerable theoretical importance with regard to chemical bonding and can sometimes be of practical significance in spectral analysis and in structure elucidation. Further discussion of this point is given in Sections 5.7 and 9.8.

5.5 Theory of Spin–Spin Coupling

The general theory of spin–spin coupling is complex, and we shall not treat it in detail. For proton–proton coupling it has been shown that the spin interaction arises principally from the electron spin–electron spin interaction, not from orbital interaction of electrons. This simplifies the theory somewhat. For some other nuclei, orbital interaction may come into play also. We shall summarize a few of the conclusions applicable to spin–spin coupling without going through the details of the theory.

It is found that the interaction depends on the density of electrons at the nucleus. It is well known that only s electrons have density at the nucleus, so we expect a relation between the magnitude of the coupling and the s character of the bond. Such a relation is found, as we shall see in Section 5.7.

The coupling interaction is proportional to the product of magnetogyric ratios of the coupled nuclei. For comparison of the magnitudes of coupling constants between different nuclei, and to compensate for the negative sign introduced in some cases by negative magnetogyric ratios, a *reduced coupling constant* K_{AB} can be defined:

$$K_{AB} = \frac{2\pi}{\gamma_A \gamma_B} J_{AB}. \tag{5.1}$$

The dependence of J on γ means that the coupling constant itself (but not the reduced coupling constant) changes with isotopic substitution even though the electron distribution in the molecule is unchanged. For example, on deuterium substitution,

$$J_{HX}/J_{DX} = \gamma_H/\gamma_D \approx 6.51. \tag{5.2}$$

The theory of spin coupling is in accord with an interaction of the form $\mathbf{I}_1 \cdot \mathbf{I}_2$, which depends on the orientation of one spin with respect to the other but does *not* depend on the orientation of the spins with respect to the magnetic field.

The theoretical development of the electron-coupled spin–spin interaction has been carried out principally by second-order perturbation theory. The resulting expression contains in the denominator terms involving the excitation energy for an electron going from the ground electronic state to excited triplet electronic states. Exact calculation of coupling constants is virtually impossible with the present limited knowledge of electronic excitation energies and wave functions, but both valence bond and molecular orbital treatments have been applied successfully to small molecules or molecular fragments in predicting the general magnitude of coupling constants and their dependence on various parameters. We shall return to a consideration of some of these predictions and correlations in Section 5.7.

5.6 Some Observed Coupling Constants

Since theory is unable to predict accurate values for coupling constants, our knowledge of the range of coupling constants found for dif-

Table 5.2

TYPICAL PROTON–PROTON SPIN
COUPLING CONSTANTS[a]

Type	J_{HH} (Hz)
H—H	280
(C)(C)C(H)(H) geminal	−12 to −15
H—C—C—H (free rotation)	7
H—C—C—C—H	~0[b]
cyclohexane ax/eq H	ax–ax 8–10 ax–eq 2–3 eq–eq 2–3
cyclopentane H,H (*cis* or *trans*)	4–5
cyclobutane H,H (*cis* or *trans*)	8
cyclopropane H,H (*cis*) (*trans*)	8–10 4–6
H—C—O—H	5
H—C—C(=O)—H	+3
H—C(=O)—C(=O)—H	8

Table 5.2— Continued

Type	J_{HH} (Hz)
	12 to 19
	−3 to +2
	7 to 11
	1–2
	7
	−1.5
	−2
	10
	±1
	7–17
	42
H—C—C≡C—H	−2

Table 5.2— Continued

Type		J_{HH} (Hz)
H—C—C≡C—C—H		2
(ring) C=C with H's	5 mem.	6
	6 mem.	10
	7 mem.	12
X=C with C—H, C—H (X = C, O)		0 to ±2
benzene H, H	$J(ortho)$	8
	$J(meta)$	2
	$J(para)$	~0.5
pyridine	$J_{(2-3)}$	5
	$J_{(3-4)}$	8
	$J_{(2-4)}$	0–3
	$J_{(3-5)}$	1.5
	$J_{(2-5)}$	1
	$J_{(2-6)}$	~0
furan	$J_{(2-3)}$	2
	$J_{(3-4)}$	4
	$J_{(2-4)}$	1
	$J_{(2-5)}$	±1.5
thiophene	$J_{(2-3)}$	5
	$J_{(3-4)}$	4
	$J_{(2-4)}$	1
	$J_{(2-5)}$	3

[a] From several sources.[71,81-84]

[b] J can be several hertz in certain cases; see discussion in text.

ferent molecular systems rests largely on observations and empirical correlation. When a spectrum can be analyzed by first-order procedure, the extraction of the value for J from a multiplet is trivial; in more complex spectra, formidable calculation may be required to find values of J's from the spectra (see Chapter 7).

Table 5.2 lists typical values of proton–proton coupling constants for various molecular species. This tabulation is meant to be illustrative, not exhaustive, with respect to both the types of molecules included and the overall ranges listed.

We have adopted the commonly used notation for coupling constants, in which the number of bonds intervening between the coupled nuclei is given as a superscript and the identity of the coupled nuclei as a subscript (e.g., $^3J_{HH}$ for a coupling between protons on adjacent carbon atoms, sometimes called *vicinal* coupling). The superscript or the subscript, or both, are deleted when there is no ambiguity.

Proton–proton couplings through single bonds are usually attenuated rapidly, so that generally $^4J < 0.5$, and is usually unobservable. Couplings are stereospecific, as we shall see in Section 5.7, so that certain geometric arrangements of nuclei result in values of ^4J and even ^5J that are observable. Coupling through more than three bonds is called *long-range coupling* and is of considerable interest in stereochemistry.

Generally, couplings are transmitted more effectively through multiple bonds than through single bonds. For example, vicinal couplings for ethane derivatives are usually < 10 Hz, while in ethylene derivatives $^3J(\text{cis}) \approx 10$ and $^3J(\text{trans}) \approx 17$.

In aromatic systems also the coupling is transmitted more effectively than through single-bonded systems. Ortho coupling constants (^3J) are generally of the order of 5–8 Hz, meta couplings (^4J) are about 1–3 Hz, and para couplings (^5J) are quite small, often < 0.5 Hz. In general, there is no coupling observed between different rings in fused polycyclic systems, so that such long-range couplings must be < 0.5 Hz. Substituents have only small effects on the magnitudes of aromatic couplings, but introduction of hetero atoms can cause significant alterations, as indicated in Table 5.2.

Geminal proton–proton coupling constants (^2J) depend markedly on substituents, as indicated in Table 5.2. The trend of ^2J with substitution has been treated successfully by theory and will be discussed in Section 5.7.

Table 5.3 lists some representative coupling constants between protons and other atoms. ^1J is usually quite large, but other coupling constants cover a wide range. The large value of ^1J for $^{13}C–H$ often permits the observation of "^{13}C satellites" in a proton resonance spectrum due to

Table 5.3

TYPICAL H—X COUPLING CONSTANTS[a]

Type	$|J|$ (Hz)
^{13}C—H (sp^3)	125
(sp^2)	160
(sp)	240
^{13}C—C—H	−5 to +5
^{13}C—C—C—H	0 to 5
$^{15}NH_3$	61
$>^{15}N$—CH_3	1–3
C=^{14}N—CH_3	3
Ph > C=^{15}NH / Ph	51
pyridinium $N_\oplus$—H	91
^{17}O—H	80
H C H / F	45
H—C—C—F	5–20
H—C—C—C—F	1–5
(ortho)	8
—F (meta)	4–7
(para)	2

Table 5.3—Continued

Type	$\lvert J \rvert$(Hz)
$\overset{\diagdown}{\underset{H\diagup}{C}}=\overset{\diagup F}{\underset{\diagdown}{C}}$	45
$\overset{H\diagdown}{\underset{\diagup}{C}}=\overset{\diagup F}{\underset{\diagdown}{C}}$	18
$\overset{\diagdown}{\underset{\diagup}{C}}=\overset{\diagup F}{\underset{\diagdown H}{C}}$	80
$H-\overset{\overset{O}{\vert}}{\underset{\vert}{P}}{=}O$	500–700
$H-P\diagup$	200
$H-\overset{\vert}{\underset{\vert}{C}}-\overset{\vert}{\underset{\vert}{P}}{=}O$	10
$H-\overset{\vert}{\underset{\vert}{C}}-\overset{\vert}{\underset{\vert}{C}}-\overset{\vert}{\underset{\vert}{P}}{=}C$	<5

^{a} Data from various sources.[47,48,82,83,85]

proton coupling with the 1.1% ^{13}C present in natural abundance. Such satellites are often of value in determining H–H coupling constants that are otherwise inaccessible (see Section 7.25). On the other hand, these large values of $^1J(^{13}C-H)$ often lead to unwanted complexity when ^{13}C spectra are observed directly. As we shall see in Chapter 9, double resonance methods are used to reduce or eliminate this problem. Geminal and vicinal couplings between protons and other nuclei, such as ^{13}C and ^{15}N, are generally small (often 1–5 Hz), while those between H and F may be quite large. (The values of the reduced coupling constants are more comparable in value.)

Table 5.4 gives a small, illustrative selection of coupling constants not involving protons. Again couplings involving ^{19}F are noticeably large.

Table 5.4

TYPICAL COUPLING CONSTANTS NOT
INVOLVING HYDROGEN[a]

Type	J (Hz)
$^{13}C{-}F$	-280 to -350
$-\overset{\|}{\underset{\|}{^{13}C}}{-}\overset{\|}{\underset{\|}{^{13}C}}{-}$	35
$-\overset{\|}{\underset{\|}{^{13}C}}{-}^{13}C{\equiv}N$	$50{-}55$
$\overset{}{\underset{}{>}}^{13}C{=}^{13}C\overset{}{\underset{}{<}}$	70
$-^{13}C{\equiv}^{13}C-$	170
$\diagup^{15}N{=}^{15}N\diagup$	5 to 15
$-\overset{\|}{\underset{\|}{^{13}C}}{-}^{15}N\overset{}{\underset{}{<}}$	-4 to -10
$-^{13}C{\equiv}^{15}N$	-17
$\overset{\diagdown}{\underset{\diagup}{>}}C\overset{\diagup F}{\underset{\diagdown F}{}}$	160
$F{-}\overset{\|}{\underset{\|}{C}}{-}\overset{\|}{\underset{\|}{C}}{-}F$	-3 to -20
[benzene ring with F substituents]	(ortho) -17 to -22 (meta) 11 to -10 (para) 14 to -14
$\overset{\diagdown}{\underset{F\diagup}{}}C{=}C\overset{\diagup F}{\underset{\diagdown}{}}$	-120
$\overset{F\diagdown}{\underset{\diagup}{}}C{=}C\overset{\diagup F}{\underset{\diagdown}{}}$	$30{-}40$

Table 5.4—Continued

Type	J (Hz)
$\diagup$P—P$\diagdown$	100
O=P—P=O (with vertical bonds)	500
F—P=O	1000

^{a} Data from several sources.[47,48,52,83,86,87]

5.7 Correlation of Coupling Constants with Other Physical Properties

Theory suggests and experiment largely confirms that coupling constants can be related to a number of physical parameters. Among the most important are (1) hybridization, (2) dihedral bond angles, and (3) electronegativity of substituents.

The dependence of J on electron density at the nucleus (Section 5.5) suggests a relation between ^1J and amount of s character in the bond. Such a relation is indeed found for ^{13}C–H couplings in sp, sp^2, and sp^3 hybridized systems, as indicated in Table 5.2. Similar rough correlations are found for other X–H couplings (X = ^{31}P, ^{15}N, ^{119}Sn), for ^{13}C–^{13}C couplings and for some other X–X couplings. In all cases, including ^{13}C–H, however, addition of substituents may well cause large changes in effective nuclear charge or uneven hybridization in different bonds, so that exact correlations should not be expected. For example, the nominally sp^3 hybridized molecules CH_4, CH_3Cl, CH_2Cl_2, and $CHCl_3$ have values of $J(^{13}C$–H) of 125, 150, 178, and 209 Hz, respectively.

One of the most fruitful theoretical contributions to the interpretation of coupling constants has been the valence bond treatment by Karplus[88] of $^3J_{HH}$ in ethanelike fragments, H_a—C_a—C_b—H_b. The most interesting conclusion is that this coupling depends drastically on the dihedral angle Φ between the H_a—C_a and the C_b—H_b bonds. The calculated results were found to fit approximately the relation

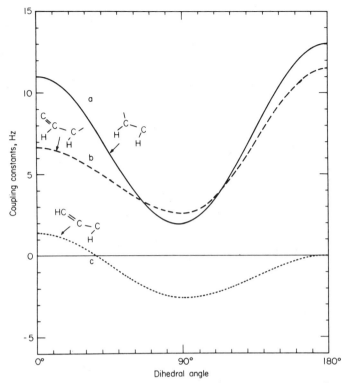

Fig. 5.6 Calculated variation of vicinal and allylic proton-proton coupling constants with dihedral angle between the C—H bonds shown in the figure. Plot[89] based on equations by Bothner-By[81] and Garbisch.[90]

$$^3J = A + B \cos \Phi + C \cos 2\Phi, \qquad (5.3)$$

with $A = 4$, $B = -0.5$, and $C = 4.5$ Hz. From empirical studies a better set of parameters, chosen by Bothner-By,[81] is $A = 7$, $B = -1$, and $C = 5$ Hz. Equation 5.3 with the latter set of parameters is plotted in Fig. 5.6. It is apparent that large values of J are predicted for cis (0°) and trans (180°) conformations but small values for gauche (60° and 120°) conformations. These predictions have been amply verified, and the Karplus relation is of great practical utility in structure determinations. It must be realized, however, that the Karplus relation contains inherent limitations due to the necessary approximations made in the quantum mechanical treatment, and also that it was derived strictly for ethane. Substitution, especially with strongly electronegative atoms such as oxygen, can cause substantial changes in coupling. Empirical modifications of the Karplus curve to take substituent effects into account have met with some success, but even with modifications one cannot justify use of the relation to determine bond angles to within a few degrees.

The success of the theoretical treatment of the angular dependence of ^3J in an ethane fragment has served as the rationale for empirical calculations of the angular dependence of $^3J_{ab}$ and $^4J_{ac}$ in the allylic system (III).

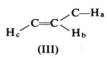

(III)

Plots of $^3J_{ab}$ and $^4J_{ac}$ as functions of dihedral angles between C—H_a and C—H_b or C—H_c are given in Fig. 5.6. It must be emphasized that these relationships are only approximate, but there is generally rather good agreement with experimental data. While all three curves in Fig. 5.6 have a similar angular dependence, the change in sign for ^4J leads to a maximum in $|J|$ near 90°, rather than a minimum as in the two curves for ^3J.

H–H couplings of 1 Hz or more through four single bonds have been

$$H{\diagdown}C{\diagup}C{\diagdown}C{\diagup}H$$

(IV)

observed almost exclusively in systems of the sort where the four bonds are in a planar "W" conformation. Usually this fragment is part of a cyclic or polycyclic system, which determines the conformation. Coupling constants of 1–2 Hz are commonly found in such cases, but larger values (3–4 Hz) have been observed in bicyclo[2.2.1]heptanes and related molecules. Conjugated systems in a planar zig-zag configuration display

$$H{\diagdown}C{\diagup\!\!=}C{\diagdown}C{\diagup\!\!=}C{\diagdown}H$$

(V)

five-bond H–H coupling constants of 0.4–2 Hz. Again, the fragment is often part of a cyclic molecule, and one of the carbon atoms may be substituted by nitrogen. The values of ^4J and ^5J in these two types of systems fall off rapidly with departures from planarity. Many examples of long-range couplings are given by Jackman and Sternhell.[71]

Vicinal *heteronuclear* coupling constants also depend on dihedral angles. Curves similar to those in Fig. 5.6 have been developed for ^3J (HCOP) and ^3J (NCCH).[91,92] In molecules where unshared (nonbonding) electrons are localized in one direction, as in some nitrogen or phosphorus-containing compounds, geminal couplings are also dependent on angular orientation. For example, in oxaziridines ^2J (H-^{15}N) $= -5$ Hz in VI but is nearly zero in VII.[93]

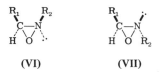

(VI)   (VII)

The molecular orbital approach has also been used to calculate proton–proton coupling constants. It has had probably its greatest success in Bothner-By and Pople's treatment of $^2J_{HH}$ in both sp^2 and sp^3 systems.[94] They did not attempt to calculate numerical values for coupling constants, but rather showed the direction and approximate magnitude of the change in $^2J_{HH}$ with change of substituents. The most interesting feature of the theory is that electronegative substituents which remove electrons from the *symmetric* bonding orbital of the CH$_2$ fragment cause increases in $^2J_{HH}$, while substituents that remove electrons from the *antisymmetric* bonding orbital cause decreases in ^2J. The former corresponds to inductive withdrawal of electrons, while the latter arises from hyperconjugation. In some cases the effects add, rather than oppose each other. In formaldehyde, for example, the electronegative oxygen causes an *inductive withdrawal* of electrons from the C—H bonds, thus increasing $^2J_{HH}$; but the two pairs of *nonbonding* electrons of the oxygen are *donated* to the C—H orbitals (hyperconjugation), thus further increasing $^2J_{HH}$. Formaldehyde is predicted, then, to have a large, positive $^2J_{HH}$; the measured value, $+42$ Hz (see Table 5.2), is the largest H–H coupling constant known, except for the directly bonded protons in the hydrogen molecule.

5.8 Tabulations of Coupling Constants

Many collections of coupling constants have been compiled in tabular form, and many other approximate values can be obtained from inspection of first-order splittings in published spectra. Most of the sources of chemical shift data listed in Section 4.12 provide useful information on coupling constants also. In addition, the following compilations are valuable:

1. A. A. Bothner-By, *Adv. Magn. Resonance* **1**, 195 (1965). A tabulation of critically evaluated geminal and vicinal H–H coupling constants.

2. R. E. Wasylishen, *Ann. Rep. NMR Spectrosc.* **7**, 246 (1977). A tabulation and discussion of $^{13}C–X$ coupling constants, where X consists of first-row elements.

3. L. M. Jackman and S. Sternhell, "Applications of NMR Spectroscopy in Organic Chemistry," pp. 269–356. Pergamon, Oxford, 1969. A good compendium of H–H coupling constants, especially valuable for long-range couplings.

4. E. G. Finer and R. K. Harris, *Prog. NMR Spectrosc.* **6**, 61 (1970). A compilation and discussion of $^{31}P–^{31}P$ coupling constants.

5. J. W. Emsley, L. Phillips and V. Wray, *Prog. NMR Spectrosc.* **10**, 83 (1976). Extensive tabulations of ^{19}F coupling constants in a wide variety of organic and inorganic compounds.

Problems

1. The spectrum at 60 MHz from two spin-coupled protons consists of four lines of equal intensity at 72, 80, 350, and 358 Hz, measured with respect to TMS. Predict the spectrum at 100 MHz and state the values of δ (in parts per million) and J.

2. The geminal H–H coupling in CH_4 has been determined as 12.4 Hz (probably negative). Since the spectrum of CH_4 consists of only a single line, how could this figure have been obtained?

3. If $J(^{14}N–H)$ for NH_3 is $+40$ Hz, what is the sign and magnitude of $J(^{15}N–H)$ in $^{15}NH_3$? Of $J(^{15}N–D)$ in $^{15}ND_3$?

4. Using reduced coupling constants, compare the magnitudes of the X–H coupling in CH_4 (125 Hz) and $^{14}NH_4^+$ (55 Hz).

5. Sketch the ^1H and ^{31}P spectra of $(CH_3O)_3P{=}O$, given that $J_{PH} = 12$ Hz.

6. Sketch the ^1H and $^2H(D)$ spectra of acetone-d_5, given that $^2J_{HD} \approx -2$ Hz. What is $^2J_{HH}$ in acetone?

7. Sketch the spectrum of the CH proton in an isopropyl group, $—CH(CH_3)_2$, assuming that first-order analysis is applicable and that $^3J_{HH} = 6$ Hz. Show that the same result is obtained by (a) considering the CH equally coupled to the six CH_3 protons as through they were a single group of magnetically equivalent nuclei, or by (b) using repetitive first-order analysis with the two CH_3 groups separately.

8. Sketch the ^1H and ^{19}F spectra of CF_3CFH_2, given $^2J_{HF} = 45.5$, $^3J_{HF} = 8.0$, and $^3J_{FF} = 15.5$ Hz.

9. Construct a table analogous to Table 5.1 for coupling to n nuclei of spin 1 for $n = 1$, 2, and 3.

10. Deduce the structures of the molecules giving Spectra 7–12, Appendix C.

Chapter 6

The Use of NMR in Structure Elucidation

Of the many applications of high resolution NMR in all branches of chemistry, the most widespread is its use in the elucidation of the structure of organic and inorganic compounds. With the background that we have presented in the preceding chapters it is now profitable to consider the way in which the NMR spectrum of a new compound might best be approached. We shall find some features of spectra that can be understood only in terms of the additional background developed in the succeeding chapters, but the examples given in this chapter and in the problems at the end demonstrate that in many cases we can obtain valuable structural information with what we have learned thus far.

6.1 A Systematic Approach to the Interpretation of NMR Spectra

In problems of structure elucidation an NMR spectrum may provide useful, even vital data, but it is seldom the sole piece of information available. A knowledge of the source of the compound or its method of synthesis is frequently the single most important fact. In addition, the interpretation of the NMR spectrum is carried out with concurrent knowledge of other physical properties, such as elemental analysis from combustion or mass spectral studies; molecular weight; and the presence or absence of structural features, as indicated by infrared or ultraviolet spectra or by chemical tests. Obviously the procedure used for analyzing the NMR spectrum is highly dependent on such ancillary knowledge. The following procedure is suggested, however, as a systematic method for extracting the information from most NMR spectra of new compounds. This procedure is equally applicable to the assignment of NMR features to given nuclei in the spectrum of a compound of known structure. Since proton resonance has thus far accounted for the vast majority of NMR studies, the procedure is aimed principally at proton NMR spectra, and the examples are drawn from such spectra. Some aspects of the interpretation of ^{13}C spectra, which are of increasing importance in organic applications, are given in Section 6.2. The use of NMR data for other nuclei in structure elucidation is also increasing, but must be regarded as sufficiently specialized to preclude our providing a general summary of the best way to approach such data.

1. Before attempting to interpret an NMR spectrum, it is wise to ascertain whether the spectrum has been obtained under suitable experimental conditions so that it is a meaningful spectrum. The appearance of the line due to TMS (or other reference) should be examined for symmetry and sharpness (as indicated by adequate ringing in spectra obtained with cw spectrometers). Any erratic behavior in the base line or appearance of very broad resonance lines should be noted as possibly indicating the presence of ferromagnetic particles (see Fig. 3.11). The trace of the integral should be consistent with proper adjustments of phasing and drift controls (see Fig. 3.9). If these criteria are not met, it is usually desirable to rerun the spectrum under better experimental conditions.

It is important to check the calibration of the spectrometer, as indicated by the spectrum. If the TMS line does not appear exactly at zero, a simple additive correction to all observed lines is sufficient, provided that the overall calibration of the sweep width is correct. Often an indication of *gross* errors can be obtained from the observed frequency of solvent lines compared with those in Table 3.1 or other sources. It should be noted, however, that small changes (a few hertz) are often found with different samples due to molecular interactions.

Usually the largest value of H_0 consistent with adequate resolution is used to maximize chemical shifts. Sometimes spectra at two field strengths provide additional information (see Chapter 7).

2. The presence of any known "impurity" lines should be noted. This includes lines due to the solvent itself or to a small amount of undeuterated solvent. If a proton-containing solvent is used, ^{13}C satellites* and spinning sidebands of the solvent peaks may be prominent. Water is often present in solvents, its resonance frequency depending on the extent of hydrogen bonding to the solvent or the sample and on the concentration of water.

---

* Proton resonance lines from molecules containing ^{13}C (natural abundance, 1.1%); see Section 7.25 for further details, and Table 5.3 for values of $^{13}C-H$ coupling constants.

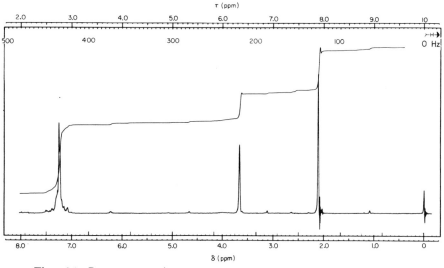

Fig. 6.1 Proton magnetic resonance spectrum (60 MHz) of phenylacetone, $C_6H_5CH_2C(O)CH_3$ in $CDCl_3$.

3. An examination of the relative areas of the NMR lines or multiplets (resolved or unresolved) is usually the best starting point for the interpretation of the spectrum. One should remember that the accuracy of a single integral trace is seldom better than 2% of the full scale value, so that the measured area of a small peak in the presence of several larger ones may be appreciably in error. If the total number of protons in the molecule is known, the total area can be equated to it, and the numbers of hydrogen atoms in each portion of the spectrum established. The opposite procedure of assigning the smallest area to one or two protons and comparing other areas with this one is sometimes helpful, but should be used with caution since appreciable error can be introduced in this way. Occasionally lines so broad that they are unobservable in the spectrum itself can be detected in the integral trace.

4. The positions of strong, relatively sharp lines should be noted and correlated with expected chemical shifts. This correlation, together with the area measurements, frequently permits the establishment of a number of methyl and methylene groups, aromatic protons (especially in some types of monosubstituted benzene), and in some instances exchangeable protons such as OH and COOH. For example, in Fig. 6.1 the line due to three protons at $\delta = 2.10$ is readily identified as a $CH_3C{=}O$, while that at $\delta = 3.67$ is a methylene group deshielded by nearby substituents. Although the aromatic protons are not precisely equivalent, the dif-

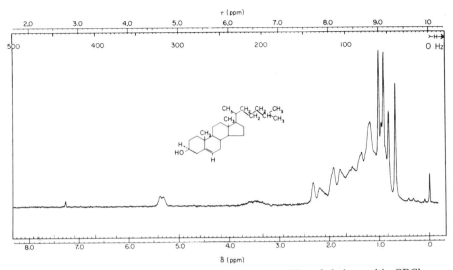

Fig. 6.2 Proton magnetic resonance spectrum (60 MHz) of cholesterol in CDCl$_3$.

ferences in their chemical shifts are so small that they give rise to a relatively sharp single line.

Figure 6.2 shows a spectrum typical of a steroid. The protons of the many CH and CH$_2$ groups in the condensed ring system are so nearly chemically equivalent that they give rise to a broad, almost featureless "hump." The angular and side-chain methyl groups, however, show very pronounced sharp lines, the positions of which can provide valuable information on molecular structure (cf. Chapter 4).

5. The approximate centers of all multiplets, broad peaks, and unresolved multiplets should be noted and correlated with functional groups. For example, in Fig. 6.2 the presence of the vinyl proton ($\sim \delta = 5.4$ ppm) and the proton adjacent to the 3-hydroxyl group ($\delta = 3.5$ ppm) can be identified. At this stage it is unnecessary to worry about *exact* chemical shifts for complex and unresolved multiplets. The *absence* of lines in characteristic regions often furnishes important data. For example, the molecule whose spectrum is given in Fig. 6.2 clearly has no aromatic protons.

6. First-order splitting in multiplets should be identified, and values of J deduced directly from the splittings. As noted in Chapter 5, the first-order criterion ($\nu_A - \nu_B$) $\gg J_{AB}$ often is not strictly obeyed, resulting in a distortion or "slanting" of the expected first-order intensity distribution. This effect can be seen in Fig. 6.3; note that the "slanting" of the intensities in one group *always* increases toward the other group with which it

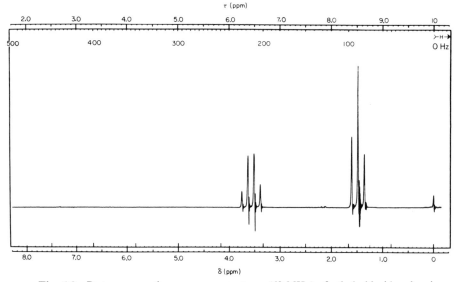

Fig. 6.3 Proton magnetic resonance spectrum (60 MHz) of ethyl chloride, showing the almost first-order splitting of the CH_3 and CH_2 resonances.

is coupled, as we shall see in Chapter 7. The number of components, their relative intensities, the value of J, and the area of the multiplet together provide much valuable information on molecular structure. Commonly occurring, nearly first-order patterns, such as that in Fig. 6.3 due to CH_3CH_2X, where X is an electronegative substituent, should be recognized immediately with a little practice. Other patterns that are actually not first order, such as that due to the magnetically nonequivalent protons in p-chloronitrobenzene (Fig. 6.4) are also characteristic and should be easily identified. Para-substituted benzene rings usually display a pattern characterized by four lines symmetrically placed with a weak, strong, strong, weak intensity relation, and a separation between the outer components of about 8 Hz [approximately J_{HH}(ortho)]. There are, however, many less intense lines, as shown in the expanded trace of Fig. 6.4. (Spectra of this type are considered in Section 7.22.)

The magnitudes of coupling constants are often definitive in establishing the relative positions of substituents. For example, Fig. 6.5 shows that the three aromatic protons of 2,4-dinitrophenol give rise to a spectrum that is almost first order in appearance. The magnitudes of the splittings suggest that two protons ortho to each other give rise to the peaks in the regions of 450 and 510 Hz, and that the latter proton is meta to the one resulting in the lines near 530 Hz. From the known effects of electron-

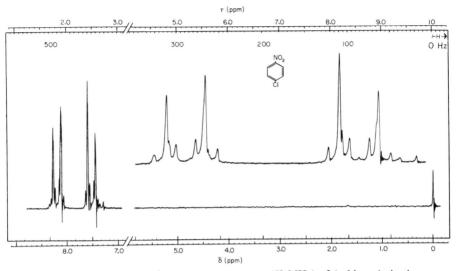

Fig. 6.4 Proton magnetic resonance spectrum (60 MHz) of 1-chloro-4-nitrobenzene in CDCl₃. Inset shows spectrum with abscissa scale expanded fivefold.

withdrawing and electron-donating substituents (Chapter 4) it is clear that the lowest field protons must be adjacent to the NO₂ groups. (Note that the slanting of intensities in this spectrum is in accord with the rule mentioned previously.)

When magnetic nuclei other than protons are present, it should be recalled that some values of J might be as large as many proton chemical shifts. For example, in Fig. 6.6, $^2J_{HF} = 48$ Hz, accounting for the widely spaced $1:3:3:1$ quartets due to the CH that is coupled to both the fluorine

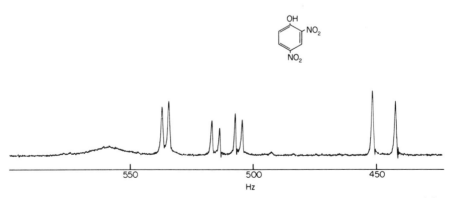

Fig. 6.5 Proton magnetic resonance spectrum (60 MHz) of 2,4-dinitrophenol in CDCl₃, with TMS as internal reference.

and the adjacent methyl group. Since $^3J_{HF} = 21$ Hz and $^3J_{HH} = 7$ Hz, the CH_3 resonance is a doublet of doublets.

7. Exchangeable protons (OH, NH, or activated CH) can often be identified by addition of a drop of D_2O to the sample (see Section 3.13), and resultant disappearance of peaks.

8. Rerunning the spectrum with the sample dissolved in another solvent is often good practice in order to resolve ambiguities arising from accidental coincidence or overlapping of peaks (see Fig. 6.7). In addition, specific information on configuration or conformation can sometimes be obtained, as we shall see in Chapter 12.

9. Some complex multiplets (nonfirst-order patterns) can be analyzed by simple procedures that we shall develop in Chapter 7. Frequently, coupling constants (or less often, chemical shifts) derived from such analyses can provide key pieces of information in the elucidation of structure.

10. If there are still ambiguities to be resolved, the technique of spin decoupling is often helpful. By selectively collapsing the splittings of multiplets to single lines, one can often determine unambiguously the origins of certain spin couplings. This important technique will be covered in Chapter 9.

No amount of discussion of the procedure for analyzing spectra can substitute for practice. The collections of spectra in the Varian catalog[95] and in the Sadtler compilation[96] provide some excellent readily available examples that should be studied in detail. In addition, there are a number of spectra of "unknowns" assigned at the end of this chapter. Other NMR

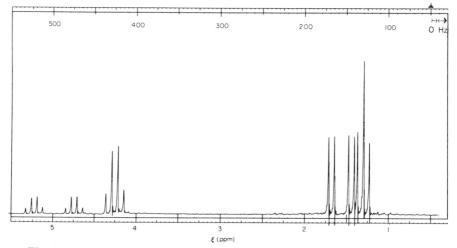

Fig. 6.6 Proton magnetic resonance spectrum (100 MHz) of $CH_3CHFCOOCH_2CH_3$.

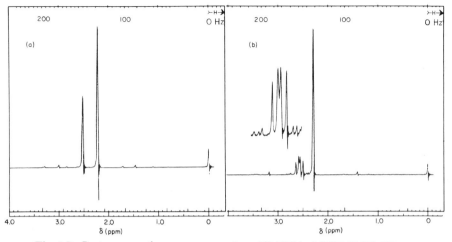

Fig. 6.7 Proton magnetic resonance spectrum (60 MHz) of $(CH_3)_2NCH_2CH_2C{\equiv}N$ (a) as neat liquid and (b) in $CDCl_3$. Note accidental coincidence of chemical shifts of protons in the two CH_2 groups in (a) and their separation into a complex nonfirst-order multiplet in (b).

books directed at structure elucidation furnish additional suggestions for procedures and many sample spectra.[97−100]

6.2 Some Features of Carbon-13 Spectra

Carbon-13 NMR spectra are becoming increasingly valuable in the structure elucidation of organic compounds. Since proton NMR generally requires smaller amounts of sample and less complex instrumentation, a proton NMR spectrum is almost always obtained prior to a ^{13}C spectrum, and the ^{13}C data are interpreted in the light of the proton spectrum, as well as non-NMR information.

Most of the general comments in Section 6.1 apply, *mutatis mutandis,* to the interpretation of ^{13}C spectra. Regarding the 10 specific steps recommended there, items 1, 2, and 4 are equally applicable to ^{13}C spectra. As pointed out in Chapter 5, most ^{13}C spectra are obtained initially with complete proton decoupling, so that (in the absence of ^{19}F, ^{31}P, or other nuclei that might couple to carbon) the spectrum consists of a single line for each chemically different carbon atom in the molecule. Thus, the recommendations regarding spin multiplets are not directly applicable. However, as we point out in more detail in Chapter 9, *off-resonance decoupling* leads to simple multiplets in a ^{13}C spectrum and provides information of great interpretive value.

For reasons that we shall discuss in Chapters 9 and 10, ^{13}C spectra are usually obtained under conditions where the areas of chemically shifted lines are *not* proportional to the numbers of carbon nuclei contributing to the lines. There is no fundamental reason why the theoretically predicted proportionality cannot be obtained, but additional experimental time is required, and in many instances it is simply not efficient to spend the time in this way.

Since most ^{13}C spectra are obtained by pulse Fourier transform methods, the means exist for determination of spin–lattice relaxation times, T_1 (see Chapters 8 and 10), which are occasionally valuable in assigning ^{13}C lines to specific carbon atoms. However, the additional experimental time required to obtain T_1 data detracts from the utility of this approach as a routine matter.

Some ^{13}C spectra of "unknowns" are included in Appendix C. Collections of ^{13}C spectra have been published, and several compilations of ^{13}C chemical shifts and coupling constants have been mentioned in Sections 4.12 and 5.8.

6.3 Structure Elucidation of Polymers

Even when the structures of the individual monomer units in a polymer are known, the determination of their sequence and of the geometrical arrangement, configuration and conformation of the entire polymer presents challenging problems. We can comment on only a few aspects here.

The NMR spectrum of a homopolymer may be very simple if the monomeric unit repeats regularly. On the other hand, irregularities, such as head-to-head junctions mixed with head-to-tail junctions, in such cases as vinyl polymers, for example, introduce additional lines that can often be valuable in structure elucidation. When the repeating unit possesses a center of asymmetry, further complexity is introduced into the spectrum. This feature will be taken up in Section 7.27.

A synthetic copolymer provides additional degrees of freedom in the arrangement of the repeating units. For example, the spectrum of a copolymer of vinylidine chloride and isobutylene, shown in Fig. 6.8, indicates that various tetrad sequences (sequences of four monomer units) display significantly different spectra. Copolymers composed of more than two monomer types, including biopolymers, may have much more complex spectra. Bovey has provided excellent discussions of the use of NMR in studies of polymers.[83,101]

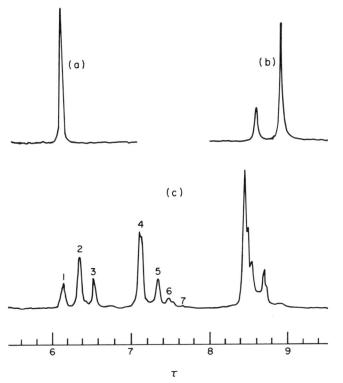

Fig. 6.8 Proton NMR spectrum (60 MHz) of (a) polyvinylidene chloride, (b) polyisobutylene, and (c) a copolymer of 70 mole % vinylidene chloride (A) and 30 mole % isobutylene (B). Peaks in (c) can be assigned to various tetrad sequences: (1) AAAA, (2) AAAB, (3) BAAB, (4) AABA, (5) BABA, (6) AABB, (7) BABB (Bovey[101]).

With the advent of NMR spectrometers operating at higher frequencies (>200 MHz) and with recent improvements in sensitivity (see Chapter 3), it has become possible to study biopolymers, such as small proteins, nucleic acids, carbohydrates, and lipids. Usually NMR is not a method of choice for analysis of the nature or sequence of monomer units, but it is of great value in providing detailed information on polymer conformation. For example, the four histidine residues in the enzyme S-nuclease have substantially different environments in the native protein and have readily distinguishable chemical shifts, as illustrated in Fig. 6.9. On acid denaturation, however, the secondary structure of the protein is lost, and all four histidines shows the same chemical shift in the random coil form of the protein. There are a number of excellent reviews of the use of NMR in studies of biopolymer structure.[103]

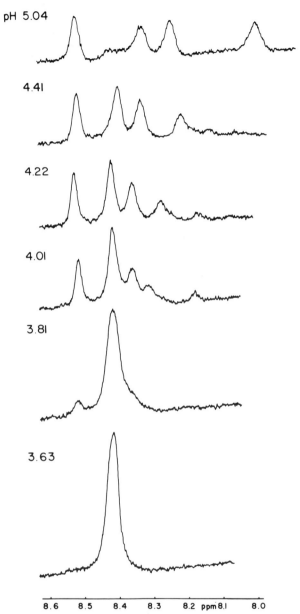

Fig. 6.9 Proton NMR spectra (220 MHz) of the C2 proton of histidines in staphylo-
coccal nuclease. The native protein at pH 5.07 was denatured by addition of DCl and rena-
tured by addition of NaOD (Epstein *et al.*[102]).

Problems

1. Determine the structural formulas of the molecules giving Spectra 13–20, Appendix C.

2. The proton NMR spectrum of a complex organic molecule has an integral with steps of 6.1, 14.0, 22.0, 28.6, 31.9, 43.0, and 61.7 units on the chart paper. (a) Assume that the smallest peak corresponds to one proton. Compute the number of protons giving rise to each peak and the total number in the molecule. (b) Suppose, alternatively, that evidence is available from mass spectroscopy that there are 44 protons in the molecule. Compute the number of protons causing each peak. How large an experimental error is required for the discrepancy between the results of parts (a) and (b)?

Chapter 7

Analysis of Complex Spectra

Thus far in discussing the various types of splitting patterns found in NMR spectra, we have used only the first-order treatment, which was derived from a qualitative examination of the interactions between nuclei due to polarization of the bonding electrons. The first-order rules regarding number of lines and their relative intensities (Section 5.3) are valid only when chemical shifts are large relative to coupling constants *and* when only groups of magnetically equivalent nuclei are involved.

In this chapter we shall consider a more general treatment of the appearance of complex spectra and of the ways in which nuclei interact with the applied magnetic field and with each other. We shall develop a certain amount of formalism using quantum mechanics, and shall find that there are very simple rules for determining in terms of chemical shifts and coupling constants what a spectrum should look like without any regard to the approximations required by the first-order treatment. The first-order treatment will then follow as a special case. In some situations where first-order treatment is inapplicable, we shall find that simple algebraic equations can be derived for treating the complex spectra. Many of the details of the derivations are omitted, but can be found in the books by Corio,[104] Emsley *et al.*,[105] and Pople *et al.*[106]

7.1 Notation

Our treatment of complex spectra is considerably simplified by restricting ourselves only to nuclei with $I = \frac{1}{2}$. This limitation permits us to cover most of the general principles without tedious algebraic manipulations. In addition, nuclei with $I = \frac{1}{2}$ are studied far more extensively than others.

120

It will be helpful to use the widely employed system of notation in which each nucleus of spin $\frac{1}{2}$ is denoted by some letter of the alphabet: A, B, X, and so forth. We shall choose letters of the alphabet representative of relative chemical shifts; that is, for two nuclei that have a small chemical shift relative to each other, we choose two letters of the alphabet that are close to each other, and for nuclei that have large relative chemical shifts, we use letters from opposite ends of the alphabet. If there are several *magnetically* equivalent nuclei, we denote this fact by a subscript (e.g., A_2X). If there are several nuclei that are *chemically* equivalent, but not magnetically equivalent, we denote this fact by repeating the letter with a prime or a double prime (e.g., $AA'X$). Nuclei with $I = 0$, as well as those that relax so rapidly that they behave as though they are not magnetic, such as Cl, Br, and I (cf. Section 2.5), are not given any designations. The few examples shown should suffice to clarify this notation.

(a) $CH_2\!=\!CCl_2$ A_2

(b) CH_2F_2 A_2X_2

(c) $CH_2\!=\!CF_2$ $AA'XX'$

(d) CH_3OH A_3B or A_3X (depending on hydrogen bonding effect on δ_{OH}; see Chapter 12)

(e) $CH_2\!=\!CHCl$ ABX or ABC

(f) $AA'BB'$ (assuming no coupling between rings)

(g) $CH_3CH\!=\!CH_2$ A_3MXY

(h) $^{13}CH_2F_2$ A_2M_2X

(i) $CH_3CH_2NO_2$ A_3X_2

(j)   ABCC'DD'EE'FF'X (for the single conformation shown; see Chapter 11 for effects of rapid interconversion of conformers)

Generally, no significance is attached to the order in which the nuclei are given. For example, in (b) either the hydrogen or fluorine nuclei could be designated A; likewise, (d) could be called an A_3B or AB_3 system. Nuclei with different magnetogyric ratio (e.g., ^1H and ^{13}C) clearly should be denoted by letters far apart in the alphabet, but for those of the same species the notation depends on relative chemical shifts, which are field dependent. When $\delta \gg J$, letters far apart in the alphabet are employed, and even systems that might be classified as, say, ABC are sometimes approximated by, say, ABX. Numerous examples will appear later in this chapter and in the problems at the end of the chapter.

The terminology *weakly coupled* or *strongly coupled* is used to denote nuclei for which $\delta \gg J$ or $\delta \approx J$, respectively. Thus in an ABX system, nuclei A and B are strongly coupled, while A and X are weakly coupled.

7.2 Energy Levels and Transitions in an AX System

Before deriving quantitative expressions for general spin systems, we shall examine qualitatively the energy levels and transitions arising from two nuclei that are not coupled or are only weakly coupled (AX system). We customarily take the static imposed magnetic field to lie along the z axis, and we express the orientation of the z component of nuclear spin, I_z, as α or β for $I_z = \frac{1}{2}$ or $-\frac{1}{2}$, respectively. This notation clearly applies only to nuclei with $I = \frac{1}{2}$. We shall define α and β more precisely in Section 7.4.

When we consider systems containing N nuclei of spin $\frac{1}{2}$ it is convenient to define a quantity F_z as the sum of the z components of all nuclear spins:

$$F_z = (I_z)_1 + (I_z)_2 + \cdots = \sum_{i=1}^{N} (I_z)_i. \tag{7.1}$$

For the AX system we can distinguish four states of spin orientations:

$$
\begin{array}{lll}
(1) & \alpha(A)\alpha(X) \equiv \alpha\alpha & F_z = 1; \\
(2) & \alpha(A)\beta(X) \equiv \alpha\beta & F_z = 0; \\
(3) & \beta(A)\alpha(X) \equiv \beta\alpha & F_z = 0; \\
(4) & \beta(A)\beta(X) \equiv \beta\beta & F_z = -1.
\end{array} \tag{7.2}
$$

The shorter notation $\alpha\alpha$ is often used in place of the notation $\alpha(A)\alpha(X)$ when there is no chance of ambiguity. In such cases it is understood that the nuclei are always given in the same order (e.g., A, X in this case).

Case I. No Coupling between A and X. If we consider the imposed magnetic field as lying in the *negative* z direction, so that a β state has lower energy than an α state, the energies of the four states in Eq. (7.2) will lie in the order given in the center portion of Fig. 7.1, provided there is no spin–spin coupling between the nuclei. States (1) and (3) differ in the "flipping" of spin A, as do states (2) and (4). Likewise, states (1) and (2) and states (3) and (4) differ in the spin orientation of X. Thus transitions between the states may be labeled as A or X transitions, and these correspond to the NMR transitions discussed in Chapter 2. (We shall see later

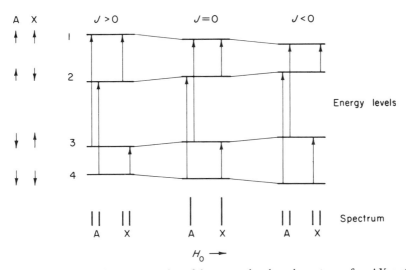

Fig. 7.1 Schematic representation of the energy levels and spectrum of an AX system under different conditions of coupling.

that the "double flip" transitions between states (2) and (3) and between states (1) and (4) are forbidden by selection rules.) If we arbitrarily take the resonance frequency of A to be greater than that of X (at constant field), then the energy levels are spaced as indicated in Fig. 7.1, and the spectrum consists of two lines, as shown. If magnetic field is held constant, frequency increases to the left; if radio frequency is held constant, magnetic field increases to the right (cf. Section 4.3).

Case II. Weak Coupling. Suppose there is a coupling $J > 0$ between A and X. Since a positive coupling constant implies that antiparallel spin orientations possess less energy than parallel orientations (cf. Section 5.4), states (2) and (3) are lower in energy than they were in the absence of coupling, while states (1) and (4) are higher in energy. This situation is depicted in the left portion of Fig. 7.1, but the magnitudes of the changes in energy relative to the separations between levels is exaggerated. The two A transitions no longer have the same energy, and a doublet results in the A portion of the spectrum; a similar doublet appears in the X portion. If $J < 0$, the energy levels change as shown in the right portion of Fig. 7.1. Again two doublets appear in the spectrum. In this case the lower-frequency A line results from a transition between states (3) and (1), rather than between states (4) and (2) as in the case where $J > 0$, but there is no *observable* difference in the spectrum with change in the sign of the coupling constant.

We cannot proceed further with this treatment, either in making it more quantitative or in extending it to strongly coupled nuclei, until we have developed some necessary quantum mechanical background.

7.3 Quantum Mechanical Formalism

A few of the elements of quantum mechanics that are required for the exact treatment of NMR spectra are presented in this section. Standard texts on quantum mechanics, as well as Corio,[104] Emsley *et al.*,[105] Pople *et al.*,[106] should be consulted for more details.

Provided the rf field is not too large (a requirement met in all high resolution NMR experiments except those involving double resonance; cf. Chapter 9), the state of a nuclear spin system may be described by a steady-state wave function arising from a time-independent Hamiltonian operator $\mathcal{H}$ satisfying the time-independent Schrödinger equation

$$\mathcal{H}\psi = E\psi. \tag{7.3}$$

(E gives the total energy, or energy level, of the system.) The Hamiltonian expresses in quantum mechanical form the potential and kinetic energies of the system. We shall examine in detail the Hamiltonian appropriate to a coupled nuclear spin system in Section 7.5. Before doing so, however, we shall review a general method of obtaining the wave functions and energy levels for any given Hamiltonian.

It is known in general that any wave function ψ may be expressed as a linear combination of some other functions Φ_n,

$$\psi = \sum_n C_n \Phi_n, \tag{7.4}$$

provided the Φ_n form an *orthonormal* and *complete set*. The first requirement is expressed in the integral equation

$$\int \Phi_m^* \Phi_n \, d\tau = \delta_{mn}, \tag{7.5}$$

where Φ_m^* is the complex conjugate of Φ_m, and the Kronecker delta is

$$\delta_{mn} = \begin{cases} 0, & m \neq n, \\ 1, & m = n. \end{cases}$$

The symbol $d\tau$ is used to denote integration over all coordinates of the system. For our purposes, we may completely neglect spatial coordinates and utilize only the x, y, and z components of the nuclear spin or nuclear magnetic moment. We shall not explore the significance of using a complete set except to note that in many quantum mechanical problems the series in Eq. (7.4) is infinite; in the case of nuclear spin interactions, however, the series contains exactly 2^N terms, where N is the number of nuclei, when the set Φ_n is formed by the simple and natural procedure described in Section 7.4.

If the expression for ψ in Eq. (7.4) is substituted in Eq. (7.3), we obtain

$$\mathcal{H} \sum_n C_n \Phi_n = E \sum_n C_n \Phi_n. \tag{7.6}$$

Multiplying by Φ_m^* and integrating, we get

$$\int \Phi_m^* \mathcal{H} \left(\sum_n C_n \Phi_n \right) d\tau = \int \Phi_m^* E \left(\sum_n C_n \Phi_n \right) d\tau. \tag{7.7}$$

E and C_n, which are constants, may be removed from the integral and the order of summing and integrating interchanged:

$$\sum_n C_n \int \Phi_m^* \mathcal{H} \Phi_n d\tau = E \sum_n C_n \int \Phi_m^* \Phi_n \, d\tau. \tag{7.8}$$

If we define

$$\mathcal{H}_{mn} = \int \Phi_m^* \mathcal{H} \Phi_n \, d\tau, \tag{7.9}$$

and make use of Eq. (7.5), we obtain

$$\sum_n C_n \mathcal{H}_{mn} = E \sum_n C_n \delta_{mn} \tag{7.10}$$

or

$$\sum_n C_n (\mathcal{H}_{mn} - E\delta_{mn}) = 0. \tag{7.11}$$

Equation (7.11) gives a set of 2^N linear, homogeneous equations in the 2^N variables C_n. The necessary and sufficient condition for nontrivial solutions of these equations is that the determinant of the coefficients of the C_n's vanish; that is,

$$|\mathcal{H}_{mn} - E\delta_{mn}| = 0. \tag{7.12}$$

In expanded form,

$$\begin{vmatrix} \mathcal{H}_{11} - E & \mathcal{H}_{12} & \mathcal{H}_{13} & \cdots \\ \mathcal{H}_{21} & \mathcal{H}_{22} - E & \mathcal{H}_{23} & \cdots \\ \mathcal{H}_{31} & \mathcal{H}_{32} & \mathcal{H}_{33} - E & \cdots \\ \cdot & \cdot & \cdot & \cdot \\ \cdot & \cdot & \cdot & \cdot \\ \cdot & \cdot & \cdot & \cdot \end{vmatrix} = 0.$$

Equation (7.12) is called the *secular equation*. The quantities $\mathcal{H}_{mn}$ may be arranged in the form of a matrix (the Hamiltonian matrix) and are called *matrix elements*. Our problem, then, reduces to developing expressions for computing these matrix elements, from which the necessary energy levels and wave functions may be calculated.

7.4 Nuclear Spin Basis Functions

In Section 7.2 we found it helpful to denote the spin orientation of a nucleus by the symbol α or β and to use the product of these functions to represent the state of spin orientation of two nuclei. We shall now define α and β more precisely and generalize to more than two nuclei.

The functions α and β must be defined in terms of the quantum mechanical operators corresponding to spin angular momentum: α and β are defined as orthonormal *eigenfunctions* of the z component of spin I_z according to the following equations:

$$I_z\alpha = \tfrac{1}{2}\alpha, \qquad I_z\beta = -\tfrac{1}{2}\beta; \tag{7.13}$$

$$\int \alpha\alpha \, d\tau = \int \beta\beta \, d\tau = 1, \qquad \int \alpha\beta \, d\tau = \int \beta\alpha \, d\tau = 0. \tag{7.14}$$

The properties of spin angular momentum operators have been worked out in detail, principally by analogy with ordinary orbital angular momentum.[104] The results as applied to the functions α and β provide several additional equations involving the x and y components of nuclear spin:

$$\begin{aligned} I_x\alpha &= \tfrac{1}{2}\beta, & I_x\beta &= \tfrac{1}{2}\alpha, \\ I_y\alpha &= \tfrac{1}{2}i\beta, & I_y\beta &= -\tfrac{1}{2}i\alpha. \end{aligned} \tag{7.15}$$

Here $i = \sqrt{-1}$. We shall employ these results in calculating matrix elements.

The preceding paragraph defined α and β for a given nucleus. For N nuclei we shall find it most convenient to define the spin state in terms of products of α's or β's for the individual nuclei. With N nuclei of spin $\tfrac{1}{2}$ there are 2^N possible product functions, or 2^N states. For example, two nuclei give the four product functions listed in Section 7.2, which would be designated as $\Phi_1, \ldots, \Phi_4$.

7.5 The Spin Hamiltonian

The Hamiltonian operator may consist of several terms describing the various contributions to the energy of the nuclear spin system. The leading term involves the interaction of each nucleus with the applied magnetic field. Suppose the field H_0 is applied in the negative z direction. We saw in Section 2.2 that the energy of interaction is given by

$$\epsilon = -\boldsymbol{\mu} \cdot \mathbf{H}_0. \qquad (2.10)$$

With H_0 in the negative z direction, this becomes

$$\epsilon = -(\mu_z)(-H_0) = \mu_z H_0. \qquad (7.16)$$

Substitution of Eqs. (2.1) and (2.2) into Eq. (7.16) gives

$$\epsilon = \frac{\gamma}{2\pi} h I_z H_0 \quad \text{(ergs)},$$

$$= \frac{\gamma}{2\pi} I_z H_0 \quad \text{(Hz)}. \qquad (7.17)$$

(Our later treatment is simplified by expressing the energy in hertz, rather than in ergs, and we shall assume henceforth that all energy terms are expressed in hertz unless otherwise stated.) If I_z is now considered to be a quantum mechanical operator, the expression for the classical interaction energy, Eq. (7.17), becomes the first term of the Hamiltonian.

Before writing the expression for this part of the Hamiltonian, we must correct for the shielding responsible for the chemical shift. Thus instead of using H_0 we use the field experienced by the nucleus itself, $H_0(1 - \sigma)$. For N nuclei we sum the contributions for each nucleus. The first term in the Hamiltonian is, then,

$$\mathscr{H}^{(0)} = \frac{1}{2\pi} H_0 \sum_{i=1}^{N} \gamma_i (1 - \sigma_i)(I_z)_i. \qquad (7.18)$$

This expression shows explicitly that the operator I_z is taken to refer only to one nucleus at a time, that more than one type of nucleus (e.g., ^{1}H and ^{19}F) may be involved, and that the shielding (chemical shift) is taken into account.

As we saw in Section 3.4, a spectrum may be obtained by applying an rf field at a fixed frequency and varying the magnetic field, or by holding the field constant and varying the ratio frequency. Our subsequent treatment will be simplified if we adopt the latter approach; the results would be precisely the same if we treated the magnetic field as a variable. To simplify the notation we substitute the Larmor relation, Eq. (2.9), in Eq. (7.18) to get

$$\mathscr{H}^{(0)} = \sum_{i=1}^{N} \nu_i (I_z)_i; \qquad (7.19)$$

ν_i is the resonance frequency (including the effect of the chemical shift) of the ith nucleus.

Whenever there are electron-coupled spin–spin interactions between two or more of the nuclei, we must add a second term to the Hamiltonian to account for the energy arising from these interactions. Following our discussion in Sections 5.1 and 7.2, the coupling depends only on the relative orientations of the nuclear spins, not on the orientations of the spins with respect to the field. This fact is expressed in the following term:

$$\mathcal{H}^{(1)} = \sum_{i<j} \sum J_{ij} \mathbf{I}_i \cdot \mathbf{I}_j$$

$$= \sum_{i<j} \sum J_{ij}[(I_x)_i(I_x)_j + (I_y)_i(I_y)_j + (I_z)_i(I_z)_j]. \qquad 7.20)$$

(The last line is just the expanded form of a vector dot product.) The summation extends over all possible pairs of nuclei, with the notation $i < j$ guaranteeing that each pair is counted only once. For N nuclei the sum includes $\frac{1}{2}N(N-1)$ terms. Thus, for a system of four nuclei there are four terms in $\mathcal{H}^{(0)}$ and six terms in $\mathcal{H}^{(1)}$.

For the types of problems we are treating, where the molecules are tumbling rapidly, the sum of $\mathcal{H}^{(0)}$ and $\mathcal{H}^{(1)}$ is adequate as the complete Hamiltonian. If there are dipole–dipole interactions to be considered, or if there are strong rf fields involved, additional terms must be added. We shall see the effect of a second rf field in Chapter 9, and the effect of dipole interactions in Section 7.28.

7.6 The Two-Spin System without Coupling

We now apply the concepts developed in the preceding sections to the system of just two nuclei, first considering the case where there is no spin coupling between them. In general, when there is no coupling (i.e., $\mathcal{H}^{(1)} = 0$), the form of $\mathcal{H}^{(0)}$ and of the basis functions Φ_n guarantees that the Φ_n are themselves the true wave functions ψ_n. We shall not demonstrate this result in general terms, but it will emerge from the following calculation of energy levels and wave functions for the case of two uncoupled spins.

We shall digress from our usual notation to call the two spins A and B, rather than A and X, since we shall later wish to use some of the present results in treating the coupled AB system. As indicated in Section 7.2, the four product basis functions are

$$\Phi_1 = \alpha_A \alpha_B = \alpha\alpha, \qquad \Phi_2 = \alpha_A \beta_B = \alpha\beta,$$
$$\Phi_3 = \beta_A \alpha_B = \beta\alpha, \qquad \Phi_4 = \beta_A \beta_B = \beta\beta. \qquad (7.21)$$

We shall now compute the matrix elements needed for the secular determinant. Since there are four basis functions, the determinant is 4×4 in size, with 16 matrix elements. Many of these will turn out to be zero. For $\mathcal{H}_{11}$ we have, from Eqs. (7.9) and (7.19) (with $\Phi_n^* = \Phi_n$ for the real functions with which we are concerned),

$$\mathcal{H}_{11} = \int \Phi_1 \mathcal{H}^{(0)} \Phi_1 \, d\tau$$

$$= \int \alpha_A \alpha_B [\nu_A (I_z)_A + \nu_B (I_z)_B] \alpha_A \alpha_B \, d\tau$$

$$= \nu_A \int \alpha_A \alpha_B (I_z)_A \alpha_A \alpha_B \, d\tau + \nu_B \int \alpha_A \alpha_B (I_z)_B \alpha_A \alpha_B \, d\tau. \quad (7.22)$$

The element $d\tau$ refers to spins A and B; thus each of the two integrals in Eq. (7.22) is really a double integral over $d\tau_A$ and $d\tau_B$. These two spins may be integrated independently:

$$\mathcal{H}_{11} = \nu_A \int \alpha_A (I_z)_A \alpha_A \, d\tau_A \int \alpha_B \alpha_B \, d\tau_B$$

$$+ \nu_B \int \alpha_A \alpha_A \, d\tau_A \int \alpha_B (I_z)_B \alpha_B \, d\tau_B. \quad (7.23)$$

Using the result of Eq. (7.13) we obtain

$$\mathcal{H}_{11} = \nu_A(\tfrac{1}{2})(1) + \nu_B(1)(\tfrac{1}{2}) = \tfrac{1}{2}(\nu_A + \nu_B). \quad (7.24)$$

By the same procedure, the other three *diagonal* matrix elements (those on the principal diagonal) may be evaluated as

$$\mathcal{H}_{22} = \tfrac{1}{2}(\nu_A - \nu_B),$$
$$\mathcal{H}_{33} = \tfrac{1}{2}(-\nu_A + \nu_B), \quad (7.25)$$
$$\mathcal{H}_{44} = -\tfrac{1}{2}(\nu_A + \nu_B).$$

In the absence of spin coupling, all *off-diagonal* elements (all those not on the principal diagonal) are zero by virtue of the orthogonality of α and β. This is a general theorem not restricted to the case of two nuclei. We shall illustrate the result for $\mathcal{H}_{12}$:

$$\mathcal{H}_{12} = \int \Phi_1 H^{(0)} \Phi_2 \, d\tau$$

$$= \int \alpha_A \alpha_B [\nu_A (I_z)_A + \nu_B (I_z)_B] \alpha_A \beta_B \, d\tau$$

$$= \nu_A \int \alpha_A (I_z)_A \alpha_A \, d\tau_A \int \alpha_B \beta_B \, d\tau_B + \nu_B \int \alpha_A \alpha_A \, d\tau_A \int \alpha_B (I_z)_B \beta_B \, d\tau_B$$

$$= \nu_A(\tfrac{1}{2})(0) + \nu_B(1)(0). \quad (7.26)$$

With all off-diagonal elements equal to zero, the secular determinant becomes

$$
\begin{vmatrix}
\mathscr{H}_{11} - E & 0 & 0 & 0 \\
0 & \mathscr{H}_{22} - E & 0 & 0 \\
0 & 0 & \mathscr{H}_{33} - E & 0 \\
0 & 0 & 0 & \mathscr{H}_{44} - E
\end{vmatrix} = 0. \quad (7.27)
$$

Thus the left-hand side of Eq. (7.27) may be written as the product of four factors and is said to be factored into four 1×1 blocks, as indicated. The four solutions of the equation are, from Eqs. (7.24) and (7.25),

$$
\begin{aligned}
E_1 &= \mathscr{H}_{11} = \tfrac{1}{2}(\nu_A + \nu_B), \\
E_2 &= \mathscr{H}_{22} = \tfrac{1}{2}(\nu_A - \nu_B), \\
E_3 &= \mathscr{H}_{33} = \tfrac{1}{2}(-\nu_A + \nu_B), \\
E_4 &= \mathscr{H}_{44} = -\tfrac{1}{2}(\nu_A + \nu_B).
\end{aligned}
\quad (7.28)
$$

These energies are easily associated with the four energy levels depicted in Fig. 7.1 in our qualitative discussion of the AX system. The allowed transitions, which are also shown in Fig. 7.1, correspond to the selection rule $\Delta F_z = \pm 1$ (F_z was defined in Eq. (7.1). The origin of this selection rule will be taken up later in Section 7.9. It is clear from the values of the energy levels in Eq. (7.28) that each observed line results from two transitions with precisely the same frequency.

The calculation of the wave function corresponding to each energy level is carried out by evaluating the ratios of the C_n's from Eq. (7.11) (for each value of E in turn) and substitution of the C_n into Eq. (7.4).[†] In this case the calculation is trivial, since for each energy level only one $C_n \neq 0$. Thus

$$
\begin{aligned}
\psi_1 &= \Phi_1 = \alpha\alpha, & \psi_2 &= \Phi_2 = \alpha\beta, \\
\psi_3 &= \Phi_3 = \beta\alpha, & \psi_4 &= \Phi_4 = \beta\beta.
\end{aligned}
\quad (7.29)
$$

This result, that the true wave functions are identical with the basis product functions, is completely general for any case where there is no spin coupling. It is only the spin coupling interaction, not the chemical shift, that causes the basis functions to *mix*. We shall see in the next section that even with coupling many of the basis functions do not mix.

---

† More efficient procedures employing matrix algebra are normally used in practice for finding the *eigenvector* that diagonalizes the Hamiltonian matrix for each *eigenvalue* E_k. There is no difference in the result, and for simple cases the algebra in the scheme used here is not too cumbersome.

7.7 Factoring the Secular Equation

Before extending our treatment to two coupled nuclei, it is helpful to consider the conditions that cause zero elements to appear in the secular equation. With this knowledge we can avoid the effort of calculating many of the elements specifically for each case we study; furthermore, the presence of zero elements usually results in the secular determinant being factored into several equations of much smaller order, the solution of which is simpler than that of a high-order equation.

Suppose the basis functions used to construct the secular equation, the Φ_n, are eigenfunctions of some operator F. Consider two of these functions, Φ_m and Φ_n, with eigenvalues f_m and f_n, respectively. Then

$$F\Phi_m = f_m\Phi_m, \qquad F\Phi_n = f_n\Phi_n. \qquad (7.30)$$

Suppose further that the operator F commutes with the Hamiltonian:

$$F\mathcal{H} - \mathcal{H}F = 0. \qquad (7.31)$$

From these premises it is shown in standard texts on quantum mechanics[107] that

$$(f_m - f_n) \int \Phi_m^* \mathcal{H} \Phi_n \, d\tau = 0,$$

$$(f_m - f_n)\mathcal{H}_{mn} = 0. \qquad (7.32)$$

Thus if $f_m \neq f_n$, $\mathcal{H}_{mn} = 0$; that is, if Φ_m and Φ_n have *different* eigenvalues of F, the matrix element connecting them in the secular equation must be zero. (If $f_m = f_n$, nothing can be said from Eq. (7.32) about the value of $\mathcal{H}_{mn}$.)

There are two types of operators F that are important in the treatment of nuclear spin systems. One is the class of operators describing the symmetry of many molecules. We shall defer a discussion of symmetry until Section 7.12.

The other operator is F_z, which was defined in Eq. (7.1). Since α and β are eigenfunctions of I_z, the product functions Φ_n are eigenfunctions of F_z with eigenvalues equal to the sum of $+\frac{1}{2}$ for each time an α appears in the product function and $-\frac{1}{2}$ for each time a β appears in the function. By using the well-established commutation rules for angular momentum, it can be shown that F_z and $\mathcal{H}$ commute, so that Eq. (7.32) is applicable.[106] (We shall not present the details of the proof here.) For example, in the general (coupled) two-spin case, Eq. (7.2) shows that the four basis functions are classified according to $F_z = 1$, 0, or -1. Only Φ_2 and Φ_3, which have the same value of F_z, can mix. Thus only $\mathcal{H}_{23}$ and $\mathcal{H}_{32}$ might be nonzero; all 10 other off-diagonal elements of the secular equation are zero.

Further factorization of the secular equation according to F_z can sometimes be accomplished to a very high degree of approximation, even when several functions have the same value of F_z. We shall explore this point further in Sections 7.8 and 7.17.

7.8 Two Coupled Spins

We are now in position to complete our calculation of the AB system in general, with no restrictions whatsoever regarding the magnitude of the coupling constant J_{AB}. By virtue of the factoring due to F_z the secular equation is

$$\begin{vmatrix} \mathcal{H}_{11} - E & 0 & 0 & 0 \\ 0 & \mathcal{H}_{22} - E & \mathcal{H}_{23} & 0 \\ 0 & \mathcal{H}_{32} & \mathcal{H}_{33} - E & 0 \\ 0 & 0 & 0 & \mathcal{H}_{44} - E \end{vmatrix} = 0. \qquad (7.33)$$

The secular equation is always symmetric about the principal diagonal; hence $\mathcal{H}_{23} = \mathcal{H}_{32}$. We thus have five matrix elements to evaluate. Since

$$\mathcal{H} = \mathcal{H}^{(0)} + \mathcal{H}^{(1)},$$

we need evaluate only the portion of the matrix elements arising from $\mathcal{H}^{(1)}$ and then merely add the portion evaluated in Section 7.6 from $\mathcal{H}^{(0)}$.

For the first matrix element we find, using Eqs. (7.9) and (7.20),

$$\mathcal{H}_{11}^{(1)} = \int \Phi_1 \mathcal{H}^{(1)} \Phi_1 \, d\tau$$

$$= J_{AB} \int \alpha_A \alpha_B [(I_x)_A (I_x)_B + (I_y)_A (I_y)_B + (I_z)_A (I_z)_B] \alpha_A \alpha_B \, d\tau. \qquad (7.34)$$

As was the case in Section 7.6, the integral in Eq. (7.34) is really a double integral over the spin coordinates of A and of B. By separating the integrations we obtain

$$\mathcal{H}_{11}^{(1)} = J_{AB} \left[\int \alpha_A (I_x)_A \alpha_A \, d\tau_A \int \alpha_B (I_x)_B \alpha_B \, d\tau_B \right.$$

$$+ \int \alpha_A (I_y)_A \alpha_A \, d\tau_A \int \alpha_B (I_y)_B \alpha_B \, d\tau_B$$

$$\left. + \int \alpha_A (I_z)_A \alpha_A \, d\tau_A \int \alpha_B (I_z)_B \alpha_B \, d\tau_B \right]. \qquad (7.35)$$

Introducing values from Eqs. (7.13), (7.14), and (7.15), we obtain

$$\mathcal{H}_{11}^{(1)} = J_{AB}[0 + 0 + (\tfrac{1}{2})(\tfrac{1}{2})] = \tfrac{1}{4} J_{AB}. \qquad (7.36)$$

For $\mathcal{H}_{22}$ the equation analogous to Eq. (7.35) is

$$\mathcal{H}_{22}^{(1)} = J_{AB} \left[\int \alpha I_x \alpha \, d\tau \int \beta I_x \beta \, d\tau + \int \alpha I_y \alpha \, d\tau \int \beta I_y \beta \, d\tau \right.$$
$$\left. + \int \alpha I_z \alpha \, d\tau \int \beta I_z \beta \, d\tau \right]$$
$$= J_{AB}[0 + 0 + (\tfrac{1}{2})(-\tfrac{1}{2})]$$
$$= -\tfrac{1}{4}J_{AB}. \tag{7.37}$$

Similar computations give

$$\mathcal{H}_{33}^{(1)} = -\tfrac{1}{4}J_{AB},$$
$$\mathcal{H}_{44}^{(1)} = \tfrac{1}{4}J_{AB}. \tag{7.38}$$

The lone nonzero off-diagonal element may be evaluated in a similar manner:

$$\mathcal{H}_{23} = \int \Phi_2 \mathcal{H}^{(1)} \Phi_3 \, d\tau$$

$$= J_{AB} \int \alpha_A \beta_B [(I_x)_A (I_x)_B + (I_y)_A (I_y)_B + (I_z)_A (I_z)_B] \beta_A \alpha_B \, d\tau$$

$$= J_{AB} \left[\int \alpha_A (I_x)_A \beta_A \, d\tau_A \int \beta_B (I_x)_B \alpha_B \, d\tau_B \right.$$
$$+ \int \alpha_A (I_y)_A \beta_A \, d\tau_A \int \beta_B (I_y)_B \alpha_B \, d\tau_B$$
$$\left. + \int \alpha_A (I_z)_A \beta_A \, d\tau_A \int \beta_B (I_z)_B \alpha_B \, d\tau_B \right]$$

$$= J_{AB}[(\tfrac{1}{2})(\tfrac{1}{2}) + (\tfrac{1}{2}i)(-\tfrac{1}{2}i) + 0]$$

$$= \tfrac{1}{2}J_{AB}. \tag{7.39}$$

Adding the contribution from $\mathcal{H}^{(0)}$ from Eqs. (7.24) and (7.25), we find that the complete secular equation factors into three equations:

$$[\tfrac{1}{2}(\nu_A + \nu_B) + \tfrac{1}{4}J_{AB} - E] = 0; \tag{7.40}$$

$$\begin{vmatrix} \tfrac{1}{2}(\nu_A - \nu_B) - \tfrac{1}{4}J_{AB} - E & \tfrac{1}{2}J_{AB} \\ \tfrac{1}{2}J_{AB} & \tfrac{1}{2}(-\nu_A + \nu_B) - \tfrac{1}{4}J_{AB} - E \end{vmatrix} = 0; \tag{7.41}$$

$$[-\tfrac{1}{2}(\nu_A + \nu_B) + \tfrac{1}{4}J_{AB} - E] = 0. \tag{7.42}$$

Equations (7.40) and (7.42) immediately give the values of two energy levels, E_1 and E_4. Equation (7.41) is a quadratic equation that is readily solved to give

$$E_2 = \tfrac{1}{2}[(\nu_A - \nu_B)^2 + J_{AB}^2]^{1/2} - \tfrac{1}{4}J_{AB},$$
$$E_3 = -\tfrac{1}{2}[(\nu_A - \nu_B)^2 + J_{AB}^2]^{1/2} - \tfrac{1}{4}J_{AB}. \tag{7.43}$$

The wave functions ψ_1 and ψ_4 are identical with Φ_1 and Φ_4, respectively, as we showed in Section 7.7. The functions ψ_2 and ψ_3 are linear combinations (mixtures) of Φ_2 and Φ_3, the extent of mixing depending on the ratio of $(\nu_A - \nu_B)/J_{AB}$, as shown in the following expressions:†

$$\psi_1 = \Phi_1,$$

$$\psi_2 = \frac{1}{(1 + Q^2)^{1/2}} (\Phi_2 + Q\Phi_3),$$

$$\psi_3 = \frac{1}{(1 + Q^2)^{1/2}} (-Q\Phi_2 + \Phi_3), \qquad (7.44)$$

$$\psi_4 = \Phi_4.$$

In Eq. (7.44) Q is defined as

$$Q = \frac{J_{AB}}{(\nu_A - \nu_B) + [(\nu_A - \nu_B)^2 + J_{AB}^2]^{1/2}}. \qquad (7.45)$$

We shall discuss the spectrum arising from an AB system in detail in Section 7.10.

One point regarding the expressions in Eq. (7.43) deserves mention at this time: If $(\nu_A - \nu_B) \gg J_{AB}$ (the general AX case), J_{AB}^2 is negligible compared with $(\nu_A - \nu_B)^2$ and may be dropped. The resultant expressions for E_2 and E_3 are then considerably simplified. Note that the same result could have been achieved by dropping the off-diagonal elements $\frac{1}{2}J_{AB}$ when the secular equation was written. Dropping such small off-diagonal terms leads to a factorization of the secular equation beyond that given by the factorization according to F_z. This is a general and extremely important procedure, applicable to cases where certain differences in chemical shifts (expressed in Hz) are large compared with the corresponding J's. This is tantamount to a classification of the basis functions, not only according to the total F_z, but also according to $F_z(G)$, where G refers to only strongly coupled nuclei. The definition of $F_z(G)$ and the application to the ABX case will be taken up in Section 7.17.

7.9 Selection Rules and Intensities

We reviewed the basic concepts of absorption of radiation in Section 2.3. We now wish to derive rules governing radiative transitions in systems consisting of N spin-coupled nuclei of spin $\frac{1}{2}$. For such a system

---

† The calculation of the C_n in Eq. (7.4) is easily carried out by using minors of the secular determinant or by matrix methods. We shall not present the details of the computation.

we need to consider the interaction between the applied rf field in the x direction and the x component of magnetization, M_x

$$M_x = \frac{1}{2\pi} \sum_{i=1}^{N} \gamma_i (I_x)_i. \tag{7.46}$$

Application of time-dependent perturbation theory leads to an expression which shows that the intensity of a spectral line (transition from state p to state q) is proportional to the *square* of the following matrix element, or integral:

$$M_{pq} \equiv \int \psi_p \sum_{i=1}^{N} \gamma_i (I_x)_i \psi_q \, d\tau$$

$$= \sum_{i=1}^{N} \gamma_i \int \psi_p (I_x)_i \psi_q \, d\tau. \tag{7.47}$$

We recall that each of the ψ's is either one of the basis functions or a linear combination of basis functions. In the former case ψ is merely a product of α's and β's, one for each nucleus. If we examine the form of one of the terms of Eq. (7.47), say the one for $i = 1$, we find that the integral is really a multiple integral over all nuclear spin coordinates. The integral over nucleus 1, which involves $(I_x)_1$, is zero (cf. Eq. (7.15)) unless nucleus 1 changes its spin state from α to β or from β to α. At the same time the integrals over all nuclei other than nucleus 1 will be different from zero only if there is no change in spin state for each of these nuclei. Thus the only permitted transition has only one nucleus changing its spin state by $\Delta I_z = \pm 1$.

In general, where ψ is a linear combination of basis functions, the requirement for an allowed transition is less stringent, for the I_x operator for each of the strongly coupled nuclei may contribute to a given transition. Even so, the results of the previous paragraph may readily be generalized to give a selection rule $\Delta F_z = \pm 1$. When the system may be divided into sets of strongly coupled nuclei, with only weak coupling between sets, we define $F_z(G)$ for each set, and the more restrictive selection rules $\Delta F_z(G) = \pm 1$ must be obeyed for each G.

7.10 The AB Spectrum

We found in Section 7.2 that the values of F_z for the two-spin system are $+1$, 0, 0, and -1 for states (1), (2), (3), and (4), respectively. Thus the selection rule $\Delta F_z = \pm 1$ forbids transitions between states (2) and (3) and between states (1) and (4). There are then four allowed transitions:

(3) $\rightarrow$ (1), (4) $\rightarrow$ (2), (4) $\rightarrow$ (3), and (2) $\rightarrow$ (1); but the selection rule itself gives no indication of the relative probabilities of these transitions or the intensities of the corresponding spectral lines. However, by using the expressions derived in Section 7.8 for the energy levels and wave functions, we can calculate the frequencies and relative intensities of the four lines of the AB spectrum.

We denote by $T_{p \rightarrow q}$ the frequency of the transition from state p to state q and by $M_{p \rightarrow q}$ the matrix element of M_x (Eq. (7.47)). The relative intensities of the spectral lines are thus proportional to $(M_{p \rightarrow q})^2$. From Eqs. (7.40)–(7.43) we obtain

$$
\begin{aligned}
l_1 &= T_{3 \rightarrow 1} = \tfrac{1}{2}(\nu_A + \nu_B) + \tfrac{1}{2}[(\nu_A - \nu_B)^2 + J^2]^{1/2} + \tfrac{1}{2}J, \\
l_2 &= T_{4 \rightarrow 2} = \tfrac{1}{2}(\nu_A + \nu_B) + \tfrac{1}{2}[(\nu_A - \nu_B)^2 + J^2]^{1/2} - \tfrac{1}{2}J, \\
l_3 &= T_{2 \rightarrow 1} = \tfrac{1}{2}(\nu_A + \nu_B) - \tfrac{1}{2}[(\nu_A - \nu_B)^2 + J^2]^{1/2} + \tfrac{1}{2}J, \\
l_4 &= T_{4 \rightarrow 3} = \tfrac{1}{2}(\nu_A + \nu_B) - \tfrac{1}{2}[(\nu_A - \nu_B)^2 + J^2]^{1/2} - \tfrac{1}{2}J.
\end{aligned}
\tag{7.48}
$$

We have dropped the subscript for J since there is one coupling constant involved and have introduced the expressions l_i as a convenient notation for the frequencies of the lines. It is apparent that lines 1 and 2 are always separated by J, as are lines 3 and 4. Thus the value of J may be extracted immediately from an AB spectrum. If J^2 is negligible relative to $(\nu_A - \nu_B)^2$, then the average of l_1 and l_2 immediately gives ν_A, and the average of l_3 and l_4 gives ν_B. In general, however, the calculation of ν_A and ν_B for an AB spectrum requires slightly more effort. The separation between l_1 and l_3 gives $[(\nu_A - \nu_B)^2 + J^2]^{1/2}$, and since J has already been found, $|\nu_A - \nu_B|$ may be calculated. The sum of l_2 and l_3 gives $\nu_A + \nu_B$, so that ν_A and ν_B may be calculated.

To find the relative intensities of the four AB lines we use the wave functions from Eq. (7.44) to evaluate the matrix elements defined by Eq. (7.47). (Since we are interested in the case involving only a single nuclear species, we can drop γ from this equation.)

$$
\begin{aligned}
M_{2 \rightarrow 1} &= \int \psi_1 [(I_x)_A + (I_x)_B] \psi_2 \, d\tau \\[2mm]
&= \frac{1}{(1 + Q^2)^{1/2}} \int \alpha\alpha [(I_x)_A + (I_x)_B](\alpha\beta + Q\beta\alpha) \, d\tau \\[2mm]
&= \frac{1}{(1 + Q^2)^{1/2}} \left\{ \int \alpha\alpha(I_x)_A \alpha\beta \, d\tau + Q \int \alpha\alpha(I_x)_A \beta\alpha \, d\tau \right. \\[2mm]
&\quad \left. + \int \alpha\alpha(I_x)_B \alpha\beta \, d\tau + Q \int \alpha\alpha(I_x)_B \beta\alpha \, d\tau \right\} \\[2mm]
&= \frac{1}{(1 + Q^2)^{1/2}} \left\{ \int \alpha(I_x)_A \alpha \, d\tau_A \int \alpha\beta \, d\tau_B \right.
\end{aligned}
$$

$$+ Q \int \alpha(I_x)_A\beta \, d\tau_A \int \alpha\alpha \, d\tau_B + \int \alpha\alpha \, d\tau_A \int \alpha(I_x)_B\beta \, d\tau_B$$

$$+ Q \int \alpha\beta \, d\tau_A \int \alpha(I_x)_B\alpha \, d\tau_B \bigg\}$$

$$= \frac{1}{(1 + Q^2)^{1/2}} \{0 + \tfrac{1}{2}Q + \tfrac{1}{2} + 0\}$$

$$= \frac{1 + Q}{2(1 + Q^2)^{1/2}}. \tag{7.49}$$

Similar calculations show that

$$M_{4\to2} = M_{2\to1},$$

$$M_{4\to3} = M_{3\to1} = \frac{1 - Q}{2(1 + Q^2)^{1/2}}. \tag{7.50}$$

The relative intensities of the lines are proportional to the $(M_{p\to q})^2$

$$M_{4\to2}^2 = M_{2\to1}^2 = \frac{(1 + Q)^2}{4(1 + Q^2)} = \frac{1}{4}\left[1 + \frac{2Q}{1 + Q^2}\right],$$

$$M_{4\to3}^2 = M_{3\to1}^2 = \frac{(1 - Q)^2}{4(1 + Q^2)} = \frac{1}{4}\left[1 - \frac{2Q}{1 + Q^2}\right]. \tag{7.51}$$

In order to simplify the expressions, we introduce the notation

$$2C \equiv [(\nu_A - \nu_B)^2 + J^2]^{1/2}. \tag{7.52}$$

Table 7.1 and Fig. 7.2 give the frequencies of the lines in terms of C. By using Eqs. (7.45) and (7.52), the relative intensities of the lines given in Eq. (7.51) may be expressed in terms of C and J, and are listed in Table 7.1.* From Table 7.1 and Fig. 7.2 it is evident that an AB spectrum consists of four lines with a symmetric weak, strong, strong, weak intensity distribution. The separation between the outer lines in either half of the spectrum is equal to J. The value of $|\nu_A - \nu_B|$ is easily found from the separation between the outer and the inner lines, for Eq. (7.52) may be rewritten

---

* One convenient way of transforming from the intensity expressions in terms of Q to those in terms of C and J is by defining

$$\sin \theta \equiv Q/(1 + Q^2)^{1/2}.$$

Expressions for $\cos \theta$ and $\sin 2\theta$ are easily obtained from trigonometric identities, and the expressions of Eq. (7.51) reduce to $\tfrac{1}{4}(1 \pm \sin 2\theta)$. Relative intensities are often expressed in this way for the AB system, and we shall use a similar notation in treating the ABX spectrum. Note that the angle θ has no physical significance.

Table 7.1

TRANSITIONS, FREQUENCIES, AND RELATIVE INTENSITIES
FOR THE AB SYSTEM

Line	Transition	Frequency (Hz)[a]	Relative intensity
1	$T_{3\rightarrow1}$	$C + \frac{1}{2}J$	$1 - \dfrac{J}{2C}$
2	$T_{4\rightarrow2}$	$C - \frac{1}{2}J$	$1 + \dfrac{J}{2C}$
3	$T_{2\rightarrow1}$	$-C + \frac{1}{2}J$	$1 + \dfrac{J}{2C}$
4	$T_{4\rightarrow3}$	$-C - \frac{1}{2}J$	$1 - \dfrac{J}{2C}$

[a] Referred to the center of the four-line pattern, $\frac{1}{2}(\nu_A + \nu_B)$.

$$\begin{aligned}
(\nu_A - \nu_B)^2 &= 4C^2 - J^2, \\
|\nu_A - \nu_B| &= [4C^2 - J^2]^{1/2} \\
&= [(2C - J)(2C + J)]^{1/2} \\
&= [(l_2 - l_3)(l_1 - l_4)]^{1/2}.
\end{aligned} \tag{7.53}$$

From Table 7.1 we see that the ratio of the *intensities* of the inner to the outer lines is

$$\frac{I(\text{inner})}{I(\text{outer})} = \frac{2C + J}{2C - J} = \frac{l_1 - l_4}{l_2 - l_3}. \tag{7.54}$$

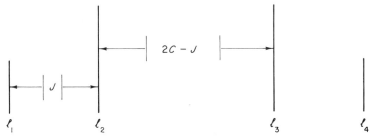

Fig. 7.2 Schematic representation of an AB spectrum.

$$\begin{aligned}
|J| &= |l_1 - l_2| = |l_3 - l_4|, \\
|\nu_A - \nu_B| &= [(l_1 - l_4)(l_2 - l_3)]^{1/2}.
\end{aligned}$$

Thus the intensities must accord with the frequencies of a set of four lines if they constitute an AB pattern.

The appearance of an AB spectrum is determined entirely by the ratio $J/(\nu_A - \nu_B)$, as was shown in Fig. 5.4. Note that the analysis of an AB spectrum gives only $|J|$ and $|\nu_A - \nu_B|$; we cannot determine from the spectrum itself the sign of J or which nucleus, A or B, is more shielded.

7.11 Spectral Contributions from Equivalent Nuclei

Let us examine the AB spectrum when $\nu_A = \nu_B$ (i.e., an A_2 system). From Eqs. (7.40)–(7.43) the energy levels are

$$E_1 = \nu_A + \tfrac{1}{4}J, \qquad E_2 = \tfrac{1}{4}J,$$
$$E_3 = -\tfrac{3}{4}J, \qquad E_4 = -\nu_A + \tfrac{1}{4}J. \qquad (7.55)$$

In principle, from the four allowed transitions we should be able to find the value of J. From Eq. (7.52), however, we find that $2C = J$. Table 7.1 shows, then, that two lines, I_2 and I_3, coincide at ν_A, while I_1 and I_4 have zero intensity.

The foregoing example is just one illustration of the lack of splitting due to spin coupling in spectra where certain types of equivalence exist among the nuclei. We shall state without proof an extremely important NMR theorem and its corollary:

1. The spectrum arising from a system of nuclear spins that includes one or more sets of *magnetically* equivalent nuclei is independent of the spin coupling between nuclei with a magnetically equivalent set. For example, the spectrum of an A_3B system does not in any way depend on J_{AA}, but the spectrum of an AA'BB' spectrum is a function of $J_{AA'}$. The proof of this theorem depends on commutation properties of angular momentum operators.[108]

2. A molecule in which *all* coupled nuclei have the same chemical shift gives a spectrum consisting of a single line. For example, CH_4, which has four magnetically equivalent protons, gives a single proton resonance line; but CH_3D does not, since here one of the set of coupled nuclei clearly has a "chemical shift" different from the rest. This corollary does not depend on the reason for the chemical equivalence, which may result from molecular symmetry or from an accidental equivalence in shielding. When *all* coupled nuclei in a molecule are chemically equivalent, they are also magnetically equivalent, since they are equally coupled ($J = 0$) to all *other* nuclei.

7.12 Symmetry of Wave Functions

The presence of symmetry in a molecule imposes severe restrictions on many chemical and spectral properties. Often the existence of symmetry permits considerable simplification in the analysis of NMR spectra.

We speak of a *symmetry operation* $\mathscr{R}$ as an operation which when applied to a molecule leaves it in a configuration that is *physically indistinguishable* from the original configuration. With regard to wave functions, it is ψ^2 that corresponds to a physically measurable quantity, not ψ itself. Hence we require that

$$\mathscr{R}\psi^2 = \psi^2. \qquad (7.56)$$

For a nondegenerate wave function, which is to remain normalized, Eq. (7.56) is valid *only* if the new ψ following the symmetry operation is either equal to the original ψ or is its negative:

$$\mathscr{R}\psi = +\psi \quad \text{or} \quad -\psi. \qquad (7.57)$$

A function that is unchanged by the operation is said to be *symmetric*; one changed into its negative is *antisymmetric*.

Let us now consider the application of this concept to a two-spin system, where the two spins are chemically equivalent (not necessarily magnetically equivalent) by virtue of their positions in a symmetric molecule. (Such nuclei are said to be *symmetrically equivalent*.) We shall designate the spins by the subscripts a and b, but this does not imply any difference between them. Following the procedure of Section 7.6, we can write four basis functions for this system:

$$
\begin{aligned}
\Phi_1 &= \alpha_a \alpha_b & \text{symmetric,} \\
\Phi_2 &= \alpha_a \beta_b & \text{asymmetric,} \\
\Phi_3 &= \beta_a \alpha_b & \text{asymmetric,} \\
\Phi_4 &= \beta_a \beta_b & \text{symmetric.}
\end{aligned} \qquad (7.58)
$$

If we perform the symmetry operation of interchanging these two identical nuclei, Φ_1 and Φ_4 will clearly be unchanged; hence we label them as symmetric functions. However, Φ_2 and Φ_3 are neither symmetric nor antisymmetric, since

$$
\begin{aligned}
\mathscr{R}\Phi_2 &\equiv \mathscr{R}\alpha_a\beta_b = \alpha_b\beta_a = \beta_a\alpha_b \equiv \Phi_3, \\
\mathscr{R}\Phi_2 &\neq \Phi_2 \quad \text{or} \quad -\Phi_2;
\end{aligned} \qquad (7.59)
$$

Φ_2 and Φ_3 have been labeled as *asymmetric*. They are not in themselves acceptable wave functions. We may, if we wish, use them as basis functions in a calculation involving the secular equation, and we shall ultimately obtain the correct solution to the problem. However, the calculations are greatly simplified by selecting in place of Φ_2 and Φ_3 two functions that have the proper symmetry and yet maintain the desirable properties of product basis functions. For the two-spin system this can be accomplished easily by selecting functions that are the sum and difference of Φ_2 and ϕ_3:

$$\Phi_2' = \frac{1}{\sqrt{2}} (\alpha_a\beta_b + \beta_a\alpha_b) \qquad \text{symmetric;}$$

$$\Phi_3' = \frac{1}{\sqrt{2}} (\alpha_a\beta_b - \beta_a\alpha_b) \qquad \text{antisymmetric.} \qquad (7.60)$$

The factor $1/\sqrt{2}$ maintains the normalization condition.

We know from Section 7.3 that the true wave functions $\psi_1, \ldots, \psi_4$ are linear combinations of the basis functions. If we begin with symmetrized functions, such as Φ_2' and Φ_3', then each of the ψ's can be formed exclusively from symmetric functions or exclusively from antisymmetric functions. Stated another way, *functions of different symmetry do not mix.* The result is that, like the situation with F_z, many off-diagonal elements of the secular equation must be zero, and the equation factors into several equations of lower order. We shall study an example of this factoring in Section 7.15, when we consider the A_2B system.

Simplification of the spectrum itself also results from the presence of symmetry, since transitions are permitted *only* between two symmetric or two antisymmetric states. We shall see in Section 7.15 that there is often a considerable reduction in the number of NMR lines.

For the two-spin system the only symmetry operation is the interchange of the two nuclei, and the correct linear combinations, Φ_2' and Φ_3', could be constructed by inspection. When three or more symmetrically equivalent nuclei are present, the symmetry operations consist of various permutations of the nuclei. The correct symmetrized functions can be determined systematically only by application of results from group theory. We shall not present the details of this procedure, but in Section 7.16 we shall use functions derived in this way for the A_3B system.

7.13 Summary of Rules for Calculating Spectra

In Sections 7.8 and 7.10 we derived in considerable detail the expressions for the secular equation and the transitions of the AB system. It is apparent that the calculation of each element in the secular equation from the general theory would become very tedious for systems of three or more nuclei. Fortunately, general rules have been derived to simplify the calculation.[106] We shall not derive these expressions, but we present the rules in a concise form to illustrate the simplicity of the calculation.

1. The calculation always begins with the writing of the 2^N simple product functions (e.g., $\alpha\alpha\beta\alpha$, etc.). In the absence of symmetry these serve as the basis functions.

2. If symmetry is present, suitable linear combinations of the product functions are used as basis functions Φ_n. Normally, group theory is used as an aid to selecting appropriate linear combinations.

3. The diagonal matrix elements of the Hamiltonian (or the secular equation) are

$$\mathscr{H}_{mm} = \sum_{i=1}^{N} \nu_i[(I_z)_i]_m + \tfrac{1}{4}\sum_{i<j}\sum J_{ij}T_{ij}, \tag{7.61}$$

where

$$[(I_z)_i]_m = \begin{cases} +\tfrac{1}{2} & \text{if nucleus } i \text{ has spin } \alpha \text{ in } \Phi_m, \\ -\tfrac{1}{2} & \text{if nucleus } i \text{ has spin } \beta \text{ in } \Phi_m; \end{cases}$$

and

$$T_{ij} = \begin{cases} +1 & \text{if spins } i \text{ and } j \text{ are parallel in } \Phi_m, \\ -1 & \text{if spins } i \text{ and } j \text{ are antiparallel in } \Phi_m. \end{cases}$$

For example, if $\Phi_m = \alpha_1\alpha_2\beta_3\alpha_4$, then

$$\mathscr{H}_{mm} = \tfrac{1}{2}(\nu_1 + \nu_2 - \nu_3 + \nu_4) + \tfrac{1}{4}(J_{12} - J_{13} + J_{14} - J_{23} + J_{24} - J_{34}).$$

4. The off-diagonal matrix elements are

$$\mathscr{H}_{mn} = \tfrac{1}{2}UJ_{ij} \tag{7.62}$$

where

$$U = \begin{cases} 1 & \text{if } \Phi_m \text{ and } \Phi_n \text{ differ } only \text{ in interchange of spins of } i \text{ and } j \\ 0 & \text{otherwise.} \end{cases}$$

For example, if $\Phi_m = \alpha_1\alpha_2\beta_3\alpha_4$ and $\Phi_n = \alpha_1\beta_2\alpha_3\alpha_4$, then $\mathscr{H}_{mn} = \tfrac{1}{2}J_{23}$. But if $\Phi_m = \alpha_1\alpha_2\beta_3\beta_4$ and $\Phi_n = \beta_1\beta_2\alpha_3\alpha_4$, then $\mathscr{H}_{mn} = 0$.

5. Matrix elements involving basis functions that are linear combinations of product functions are evaluated by expansion, that is, as the sum of several integrals. For example, if Φ_2' and ϕ_3' are the functions of Eq. (7.60), then

$$\mathscr{H}_{2'3'} = \tfrac{1}{2}\left(\int \alpha\beta\mathscr{H}\alpha\beta \, d\tau + \int \alpha\beta\mathscr{H}\beta\alpha \, d\tau + \int \beta\alpha\mathscr{H}\alpha\beta \, d\tau \right.$$
$$\left. + \int \beta\alpha\mathscr{H}\beta\alpha \, d\tau\right).$$

6. Many matrix elements need not be evaluated since the secular equation factors according to (a) symmetry; (b) F_z; and (c) $F_z(G)$, to a high degree of approximation.

7. The energy levels are found as the solutions to the secular equation by treating each factor separately.

8. The frequencies of the spectral lines are calculated as differences between energy levels according to the selection rules for symmetry, F_z and $F_z(G)$.

9. The wave functions are found either from the original 2^N linear equations or as the eigenvectors that diagonalize the Hamiltonian matrix.

10. The intensities of the spectral lines are found from Eq. (7.47).

When the application of step 6 results in factors no larger than 2×2, the equations may be solved readily and algebraic expressions derived for the frequencies and intensities of the spectral lines in terms of ν's and J's. The AB case served as an example of this procedure. Most other simple systems have been treated in this way; we shall consider several of them in succeeding sections.

When factors larger than 2×2 are present in the factored secular equation, general algebraic solutions are not possible, and the analysis of each spectrum must be carried out individually, usually by a trial and error procedure. First, assumed values of ν's and J's are used with the foregoing rules to calculate a spectrum, which is compared with an experimentally determined spectrum. The trial values of ν and J are then altered systematically until a suitable fit is obtained. This process is clearly adaptable to a high-speed digital computer, which can be programmed to carry out steps 1–10 for each choice of trial parameters and to check for best agreement between calculated and experimental spectra according to a least squares criterion.

The first widely used iterative programs for NMR spectral analysis were those of Castellano and Bothner-By[109] (LAOCOON) and Swalen and Reilly[110] (NMRIT), with almost all subsequently developed programs tracing their origins back to one of these two. The approach used in LAOCOON is to compute the frequencies of the transitions from assumed input parameters, permit the spectroscopist to assign the lines in the observed spectrum to these computed frequencies, and systematically modify the parameters (chemical shifts and coupling constants) in an iterative calculation that seeks convergence of computed and observed frequencies. The approach in NMRIT differs in that the spectroscopist must assign each observed spectral line to the two energy levels between which the transition occurs, and the iterative steps aim toward convergence of computed and observed energy levels. Although both programs have been widely used in their original and improved versions, the LAOCOON approach seems to be somewhat more satisfactory and simpler to use. Improvements in both programs have been introduced to take advantage of symmetry factoring of the equations due to equalities in chemical shifts, magnetic equivalence of nuclei, and effects of large chemical shift differences. As we shall see in Chapter 9, double resonance results can provide considerable help in spectral analysis. Such data can be introduced readily into NMRIT type programs or into another program, UEAITR,[111] based on the LAOCOON philosophy. There are several excellent reviews describing the structure and application of the various NMR computer programs.[112-114] Most of the programs are available from the Quantum Chemistry Program Exchange and other sources.[115]

7.14 The Three-Spin System: ABC

Let us now consider the general three-spin system with no restrictions regarding relative sizes of chemical shifts and coupling constants. Such ABC systems are frequently found, for example, among vinyl compounds, trisubstituted aromatics, and disubstituted pyridines. There are 2^3 basis functions that can be formed as products without regard to symmetry considerations. These can be classified into four sets according to the values of F_z, as indicated in Table 7.2. Application of the selection rule $\Delta F_z = \pm 1$ shows that there are 15 allowed transitions.

Because the functions for $F_z = \frac{1}{2}$ and $-\frac{1}{2}$ lead to cubic equations, it is not possible to express the transition energies in simple algebraic form as functions of the six parameters ν_A, ν_B, ν_C, J_{AB}, J_{AC}, and J_{BC}. Hence, an analysis of an ABC spectrum must be carried out for each case individually, using an iterative procedure such as that mentioned in the preceding section. Often the ABC spectrum may be roughly approximated as an ABX spectrum and analyzed by procedures we shall discuss in Section 7.17. The parameters thus obtained can be used as initial guesses for the iterative treatment of the ABC system. In favorable cases excellent agreement between observed and calculated spectra may be obtained, as indicated in Fig. 7.3. One must be extremely careful in using iterative computer analyses for ABC and other complex systems since the values of the

Table 7.2

BASIS FUNCTIONS FOR THE ABC SYSTEM

$\Phi_1 = \alpha\alpha\alpha,$			$F_z = \frac{3}{2};$
$\Phi_2 = \alpha\alpha\beta,$	$\Phi_3 = \alpha\beta\alpha,$	$\Phi_4 = \beta\alpha\alpha,$	$F_z = \frac{1}{2};$
$\Phi_5 = \beta\beta\alpha,$	$\Phi_6 = \beta\alpha\beta,$	$\Phi_7 = \alpha\beta\beta,$	$F_z = -\frac{1}{2};$
$\Phi_8 = \beta\beta\beta,$			$F_z = -\frac{3}{2}.$

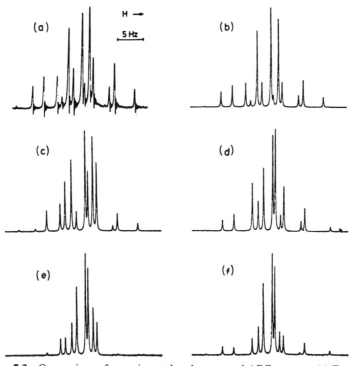

Fig. 7.3 Comparison of experimental and computed ABC spectra. (a) Experimental spectrum of 2-chlorothiophene; (b)–(f) Five of the eight different spectra that agree in all frequencies with the experimental spectrum but are computed from entirely different sets of parameters. Note that the intensities of the lines in (b) agree with those of the experimental spectrum, while the other computed spectra have very different intensity distributions (Diehl *et al.*[114]).

parameters selected to provide the "best" agreement with the experimental frequencies may not be unique. In fact, if the criteria of agreement for both frequencies and intensities are not sufficiently stringent, these parameters may not even represent one correct solution.

For the ABC system, alternative mathematical treatments have been developed which provide all possible sets of parameters consistent with the observed spectrum.[116] The complexity of the mathematics virtually limits its application to the three-spin case.

7.15 The A₂B System

When two of three strongly coupled nuclei are magnetically equivalent, the presence of symmetry results in considerable simplification of the calculation. In the first place, the spectrum is now determined by only

Table 7.3

Basis Functions for the A_2B System

Function	A_2	B	F_z	Symmetry
Φ_1	$\alpha\alpha$	α	$\frac{3}{2}$	s
Φ_2	$(1/\sqrt{2})(\alpha\beta + \beta\alpha)$	α	$\frac{1}{2}$	s
Φ_3	$(1/\sqrt{2})(\alpha\beta - \beta\alpha)$	α	$\frac{1}{2}$	a
Φ_4	$\beta\beta$	α	$-\frac{1}{2}$	s
Φ_5	$\alpha\alpha$	β	$\frac{1}{2}$	s
Φ_6	$(1/\sqrt{2})(\alpha\beta + \beta\alpha)$	β	$-\frac{1}{2}$	s
Φ_7	$(1/\sqrt{2})(\alpha\beta - \beta\alpha)$	β	$-\frac{1}{2}$	a
Φ_8	$\beta\beta$	β	$-\frac{3}{2}$	s

two chemical shifts, ν_A and ν_B, and one coupling constant, J_{AB}. We saw in Section 7.11 that the appearance of the spectrum does not depend on J_{AA}, and there is no way that this parameter can be derived from the observed spectrum.*

Each of the eight basis functions is now formed as the product of the spin function of B with one of the symmetrized functions given in Eqs. (7.58) and (7.60). These basis functions are given in Table 7.3.

As with the ABC case, the basis functions divide into four sets according to F_z, with 1, 3, 3, and 1 functions in each set. However, of the three functions in the set with $F_z = \frac{1}{2}$ or $-\frac{1}{2}$, two are symmetric and one antisymmetric. Hence each of the two 3×3 blocks of the secular equation factors into a 2×2 block and a 1×1 block. Algebraic solutions are thus possible. Furthermore, the presence of symmetry reduces the number of allowed transitions from 15 to 9, since no transitions are allowed between states of different symmetry.

The computation of transition frequencies and intensities can be carried out according to the rules of Section 7.13. We shall not reproduce the expressions thus derived, but rather we can illustrate the behavior of the spectrum in Fig. 7.4. It may be noted from this figure that the frequency of line 3 always gives ν_B. (This is the single transition allowed between the antisymmetric states ψ_3 and ψ_7.) The value of ν_A may also be found readily; it is the average of the frequencies of lines 5 and 7. From the detailed expressions for the line frequencies it is easily shown that

$$J = \tfrac{1}{3}|l_1 - l_4 + l_6 - l_8|, \tag{7.63}$$

---

* J_{AA} might be determined, however, from studies of an isotopic derivative (Chapter 5) or from oriented molecules (Section 7.28).

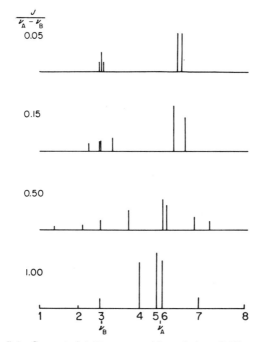

Fig. 7.4 Computed A₂B spectra with variation of $J/(\nu_A - \nu_B)$.

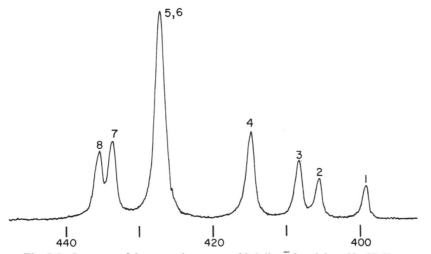

Fig. 7.5 Spectrum of the aromatic protons of 2,6-di-*tert*-butylphenol in CDCl₃, an example of an A₂B spectrum.

where the l's are the line frequencies. Detailed tables of frequencies and relative intensities for various ratios of $|J/(\nu_A - \nu_B)|$ are available[104] and may be used to construct spectra for comparison with one being analyzed.

Of the nine transitions expected for an A_2B spectrum, a maximum of only eight are observed, since the ninth is a *combination* line, corresponding to the simultaneous "flipping" of all three nuclei, and has extremely low intensity. Sometimes other lines are not resolved; for example, in the A_2B spectrum shown in Fig. 7.5 lines 5 and 6 are unresolved.

The sign of J cannot be determined from the spectrum, but, unlike the AB case, it *is* possible simply from the areas under the lines to determine which nucleus, A or B, is more shielded.

7.16 The A_3B System; Subspectral Analysis

The symmetry inherent in this four-spin system can be employed to simplify the treatment. The 16th-order secular determinant factors by symmetry and F_z into blocks no larger than 2×2, so that again algebraic solution is possible. Here too the appearance of the spectrum depends entirely on the ratio $J_{AB}/(\nu_A - \nu_B)$, and the analysis of an A_3B spectrum can be carried out using a table analogous to that used for the A_2B system (see, e.g., Corio[104]).

An alternative procedure for analyzing an A_3B spectrum, as well as many more complex spectra, is the method of *subspectral analysis*. Many complex spectra can be shown to contain one or more simpler subspectra, which may be analyzed separately provided the observed spectral lines

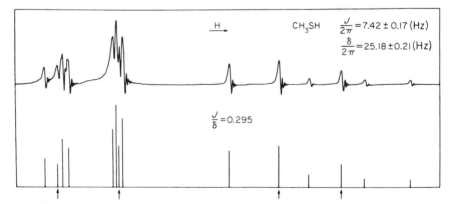

Fig. 7.6 Application of subspectral analysis to the proton resonance spectrum of methyl mercaptan (40 MHz). The calculated spectrum is shown for $J/(\nu_A - \nu_B) = 0.295$ with the ab subspectrum indicated. Spectrum and analysis from Corio.[104]

can be properly assigned to the correct subspectra. For example, the A_3B spectrum is composed of an ab subspectrum (small letters are used to denote subspectra), as well as other lines not belonging to a subspectrum of a simpler system. However, since both the A_3B spectrum and the ab subspectrum are determined completely by the parameters $(\nu_A - \nu_B)$ and J_{AB}, analysis of the ab subspectrum provides all the information that could be extracted from the more complex A_3B spectrum. An example of this analysis is shown in Fig. 7.6. The essential point, of course, is the selection of the lines belonging to the ab subspectrum. This must be done in accordance with the spacing and intensity relationships given in Section 7.10.

The method of subspectral analysis has its greatest utility with more complex spectra, such as $ABB'XX'$. A discussion of such applications is beyond the scope of this book, but is given in a recent review article.[117]

7.17 The ABX System

Intermediate in complexity between the AMX system, which can be analyzed by first-order procedures, and the completely strongly coupled ABC system, which must be treated by individual computer-aided analysis, is the ABX system. The presence of one nucleus only weakly coupled to the others permits factoring of the secular equation so that algebraic solutions are possible. We shall summarize the results of the solution of the secular equation and shall devote considerable attention to the application of the resulting equations to the analysis of observed ABX spectra. There are two reasons for this emphasis on ABX spectra. First, ABX spectra occur frequently, for example in trisubstituted aromatics and in vinyl systems. Second, the ABX system is the simplest one in which many important concepts common to more complex systems can be demonstrated. These include the effects of the signs of coupling constants and the "deceptive simplicity" often found in complex spectra.

The basis functions for the ABX system are just those used in Eq. (7.63) for the general three-spin system. However, because $(\nu_A - \nu_X)$ and $(\nu_B - \nu_X)$ are much larger than J_{AX} and J_{BX}, we can define an F_z for the AB nuclei separately from F_z for the X nucleus (cf. Section 7.8). The basis functions classified in this way are given in Table 7.4. Of the three functions with $F_z = \frac{1}{2}$, Φ_3 and Φ_4 have the same values of $F_z(AB)$ and $F_z(X)$, but Φ_2 is in a separate class and does not mix with Φ_3 and Φ_4. Thus the 3×3 block of the secular equation factors into a 2×2 block and a 1×1 block. Analogous factoring occurs for the 3×3 block arising from the three functions with $F_z = -\frac{1}{2}$.

Table 7.4

BASIS FUNCTIONS FOR THE ABX SYSTEM

Function	AB	X	$F_z(AB)$	$F_z(X)$	F_z
Φ_1	$\alpha\alpha$	α	1	$\frac{1}{2}$	$\frac{3}{2}$
Φ_2	$\alpha\alpha$	β	1	$-\frac{1}{2}$	$\frac{1}{2}$
Φ_3	$\alpha\beta$	α	0	$\frac{1}{2}$	$\frac{1}{2}$
Φ_4	$\beta\alpha$	α	0	$\frac{1}{2}$	$\frac{1}{2}$
Φ_5	$\beta\beta$	α	-1	$\frac{1}{2}$	$-\frac{1}{2}$
Φ_6	$\beta\alpha$	β	0	$-\frac{1}{2}$	$-\frac{1}{2}$
Φ_7	$\alpha\beta$	β	0	$-\frac{1}{2}$	$-\frac{1}{2}$
Φ_8	$\beta\beta$	β	-1	$-\frac{1}{2}$	$-\frac{3}{2}$

The computation of the matrix elements, the solution of the factors of the secular equation, and the calculation of the transition frequencies and intensities are readily carried out using the procedure outlined in Section 7.13. The results are summarized in Table 7.5 in terms of the following commonly employed notation.

$$2D_+ \cos 2\theta_+ = (\nu_A - \nu_B) + \tfrac{1}{2}(J_{AX} - J_{BX}); \tag{7.64a}$$

Table 7.5

ABX SPECTRUM: FREQUENCIES AND RELATIVE INTENSITIES[a,b]

Line	Origin	Energy	Relative intensity
1	B	$\nu_{AB} + \tfrac{1}{4}(-2J_{AB} - J_{AX} - J_{BX}) - D_-$	$1 - \sin 2\theta_-$
2	B	$\nu_{AB} + \tfrac{1}{4}(-2J_{AB} + J_{AX} + J_{BX}) - D_+$	$1 - \sin 2\theta_+$
3	B	$\nu_{AB} + \tfrac{1}{4}(2J_{AB} - J_{AX} - J_{BX}) - D_-$	$1 + \sin 2\theta_-$
4	B	$\nu_{AB} + \tfrac{1}{4}(2J_{AB} + J_{AX} + J_{BX}) - D_+$	$1 + \sin 2\theta_+$
5	A	$\nu_{AB} + \tfrac{1}{4}(-2J_{AB} - J_{AX} - J_{BX}) + D_-$	$1 + \sin 2\theta_-$
6	A	$\nu_{AB} + \tfrac{1}{4}(-2J_{AB} + J_{AX} + J_{BX}) + D_+$	$1 + \sin 2\theta_+$
7	A	$\nu_{AB} + \tfrac{1}{4}(2J_{AB} - J_{AX} - J_{BX}) + D_-$	$1 - \sin 2\theta_-$
8	A	$\nu_{AB} + \tfrac{1}{4}(2J_{AB} + J_{AX} + J_{BX}) + D_+$	$1 - \sin 2\theta_+$
9	X	$\nu_X - \tfrac{1}{2}(J_{AX} + J_{BX})$	1
10	X	$\nu_X + D_+ - D_-$	$\cos^2(\theta_+ - \theta_-)$
11	X	$\nu_X - D_+ + D_-$	$\cos^2(\theta_+ - \theta_-)$
12	X	$\nu_X + \tfrac{1}{2}(J_{AX} + J_{BX})$	1
13	Comb.	$2\nu_{AB} - \nu_X$	0
14	Comb.(X)	$\nu_X - D_+ - D_-$	$\sin^2(\theta_+ - \theta_-)$
15	Comb.(X)	$\nu_X + D_+ + D_-$	$\sin^2(\theta_+ - \theta_-)$

[a] Pople et al.[106]

[b] See Eqs. (7.64) and (7.65) for definition of terms.

$$2D_+ \sin 2\theta_+ = J_{AB}; \tag{7.64b}$$

$$2D_- \cos 2\theta_- = (\nu_A - \nu_B) - \tfrac{1}{2}(J_{AX} - J_{BX}); \tag{7.64c}$$

$$2D_- \sin 2\theta_- = J_{AB}; \tag{7.64d}$$

$$\nu_{AB} = \tfrac{1}{2}(\nu_A + \nu_B). \tag{7.65}$$

The four quantities D_+, D_-, θ_+, and θ_- are *defined* by Eqs. (7.64); D_+ and D_- are analogous to the quantity C utilized in the analysis of the AB spectrum (cf. Eqs. (7.52)), as we shall see below. The angles θ_+ and θ_- are analogous to the quantity θ mentioned in the AB analysis in Section 7.10. They have *no* physical significance and merely provide a convenient way of expressing the relations of the spectral parameters to the intensities of the spectral lines.

There are certain limitations that we shall impose on the quantities defined in Eqs. (7.64); D_+ and D_- are defined as *positive* quantities. There are, in principle, no restrictions on θ_+ and θ_-, but we shall see later that they can be limited without affecting the observed spectrum. Without any loss of generality we shall always label the nuclei so that

$$\nu_A \geqslant \nu_B. \tag{7.66}$$

If the right-hand side of one or more of Eqs. (7.64) is negative, no inconsistencies result, since θ_+ and θ_- can assume values such that the sine and/or cosine factor is negative. By squaring Eqs. (7.64a) and (7.64b) (or Eqs. (7.64c) and (7.64d)), adding, and taking the square root, we obtain

$$2D_+ = \{[(\nu_A - \nu_B) + \tfrac{1}{2}(J_{AX} - J_{BX})]^2 + J_{AB}^2\}^{1/2},$$
$$2D_- = \{[(\nu_A - \nu_B) - \tfrac{1}{2}(J_{AX} - J_{BX})]^2 + J_{AB}^2\}^{1/2}. \tag{7.67}$$

Equations (7.67) show the analogy between D_+ and C used in the AB analysis. Note, however, that where $(\nu_A - \nu_B)$ appeared in the expression for the AB case, Eqs. (7.67) contain $[(\nu_A - \nu_B) \pm \tfrac{1}{2}(J_{AX} - J_{BX})]$, which serve as "effective chemical shifts." We shall refer later to the concept of "effective Larmor frequencies" and effective chemical shifts.

A better appreciation of the significance of the expressions in Table 7.5 may be obtained from an examination of some calculated ABX spectra. Figure 7.7 shows ABX spectra computed for the parameters shown. All parameters remain constant through the series except ν_B. The spectrum in Fig. 7.7a may be analyzed approximately as a first-order AMX case, as shown. The frequencies calculated from the first-order treatment are nearly identical with those found in the ABX calculation, while the intensities are only slightly in error. As $(\nu_A - \nu_B)$ decreases (Fig. 7.7b), there is a greater departure from the first-order calculation. In Fig. 7.7c the lines originating with nucleus A and those originating with

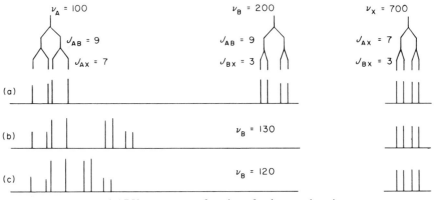

Fig. 7.7 Computed ABX spectra as a function of only one changing parameter, ν_B.

nucleus B can still be recognized by comparison with the spectrum in Fig. 7.7b; but the wave functions are now mixtures of A and B functions, and the transitions cannot strictly be called A or B transitions. The labeling "A" or "B" in the second column of Table 7.5 is convenient but is strictly applicable only in the limiting case of large $(\nu_A - \nu_B)$. The "combination" transition 13 involves simultaneous flipping of all spins and is forbidden in the ABX case. Transitions 14 and 15 also involve change in spin of all nuclei, but may have considerable intensity when $(\nu_A - \nu_B)$ is small. They appear in the X region and are discussed with the four X lines.

7.18 Analysis of an ABX Spectrum

The "complete" analysis of an ABX spectrum would require the determination from the spectrum of nine quantities: the three chemical shifts ν_A, ν_B, and ν_X; the magnitudes of the three coupling constants $|J_{AB}|$, $|J_{AX}|$, and $|J_{BX}|$; and the signs of the coupling constants. As we shall see, some of these quantities cannot be determined.

From Table 7.5 and Fig. 7.8 it is apparent that the X lines are *symmetrically* arranged about ν_X, so that ν_X is immediately determined. It is seen from Table 7.5 that the average of the frequencies of all eight AB lines gives $\nu_{AB} = \frac{1}{2}(\nu_A + \nu_B)$; a determination of ν_A and ν_B then requires only the additional value of $(\nu_A - \nu_B)$ (see below).

Consideration of Table 7.5 shows that a change in sign of J_{AB} has no effect on either the observed frequencies or intensities. The AB lines

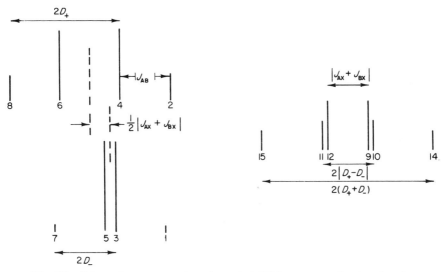

Fig. 7.8 Schematic representation of a typical ABX spectrum, showing the (ab)$_+$ and (ab)$_-$ quartets and the parameters that may be extracted readily from the spectrum. The numbering of the lines applies to the case $\nu_A > \nu_B$ and $J_{AX} > J_{BX} > 0$.

would have different labels (e.g., lines 1 and 3 would be interchanged), but no *observable* change would take place.

Likewise, the absolute signs of J_{AX} and J_{BX} cannot be determined. Examination of Eq. (7.67) shows that a change in sign of *both* J_{AX} and J_{BX} corresponds to an interchange of D_+ and D_-. Interchange of these quantities and of the signs of J_{AX} and J_{BX} in Table 7.5 shows that the spectrum is again unchanged in both frequency and intensity. A change in sign of only *one* of these two coupling constants may well change some features in the spectrum, so that it is often possible to determine the *relative* signs of J_{AX} and J_{BX}. This point will be explored later.

Figure 7.8 shows that the AB portion of the spectrum may be divided into two AB-type quartets, or subspectra, which we shall designate as (ab)$_+$ and (ab)$_-$. From the expressions for the line frequencies in Table 7.5, we see that such a division may always be made. The difference in frequency between lines 1 and 3, lines 2 and 4, lines 5 and 7, and lines 6 and 8 is equal to J_{AB}. As indicated in Fig. 7.8, the difference between the first and third lines of each quartet (between lines 2 and 6 and between lines 1 and 5), gives $2D_+$ and $2D_-$, respectively, so that the effective chemical shifts for the (ab)$_+$ and (ab)$_-$ subspectra are

$$\delta_+ = (\nu_A - \nu_B) + \tfrac{1}{2}(J_{AX} - J_{BX}),$$
$$\delta_- = (\nu_A - \nu_B) - \tfrac{1}{2}(J_{AX} - J_{BX}), \qquad (7.68)$$

as pointed out previously. The relative intensities are also correct for AB spectra. The centers of the quartets are separated by $\frac{1}{2}|J_{AX} + J_{BX}|$ (difference between the average of lines 3 and 5 and the average of lines 4 and 6). The absolute value symbol is used since we cannot know the labeling of the observed lines.

We have no way of telling *a priori* which of the two quartets to associate with $(ab)_+$ and which with $(ab)_-$. The later calculations will be considerably simplified if we arbitrarily choose

$$D_+ > D_-. \tag{7.69}$$

From Eq. (7.67) this choice is equivalent to taking the quantities $(\nu_A - \nu_B)$ and $(J_{AX} - J_{BX})$ to be of the same sign. This relation, together with our previous choice of $\nu_A > \nu_B$ (Eq. (7.66)), requires* that

$$J_{AX} > J_{BX}. \tag{7.70}$$

The X portion of the spectrum (Fig. 7.8 and Table 7.5) consists of three pairs of lines symmetrically placed around ν_X. The two strongest lines (9 and 12) are separated by $|J_{AX} + J_{BX}|$, which is just twice the separation of the centers of the $(ab)_+$ and $(ab)_-$ quartets. A pair of lines (10 and 11) is separated by $2(D_+ + D_-)$, while another pair (14 and 15) is separated by $2(D_+ + D_-)$. Lines 14 and 15 must lie outside lines 10 and 11, but the relation to lines 9 and 12 is variable. Frequently lines 14 and 15 (the combination lines) have so little intensity that they are not observed; that is, $\sin^2(\theta_+ - \theta_-) \approx 0$. In that case $\cos^2(\theta_+ - \theta_-) \approx 1$ by trigonometric identity, so that lines 10 and 11 are essentially equal in intensity to lines 9 and 12. In some cases the converse occurs, with lines 10 and 11 virtually disappearing and lines 14 and 15 becoming intense.

The analysis of an ABX spectrum to extract the magnitudes of the three chemical shifts and the three coupling constants is in principle straightforward, but in practice a number of ambiguities can occur. We shall summarize insofar as possible a general approach to the analysis of an ABX spectrum and indicate the limitations that might occur in practical examples.

1. The starting point for analysis is usually the identification of the two quartets $(ab)_+$ and $(ab)_-$; J_{AB} can usually be found with no difficulty.

---

* This requirement may appear to impose an unacceptable restriction, since physically there is no reason why J_{AX} might not be smaller in magnitude than J_{BX}. However, if we recall that both J_{AX} and J_{BX} could be negative or both positive with no *observable* change in the spectrum, then we note that a *mathematical* solution of, for example, $J_{AX} = -3$ and $J_{BX} = -10$ meets the requirement that $J_{AX} > J_{BX}$, yet is physically indistinguishable from the solution $J_{AX} = +3$ and $J_{BX} = +10$. (If $D_+ = D_-$, then $J_{AX} = J_{BX}$, and the magnitudes of both are immediately found from the known value of $|J_{AX} + J_{BX}|$.)

The association of the left and right halves of the quartets can often be done in two ways consistent with the frequencies but *usually* only in one way that is consistent with AB intensity relations. Sometimes, however, there is ambiguity in this selection, partly because of experimental inaccuracies or overlapping lines. For the present we shall assume that an unequivocal selection of the $(ab)_+$ and $(ab)_-$ quartets can be made. In Section 7.19 we shall examine the consequences of an incorrect selection of the two quartets.

2. The centers of the $(ab)_+$ and $(ab)_-$ quartets are separated by $\frac{1}{2}|J_{AX} + J_{BX}|$, so this quantity is now readily determined.

3. Turning to the X region, we can now identify lines 9 and 12, since they are separated by $|J_{AX} + J_{BX}|$, and thus confirm the value of this quantity.

4. D_+ and D_- can be found from the $(ab)_+$ and $(ab)_-$ quartets, and the values checked by the separations in the X region. (In accordance with our assumption (Eq. (7.69)), D_+ is taken to be the larger of these two quantities.) Sometimes overlap of lines renders the values obtained from one or the other region of the spectrum less accurate.

5. By squaring, rearranging, and taking square roots in Eq. (7.67), we obtain the relations

$$\delta_+ = (\nu_A - \nu_B) + \tfrac{1}{2}(J_{AX} - J_{BX}) = \pm[4D_+^2 - J_{AB}^2]^{1/2} \equiv \pm 2M,$$
$$\delta_- = (\nu_A - \nu_B) - \tfrac{1}{2}(J_{AX} - J_{BX}) = \pm[4D_-^2 - J_{AB}^2]^{1/2} \equiv \pm 2N; \quad (7.71)$$

M and N are defined as *positive* numbers by the relations at the right of Eq. (7.71) and are introduced here to simplify the notation. Since D_+, D_- and J_{AB} have all been found from the observed spectrum, M and N now represent known quantities. Mathematically, the following four solutions result from the possible choices of sign in Eqs. (7.71):

	①	②	③	④
$\nu_A - \nu_B$	$M + N$	$M - N$	$-M + N$	$-M - N$;
$\tfrac{1}{2}(J_{AX} - J_{BX})$	$M - N$	$M + N$	$-M - N$	$-M + N$.

From the restrictions expressed in Eqs. (7.66) and (7.70), both $(\nu_A - \nu_B)$ and $(J_{AX} - J_{BX})$ must be positive; hence solutions ③ and ④ can be disregarded. But since $M > N$, as a result of $D_+ > D_-$ (Eq. (7.69)), ① and ② are both valid so far as our present equations are concerned. Thus one of the two quantities $(M + N)$ and $(M - N)$ gives $(\nu_A - \nu_B)$, but we cannot tell at this point which one. The resolution of this ambiguity normally requires a consideration of the intensity distribution in the X region. In some cases, however, we have enough information from prior studies of similar molecules to be able to reject one of the two solutions as including physically unreasonable parameters. In such instances the computations in paragraph 6 can be omitted.

6. The relative intensities of the X lines depend on the angles θ_+ and θ_-, as shown in Table 7.5. Equation (7.64) defined these angles with no restrictions on their magnitudes. For ease of calculation, however, we shall restrict θ_+ to lie between 0° and 45° and θ_- to lie between 0° and 90°; that is, $0° \leqslant 2\theta_+ \leqslant 90°$, and $0° \leqslant 2\theta_- \leqslant 180°$. This restriction is equivalent to requiring that J_{AB} be positive, but as we have seen, the entire spectrum is independent of the sign of J_{AB}. The restriction is also consistent with our choices of $\nu_A > \nu_B$ and $J_{AX} > J_{BX}$.

From the values already found for J_{AB}, $2D_+$, and $2D_-$, the values of $\sin 2\theta_+$ and $\sin 2\theta_-$ can be calculated from Eqs. (7.64b) and (7.64d). As indicated in Fig. 7.9, there are two values of the angle $2\theta_-$, r and $(180° - r)$, consistent with this value of $\sin 2\theta_-$. The two angles correspond to positive and negative values of $\cos 2\theta_-$, hence, from Eqs. (7.64a) and (7.64c), to interchange of $(\nu_A - \nu_B)$ and $\frac{1}{2}(J_{AX} - J_{BX})$. Thus each of the two possible values of $2\theta_-$ is associated with one of the two possible solutions in paragraph 5. Since M and N are positive, solution ① gives $(\nu_A - \nu_B) > \frac{1}{2}(J_{AX} - J_{BX})$, and thus must have a value of $\cos 2\theta_- > 0$, or $0° \leqslant 2\theta_- < 90°$. Solution ② has $90° < 2\theta_- \leqslant 180°$.

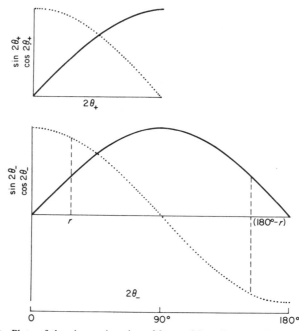

Fig. 7.9 Plots of the sine and cosine of $2\theta_+$ and $2\theta_-$. In general, two values of $2\theta_-$, r and $(180° - r)$, are consistent with the value of $\sin 2\theta_-$ found from the spectrum.

From the two possible values of $2\theta_-$, two values of $(\theta_+ - \theta_-)$ can be calculated, leading in general to quite different ratios of the intensities of lines 10 and 11 relative to lines 14 and 15. If one of the calculated intensity ratios is inconsistent with the observed intensities, then that value of θ_- and the corresponding solution ① or ② can be discarded. Sometimes the difference in intensity distribution between the two solutions is too small to permit an unequivocal decision as to the correct solution, but usually the distinction is clear.

For example, an analysis of the spectrum of Fig. (7.7c) leads to the two possible solutions: ① $(\nu_A - \nu_B) = 20$ Hz and $\frac{1}{2}(J_{AX} - J_{BX}) = 2$ Hz; ② $(\nu_A - \nu_B) = 2$ Hz and $\frac{1}{2}(J_{AX} - J_{BX}) = 20$ Hz. Solution ① gives $\theta_- = 13°$, while solution ② gives $\theta_- = 77°$. In either case $\theta_+ = 11°$. The intensities of the X lines can be found from Table 7.5. Solution ① gives intensities for lines 10 and 11 of 0.999 each, essentially equal to those of lines 9 and 12, with the combination lines 14 and 15 virtually absent (intensity 0.001). Solution ②, on the other hand, gives intensities for lines 10 and 11 of 0.17 and for lines 14 and 15 of 0.83. Clearly solution ① is compatible with the four X lines of equal intensities in the actual spectrum, while solution ② is incompatible.

7. With the correct value of $(\nu_A - \nu_B)$ selected, the previously determined value of $(\nu_A + \nu_B)$ can be used to find ν_A and ν_B.

8. From the correctly chosen value of $\frac{1}{2}(J_{AX} - J_{BX})$ and the previously determined value of $|J_{AX} + J_{BX}|$, J_{AX} and J_{BX} can be found. The question arises whether $(J_{AX} + J_{BX})$ is positive or negative. It was pointed out in Section 7.18 that a change in sign of this sum does not affect the observed spectrum, since it corresponds merely to an interchange of D_+ and D_-. However, at this point in the calculation it is extremely important to recognize the distinction, since we have already for purposes of calculation made a choice in the values of D_+ and D_-. A choice in the sign of $(J_{AX} + J_{BX})$ that is inconsistent with the values of D_+ and D_- will cause an interchange in the calculated values of J_{AX} and J_{BX}, which will not fit the observed spectrum. An examination of Table 7.5 shows that the expressions for the lines of the (ab)$_+$ quartet (characterized by D_+) always contain $(J_{AX} + J_{BX})$, while the expressions for the (ab)$_-$ quartet contain $(-J_{AX} - J_{BX})$. Thus the (ab)$_+$ quartet will be centered at a higher frequency than the (ab)$_-$ quartet if the sum $(J_{AX} + J_{BX})$ is positive, but at a lower frequency if the sum is negative. Hence the sign of this sum may be chosen unambiguously from the appearance of the spectrum.

Table 7.6 summarizes the procedure suggested in the foregoing paragraphs for the analysis of an ABX spectrum. The numbered steps in the table correspond to the paragraphs in this section. Problems involving the analysis of ABX spectra are given at the end of this chapter.

Table 7.6

PROCEDURE FOR THE ANALYSIS OF AN ABX SPECTRUM

1. Identify the two ab quartets on the basis of frequency and intensity relations. Note the value of J_{AB}.
2. Find the value of $\frac{1}{2}|J_{AX} + J_{BX}|$ from the separation of the centers of the two ab quartets.
3. Check the value of $|J_{AX} + J_{BX}|$ from the separation of the two strongest X lines, and identify lines 9 and 12 (see Fig. 7.8).
4. Find $2D_+$ and $2D_-$ from the separations of the first and third lines in the $(ab)_+$ and $(ab)_-$ quartets. Choose $2D_+$ as the larger. Check the values of $2D_+$ and $2D_-$ from the separations of lines in the X region and identify lines 10, 11, 14, and 15 (see Fig. 7.8).
5. Calculate M and N, where

$$2M = (4D_+^2 - J_{AB}^2)^{1/2}; \qquad 2N = (4D_-^2 - J_{AB}^2)^{1/2}.$$

   The two solutions for $(\nu_A - \nu_B)$ and $\frac{1}{2}(J_{AX} - J_{BX})$ are

	①	②
$\nu_A - \nu_B$	$M + N$	$M - N$
$\frac{1}{2}(J_{AX} - J_{BX})$	$M - N$	$M + N$

6. Find the value of $2\theta_+$, where $0 \leqslant 2\theta_+ \leqslant 90°$, from the relation

$$\sin 2\theta_+ = J_{AB}/2D_+.$$

   Find the *two* possible values of $2\theta_-$, where $0 \leqslant 2\theta_- \leqslant 180°$, from the relation

$$\sin 2\theta_- = J_{AB}/2D_-.$$

   Calculate the two possible values of $\sin(\theta_+ - \theta_-)$ and $\cos(\theta_+ - \theta_-)$ and from Table 7.5 compute the intensities of the X lines for each solution. If the smaller value of $\theta_-(0°-45°)$ gives X intensities consistent with the observed spectrum, while the larger value $(45°-90°)$ does not, then choose solution ① as the correct solution. If the converse is true, choose solution ②.
7. Find $\frac{1}{2}(\nu_A + \nu_B)$, which is the average of the centers of the $(ab)_+$ and $(ab)_-$ quartets, or equivalently, the average of the frequencies of all eight AB lines. From this value and the correct value of $(\nu_A - \nu_B)$ determined in steps 5 and 6, calculate ν_A and ν_B.
8. Assign to the sum $\frac{1}{2}(J_{AX} - J_{BX})$, for which the absolute value was found in step 2, a positive sign if the $(ab)_+$ quartet is centered at a higher frequency than the $(ab)_-$ quartet, or a negative sign if the reverse order is true. From this value and the correct value of $\frac{1}{2}(J_{AX} - J_{BX})$ determined in steps 5 and 6, calculate J_{AX} and J_{BX}.

7.19 Relative Signs of J_{AX} and J_{BX} in an ABX Spectrum

The analysis discussed in the previous section *should* lead unambiguously to the correct set of parameters needed to describe an observed ABX spectrum. But the solution derived in this way is in some instances not unique. In this section and the following one we shall investigate these ambiguities and the conditions responsible for them.

From the procedure of Section 7.18 we derive numerical values for J_{AX} and J_{BX}. Since these numbers are signed, we know the relative signs of these two coupling constants (same or opposite). We know the spectrum would be unchanged if the signs of *both* J's were changed. However, in some cases, we shall see that the spectrum may be virtually unchanged if the sign of only *one* of the J's is changed. In an AMX spectrum, which is just an ABX spectrum with a very large value of $(\nu_A - \nu_B)$ relative to J_{AB}, first-order analysis applies, and a change in sign of one or more coupling constants leaves the observed spectrum completely unchanged, as indicated at the top of Fig. 7.10.* However, as $(\nu_A - \nu_B)/J_{AB}$ decreases, the spectrum becomes progressively more dependent on the relative signs of J_{AX} and J_{BX}, as illustrated in Fig. 7.10.

* The relative signs of the J's can be determined by double resonance, as pointed out in Section 9.8.

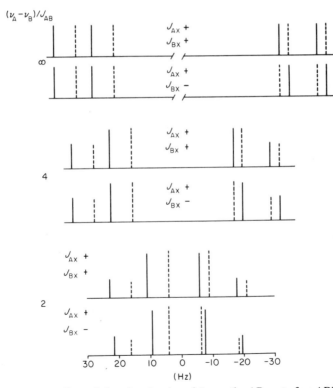

Fig. 7.10 The effect of changing the sign of J_{BX} on the AB part of an ABX spectrum. Solid lines: (ab)$_+$ subspectrum; dashed lines: (ab)$_-$ subspectrum. Parameters: $J_{AB} = 12$, $J_{AX} = 7$, $J_{BX} = \pm 3$ Hz; $(\nu_A - \nu_B)/J_{AB}$ indicated in figure.

It is apparent from Fig. 7.10 that a change in sign of J_{BX} for large $(\nu_A - \nu_B)/J_{AB}$ merely interchanges pairs of lines. In analyzing a spectrum, then, the choice of the wrong relative signs for the J's is associated with an incorrect choice of the $(ab)_+$ and $(ab)_-$ quartets. If the left half of $(ab)_+$ is mistakenly associated with the right half of $(ab)_-$ and vice versa, the centers of the two quartets thus selected are separated by $(D_+ - D_-)$, not $\frac{1}{2}|J_{AX} + J_{BX}|$. The X region may be of no help in rectifying this misassignment, since it is sometimes impossible to distinguish between lines 9 and 12, separated by $|J_{AX} + J_{BX}|$, and lines 10 and 11, separated by $2(D_+ - D_-)$. If all six X lines are observed, the intensities will permit the distinction, since lines 9 and 12 are then the most intense. The correct as-

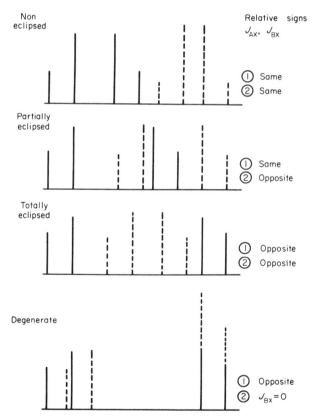

Fig. 7.11 Possible general patterns in the AB region of an ABX spectrum. Relative signs of J_{AX} and J_{BX} are given for the two solutions consistent with the spectrum in the AB region. In the totally eclipsed case the "dashed" subspectrum is contained between the first and fourth lines of the "solid" subspectrum but need not lie entirely between the second and third lines as in the illustration.

signment of the (ab)$_+$ and (ab)$_-$ quartets can be checked in principle by the relative intensities of the AB lines where each of the two quartets is typical of an AB-type system. However, the distinction may not be clear if either (a) the line intensities are perturbed because the X nucleus has a chemical shift that is not infinitely far removed from those of A and B, or (b) $(\nu_A - \nu_B)/J_{AB}$ is large enough so that the lines behave more like an AM pair than an AB pair. Generally, values of $(\nu_A - \nu_B)/J_{AB} < 2$ cause little difficulty. It should be noted that an incorrect assignment of (ab)$_+$ and (ab)$_-$ leads only to a reversal of one of the signs of J_{AX} or J_{BX} in the case of large $(\nu_A - \nu_B)/J_{AB}$, but can actually lead to the calculation of slightly different magnitudes of these J's as well, if $(\nu_A - \nu_B)/J_{AB} < \sim 2$.

We have seen that the AB region of the spectrum is in general compatible with two distinct solutions, even when the (ab)$_+$ and (ab)$_-$ quartets have been correctly assigned. The general appearance of the AB region can, however, sometimes give information on the relative signs of J_{AX} and J_{BX}.[118] If the (ab)$_+$ and (ab)$_-$ quartets are completely noneclipsed, as indicated in Fig. 7.11, then *both* possible solutions have J's with like signs. In the most general case, with partially overlapping quartets, one solution has J's of the same signs, while the other has J's of opposite signs. When one quartet lies completely inside the other, both solutions give J's of opposite signs. One final case depicted in Fig. 7.11 is the "degenerate" spectrum arising when two pairs of lines accidentally overlap. The degenerate case pictured could arise from $J_{BX} = 0$ or from the J's of opposite signs with all parameters having certain ratios. Other types of degeneracy in line positions in ABX spectra can occur; one example will be considered in the next section.

7.20 ABX Patterns; Deceptively Simple Spectra

Our analysis of an ABX spectrum was based on the assumption that all eight AB lines and either four or six X lines are observed. Frequently, however, some of the lines coincide, creating a spectrum whose appearance is not that of a typical ABX pattern. For example, Fig. 7.12 shows calculated ABX spectra as a function of only one changing parameter, ν_B. The spectrum at the top, which is the same as that at the bottom of Fig. 7.7, is readily recognized as an ABX pattern. As ν_B approaches ν_A, the appearance changes drastically, and when $\nu_B = \nu_A$, the AB region appears to be simply a doublet, while the X region is a 1:2:1 triplet. Actually there are other weak lines, as indicated, but in a practical case these would be lost in noise. Each of the strong AB lines is actually a very

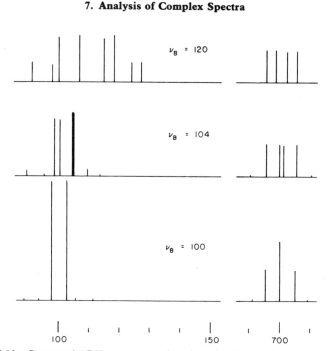

Fig. 7.12 Computed ABX spectra as a function of only one changing parameter, ν_B. All other parameters are identical to those in Fig. 7.7. The spectrum at the bottom is an AA'X spectrum.

closely spaced doublet, but in practice the splittings would very probably not be detected.

This spectrum is an excellent example of what have been termed "deceptively simple spectra." If one were confronted with such a spectrum and did not realize that it is a special case of ABX (actually AA'X, since the chemical shifts of A and B are equal), the doublet and triplet might mistakenly be interpreted as the components of a first-order A_2X spectrum, with $J_{AX} = 5$ Hz. Actually, the observed splitting is the average of J_{AX} and $J_{A'X}$. Deceptively simple spectra are widespread and are, of course, not limited to ABX systems; misinterpretations must be guarded against. An example of an AA'X deceptively simple spectrum is shown in Fig. 7.13. Note that the AA'X spectrum is independent of the value of $J_{AA'}$, which in this example is J(ortho), usually about 8 Hz.

Other pronounced departures from the "classical" ABX pattern occur for specific values of certain parameters. As pointed out in Section 7.17, the effective chemical shifts of the (ab)$_+$ and (ab)$_-$ quartets are $[(\nu_A - \nu_B) + \frac{1}{2}(J_{AX} - J_{BX})]$ and $[(\nu_A - \nu_B) - \frac{1}{2}(J_{AX} - J_{BX})]$, respectively. If one of these quantities is zero for particular values of the ν's and the J's,

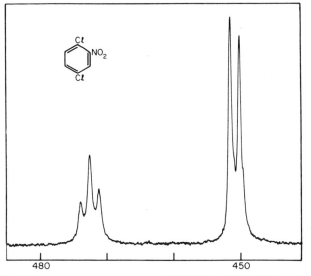

Fig. 7.13 Proton resonance spectrum (60 MHz) of 2,5-dichloronitrobenzene in CDCl$_3$. The low field triplet is due to H$_6$ and the doublet to H$_3$ and H$_4$, which are fortuitously chemically equivalent. The observed splitting of 1.6 Hz is about what would be expected as the average of J(meta) (3 Hz) and J(para) (0).

the corresponding ab subspectrum degenerates to a single line, while the other ab subspectrum remains a typical AB-type quartet.

Because of the many ways in which degeneracies or near degeneracies in line positions can occur, it is not possible to give a systematic treatment. The best ways of resolving degeneracy or deceptive simplicity are to obtain the spectrum at a different applied field or in another solvent (see Section 7.24).

7.21 "Virtual Coupling"

The ABX system provides a convenient framework for introducing the concept (actually poorly named) of *virtual coupling*. Consider the ABX system in the molecular fragment (I). The exact nature of the sub-

$$
\begin{array}{ccc}
R & R & R \\
| & | & | \\
R-C-C-C-OR \\
| & | & | \\
H_A & H_B & H_X
\end{array}
$$

(I)

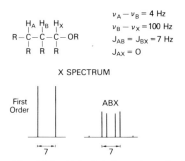

X SPECTRUM

Fig. 7.14 Calculated spectrum for H_x in the molecular fragment shown, with the parameters as indicated.

stituents is not important; it is only necessary that they contain no protons that couple with the three shown. Furthermore, let us assume that the R's are such that we have the chemical shift parameters given in Fig. 7.14. If we assume that there is free rotation of the C—C bonds, then the couplings J_{AB} and J_{BX} should be nearly equal and each of the order of 7 Hz. On the other hand, J_{AX} should be nearly zero. If we direct our attention only to the X portion of the spectrum (as might often happen in a complex molecule where the A and B portion would be overlapped by other aliphatic protons), we might well be tempted to treat the X portion of the spectrum by simple first-order analysis since $(\nu_B - \nu_X) \gg J_{BX}$. First-order analysis would then predict a simple doublet of approximately equal intensities with a splitting of 7 Hz. If, however, we recognize that this is really an ABX system and treat it accordingly, the X portion of the spectrum has 6 lines, as shown in Fig. 7.14. The two most intense lines are again separated by 7 Hz, but the other lines are quite significant in intensity.

Since the ABX calculation is certainly correct, the first-order approximation must be wrong in this case. In fact, the first-order calculation will always fail in circumstances of this sort where the proton in question (X) is coupled to one of a set of strongly coupled nuclei (i.e., a set in which J is greater than the chemical shift difference in hertz). In these circumstances we sometimes say that the X proton will "behave" as though it were coupled to both A and B, whereas in fact it is coupled to only one of the two. This *apparent but not real coupling* has been termed virtual coupling. It is important to note that virtual coupling is *not* a new and different phenomenon. It is merely a way of expressing the fact that first-order analysis is not applicable in this type of situation, and as a result the *splittings* observed in the spectrum are not necessarily equivalent to the magnitudes of certain J's.

The term virtual coupling is usually reserved for those nonfirst-order situations where a hasty examination might lead one to infer incorrectly that first-order rules are applicable. Such cases often occur in symmetric molecules where one might confuse chemical and magnetic equivalence. Two examples of systems other than ABX may help emphasize the types of circumstances where virtual coupling can occur. Fig. 7.15a shows the spectrum of 1,4-dibromobutane in which the protons attached to carbons 2 and 3 are chemically but not magnetically equivalent. They are coupled to each other with a coupling constant $J_{23} \approx 7$ Hz (a typical average value for such vicinal coupling); since $(\nu_2 - \nu_3) = 0$, the four protons on C_2 and C_3 behave as a strongly coupled group. The result, as shown in spectrum a, Fig. 7.15, is that the protons on C_1 and C_4 give rise not to a simple triplet, as might be expected from first-order analysis, but to the complex low field multiplet shown. The multiplet for the protons on C_2 and C_3 is likewise complex. This situation can be contrasted with that in 1,5-dibromopentane (Fig. 7.15b). In this case the chemical shifts of the protons on C_2 and C_3 are different; for $J \approx 7$ Hz, $J_{23}/(\nu_2 - \nu_3) \approx 0.3$, so the C_2 and C_3 protons are not strongly coupled. The C_1 and C_5 protons thus

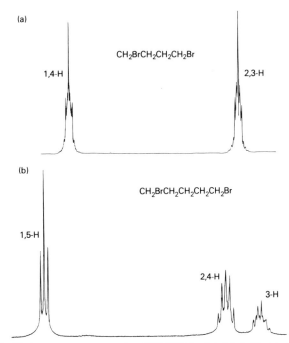

Fig. 7.15 Proton resonance spectra (100 MHz) of (a) 1,4-dibromobutane and (b) 1,5-dibromopentane. Virtual coupling occurs in (a) but not in (b).

give a first-order triplet. Protons on C_2 and C_4 are, of course, chemically equivalent, but $J_{24} \approx 0$, so these are not strongly coupled.

Another interesting example of virtual coupling is shown in Fig. 7.16. The spectrum of 2,5-dimethylquinone (a) is readily interpreted by first-order analysis; that of 2,6-dimethylquinone (b) shows additional splittings caused by the virtual coupling of the C_2 methyl protons with C_5H, and by symmetry of the C_6 methyl with C_3H. In the latter molecule the protons on C_3 and C_5 are coupled by about 2 Hz, and their chemical equivalence results in a large value of J/δ. In 2,5-dimethylquinone, however, protons on C_3 and C_6 are apparently not coupled; at least, any coupling between them is too small to permit observable effects with the available resolution.

Often the effect of virtual coupling is merely to bring about an apparent broadening of peaks when many lines fall close together. The possibility of virtual coupling should always be suspected when there are two or

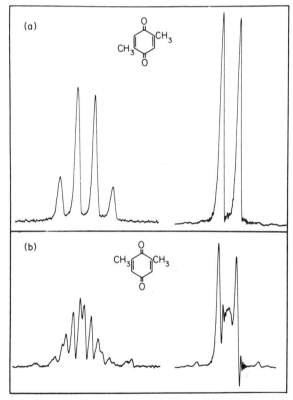

Fig. 7.16 Proton resonance spectra (60 MHz) of (a) 2,5-dimethylquinone and (b) 2,6-dimethylquinone, showing the effect of virtual coupling in (b).

more strongly coupled nuclei that are chemically but not magnetically equivalent.

7.22 The AA'BB' and AA'XX' Systems

These four-spin systems are characterized by two chemical shifts and four coupling constants, $J_{AA'}$, $J_{BB'}$, J_{AB}, and $J_{AB'}$. The last two are not equal, leading to magnetic nonequivalence.

The calculation of the energy levels and transitions is considerably simplified by inclusion of the symmetry of the system. As we saw in Section 7.12, the four basic symmetry functions for two equivalent nuclei can easily be constructed to be either symmetric or antisymmetric with respect to interchange of the nuclei. The 16 basis functions for the AA'BB' system are the products of the four symmetrized AA' functions (Eqs. (7.58) and (7.60)) with the four identical symmetrized BB' functions. The resultant secular determinant factors according to symmetry and F_z into two 1 × 1, five 2 × 2, and one 4 × 4 blocks. There are 28 transitions allowed by symmetry and the selection rule $\Delta F_z = \pm 1$, but four of these are combination transitions and are normally too weak to be observed. The remaining transitions are symmetrically arranged around the average of the A and B chemical shifts, $\frac{1}{2}(\nu_A + \nu_B)$.

Because of the presence of the 4 × 4 block in the Hamiltonian, explicit algebraic expressions for only 12 of the 24 expected transitions can be obtained. As a result, the analysis of an AA'BB' spectrum is usually tedious and requires a trial and error procedure aided by a computer. Even if this analysis is accomplished, there are some ambiguities in the signs and assignments of coupling constants. We shall return to AA'BB' spectra after we have taken up the AA'XX' system.

As in the ABX case, the larger chemical shift difference found in the AA'XX' system permits the definition of $F_z(A)$ and $F_z(X)$. As a result, the Hamiltonian factors into twelve 1 × 1 and two 2 × 2 blocks. As in the AA'BB' system, there are 24 allowed transitions of significant intensity, again arranged symmetrically around the frequency $\frac{1}{2}(\nu_A + \nu_X)$. Table 7.7 gives the frequencies and intensities of the A transitions, and Fig. 7.17 depicts the half-spectrum schematically. (The X spectrum would be identical and furnishes no additional information on the coupling constants. It does give the X chemical shift.) It is apparent that the A spectrum is also symmetric about its midpoint, which is ν_A. In Table 7.7 the frequencies are given in terms of the parameters K, L, M, N, P, and R defined in the table. Lines 1 and 2 always coincide, as do lines 3 and 4. Thus the half-spectrum has only 10 lines.

The analysis of an AA'XX' spectrum is straightforward, but a number

Table 7.7

AA′XX′ Spectrum: Frequencies and Relative
Intensities of the A Portion[a]

Line	Frequency relative to ν_A	Relative intensity
1	$\frac{1}{2}N$	1
2	$\frac{1}{2}N$	1
3	$-\frac{1}{2}N$	1
4	$-\frac{1}{2}N$	1
5	$P + \frac{1}{2}K$	$1 - K/2P$
6	$P - \frac{1}{2}K$	$1 + K/2P$
7	$-P + \frac{1}{2}K$	$1 + K/2P$
8	$-P - \frac{1}{2}K$	$1 - K/2P$
9	$R + \frac{1}{2}M$	$1 - M/2R$
10	$R - \frac{1}{2}M$	$1 + M/2R$
11	$-R + \frac{1}{2}M$	$1 + M/2R$
12	$-R - \frac{1}{2}M$	$1 - M/2R$

[a] $K = J_{AA'} + J_{XX'}$, $\quad L = J_{AX} - J_{AX'}$, $\quad M = J_{AA'} - J_{XX'}$, $N = J_{AX} + J_{AX'}$, $\quad 2P = (K^2 + L^2)^{1/2}$, $\quad 2R = (M^2 + L^2)^{1/2}$.

of ambiguities occur because of the symmetry of the spectrum. The chemical shifts ν_A and ν_X are of course easily determined as the midpoints of the respective half-spectra. The two strongest lines in the half-spectrum are separated by $|N|$. The remaining lines can be shown to arise from two ab subspectra characterized by "coupling constants" of K and M, respectively; P and R are defined by analogy to C in Eq. (7.52). Table 7.7 and Fig. 7.17 show that the frequencies and intensity ratios conform to the AB pattern. The values of $|K|$ and $|M|$ are easily found from the subspectra, but cannot be distinguished from each other; $|L|$ is easily calculated from the spectral line separations and the value of either $|K|$ or $|M|$ already found (cf. Fig. 7.2). The relative signs of J_{AX} and $J_{AX'}$ can be determined by noting whether $|N|$ is larger or smaller than $|L|$. Since K and M cannot be distinguished, we cannot ascertain the relative signs of $J_{AA'}$ and $J_{XX'}$. And finally, there is no way from the spectrum alone that we can decide which of the calculated pair is $J_{AA'}$ and which is $J_{XX'}$. The same ambiguity exists with J_{AX} and $J_{AX'}$. Often the coupling constants can be assigned to the proper nuclei on the basis of analogy to other systems. For example, the spectrum of 1,1-difluoroethylene in Fig. 5.3 is an AA′XX′ spectrum, which has been analyzed to give the absolute values $J_{HH'} = 4.8$, $J_{FF'} = 36.4$, $J_{HF}(cis) = 0.7$, $J_{HF}(trans) = 33.9$ Hz.[119] The assignments were readily made by analogy (see Tables 5.3 and 5.4).

As $(\nu_A - \nu_X)$ becomes smaller and the system converts to an AA′BB′

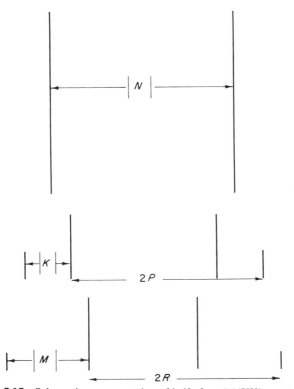

Fig. 7.17 Schematic representation of half of an AA′XX′ spectrum.

system, the lines toward the center of the complete spectrum grow in intensity, while those toward the outside diminish in intensity (cf. AX → AB, Fig. 5.4). Explicit algebraic expressions can be written for 12 of the 24 lines, as indicated in Table 7.8. The frequencies of lines 2 and 4,

Table 7.8

AA′BB′ SPECTRUM: FREQUENCIES OF
THE A PORTION

Line	Frequency relative to $\frac{1}{2}(\nu_A + \nu_B)$
1	$\frac{1}{2}N + \frac{1}{2}[(\nu_A - \nu_B)^2 + N^2]^{1/2}$
3	$-\frac{1}{2}N + \frac{1}{2}[(\nu_A - \nu_B)^2 + N^2]^{1/2}$
9	$\frac{1}{2}\{[(\nu_A - \nu_B + M)^2 + L^2]^{1/2} + [M^2 + L^2]^{1/2}\}$
10	$\frac{1}{2}\{[(\nu_A - \nu_B - M)^2 + L^2]^{1/2} + [M^2 + L^2]^{1/2}\}$
11	$\frac{1}{2}\{[(\nu_A - \nu_B + M)^2 + L^2]^{1/2} - [M^2 + L^2]^{1/2}\}$
12	$\frac{1}{2}\{[(\nu_A - \nu_B - M)^2 + L^2]^{1/2} - [M^2 + L^2]^{1/2}\}$

which coincided with lines 1 and 3, respectively, in the AA'XX' spectrum, cannot be expressed algebraically. However, it is found that these lines often remain close to their previous degenerate partners and thus provide a concentration of intensity that aids in the identification of these four lines. It is then possible to determine the value of N from the frequency difference between lines 1 and 3. (This value may be only approximate if lines 1 and 3 are confused with 2 and 4.) With N determined, $(\nu_A - \nu_B)$ can be found from the sum of the frequencies of lines 1 and 3, measured relative to $\frac{1}{2}(\nu_A + \nu_B)$, the center of the entire AA'BB' spectrum. Lines 9–12, which originated in one of the ab-type subspectra in the AA'XX' system, are determined by the three parameters L, M, and $(\nu_A - \nu_B)$. In principle these quantities can be found from the four line frequencies *if* the appropriate lines can be identified. When $(\nu_A - \nu_B)$ becomes small, the original ab-type quartet is greatly distorted, as indicated in Fig. 7.18, and line assignments may be quite ambiguous. The separation between lines 9 and 11 always remains equal to that between lines 10 and 12, which provides a point of reference.

The partial line assignment and resultant algebraic manipulations can at best provide only a reasonable starting point for a computer-aided iterative analysis of an unknown spectrum. The parameter $K(J_{AA'} + J_{BB'})$ does not appear in Table 7.8, so it must be found from this more detailed analysis. Double resonance methods (see Chapter 9) have been found to be quite useful in assigning all of the lines in an AA'BB' spectrum.[120]

AA'BB' spectra can vary markedly in appearance, depending on the specific values of the five parameters of the system, but the pattern is always symmetric about the midpoint. Bovey[83] has computed a number of examples, which can be of great value in approaching the analysis of an unknown spectrum. We shall describe here two examples of AA'BB' spectra commonly encountered in aromatic systems.

One situation occurs when $|J_{AB}|$ is much greater than $|J_{AA'}|$, $|J_{BB'}|$, and $|J_{AB'}|$, as in *p*-disubstituted benzenes (cf. Table 5.2) (II). The dominance of

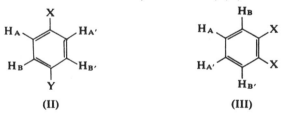

(II)   (III)

one coupling constant causes the spectrum to resemble roughly an AB quartet, but closer examination reveals several small peaks, which may be used in analyzing the spectrum. One example was given in Fig. 6.4, and many others are included among the spectra in Appendix *C*.

A different relation among the coupling constants, $|J_{AB}| \gg |J_{AB'}|$ and

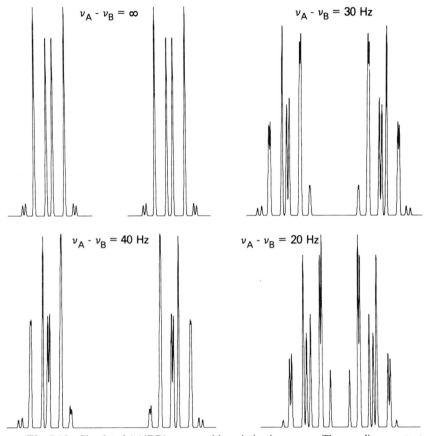

Fig. 7.18   Simulated AA'BB' spectra with variation in $\nu_A - \nu_B$. The coupling constants are fixed at $J_{AB} = 7.5$, $J_{AA'} = 0.5$, $J_{AB'} = 2.0$, and $J_{BB'} = 7.0$ (all in Hz). These values are typical of proton–proton coupling constants in an ortho-disubstituted aromatic ring. (The limiting AA'XX' case fortuitously has two pairs of lines too close to be resolved.)

simultaneously $|J_{AA'}| \gg |J_{BB'}|$, occurs, for example, in symmetrically *o*-disubstituted benzenes (III). In this case the spectrum is complex, with lines frequently tending to appear as very closely spaced doublets, as shown in Fig. 7.19. Many other AA'BB' patterns are possible, depending on the relations among the various parameters.

7.23 Other Complex Spectra

A large number of spin systems have been studied in detail and are discussed elsewhere. Those involving groups of magnetically equivalent nuclei, such as $A_m B_n X_p$, can usually be treated algebraically, while others

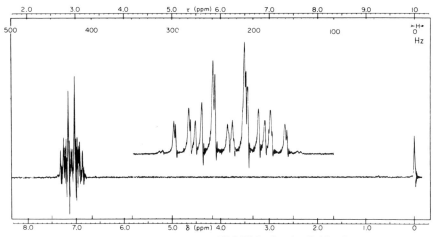

Fig. 7.19 Proton resonance spectrum (60 MHz) of *ortho*-dichlorobenzene, one type of AA'BB' spectrum.

possessing fewer spins, such as ABXY, can be treated analytically only in certain limiting cases.[121]

7.24 Aids in the Analysis of Complex Spectra

Variation of H_0. A number of procedures are available to simplify the analysis of complex spectra. Probably the most useful is observation of the spectrum at two or more values of the applied magnetic field. A spectrum that defies analysis at one field may be simplified greatly at higher field. The present general availability of proton resonance spectrometers operating at 60 and at 100 MHz and the increasing number of instruments operating above 200 MHz make field-dependent studies feasible. The field dependence is sometimes turned to advantage in another way by deliberately reducing the field to convert a first-order spectrum that is independent of certain parameters, such as signs of J's, to a more complex pattern from which the desired information can be extracted by iterative computer-aided analysis.[122]

Isotopic Substitution. Isotopic substitution, especially involving the specific interchange of magnetic and nonmagnetic nuclei, frequently aids analysis of complex spectra. Deuterium substitution is most common, with the smaller magnetogyric ratio of deuterium rendering many spin couplings negligibly small. If necessary, the deuterium coupling effects can be eliminated by spin decoupling (see Chapter 9). Nitrogen-15 substitution is occasionally used to avoid the broadening effects of the quadrupole containing ^{14}N.

Double Quantum Transitions. When very high rf power is used in scanning a spectrum, it is possible that two quanta of the *same* frequency will be absorbed to produce flipping of two coupled spins. These "double quantum transitions" appear as sharp lines under conditions where the normal spectrum is almost completely saturated, as indicated in Fig. 7.20. Theory shows that from the frequencies of the double quantum lines we can often obtain information on coupling constants that differs from the information obtainable from the ordinary spectrum. For example, in the ABC system the double quantum spectrum, together with the ordinary spectrum, provides enough information to determine from the frequencies the relative signs of all three J's. Double quantum transitions occur for only a narrow range of rf power. They are sometimes observed inadvertently when a normal scan is conducted with excess rf power.

Double resonance and *solvent effects* may be useful aids to analysis. They are discussed in Chapters 9 and 12, respectively.

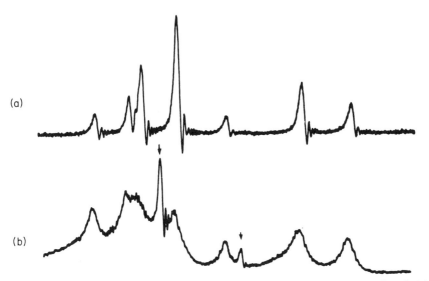

Fig. 7.20 Proton resonance spectrum (60 MHz) of 2-fluoro-4-6-dichlorophenol. (a) Spectrum with small H_1, showing AB portion of an ABX spectrum (from the aromatic protons and fluorine). (b) Spectrum with larger H_1, showing double quantum transitions (arrows) at the centers of the $(ab)_+$ and $(ab)_-$ quartets (Corio[104]).

7.25 Carbon-13 Satellites

One particularly useful isotopic substitution that is always available in organic molecules involves ^{13}C. This isotope is present at a natural abundance of 1.1%. Carbon-13 has a spin of $\frac{1}{2}$, and as we saw in Chapter 5, $^1J(^{13}C—H)$ is normally 100–200 Hz. Hence a resonance line from a proton attached to a carbon atom (^{12}C) will be accompanied by weak ^{13}C satellites symmetrically placed about it at a frequency separation of $\frac{1}{2}J(^{13}C—H)$.* For example, the proton resonance of chloroform in Fig. 7.21a shows ^{13}C satellites (or ^{13}C sidebands).

When the molecule in question contains more than one carbon atom, the ^{13}C satellites often become much more complex. Consider, for example, the molecule $CHCl_2CHCl_2$, the proton resonance of which is shown in Fig. 7.21b. The ordinary spectrum is a single line because of the magnetic equivalence of the two protons. On the other hand, the approximately 2.2% of the molecules that contain one ^{13}C and one ^{12}C have protons that are not magnetically equivalent. In fact, the proton resonance spectrum of these molecules should be an ABX spectrum in which $\nu_A \approx \nu_B$. The ABX analysis is also shown in Fig. 7.21. Note that because $(\nu_A - \nu_B) \approx 0$ and $J_{AX} \gg J_{BX}$, the protons behave as though the effective chemical shift of H_A is about 91 Hz ($\frac{1}{2}J_{AX}$) away from H_B. Since this value is much greater than J_{AB} (3 Hz), the A—B coupling can then be interpreted to a high degree of approximation on a first-order basis. This is a general result with ^{13}C satellites since $^1J(^{13}C—H)$ normally is much greater than $^2J(^{13}C—H)$ and $^3J((H—H)$.

With the recent marked improvement in NMR sensitivity (see Section 3.5) so that ^{13}C satellites may be readily observed, this technique has become a powerful tool in determining H—H coupling constants in symmetric molecules where this coupling normally leads to no observable splitting.

The phenomena discussed here are not, of course, restricted to ^{13}C. Other magnetic nuclei present at low abundance with the principal isotope of $I = 0$ display similar spectra. Among the best known are ^{29}Si, ^{199}Hg, and ^{183}W.

7.26 Effects of Molecular Asymmetry

The type of spectrum found in any coupled spin system can be drastically altered when elements of symmetry are present in the molecule. In many instances the extent of symmetry can be recognized readily, but in

---

* The position of the satellites about the main peak is usually not quite symmetric because of a small isotope effect on the chemical shift (see Section 4.14).

(a)

(b)

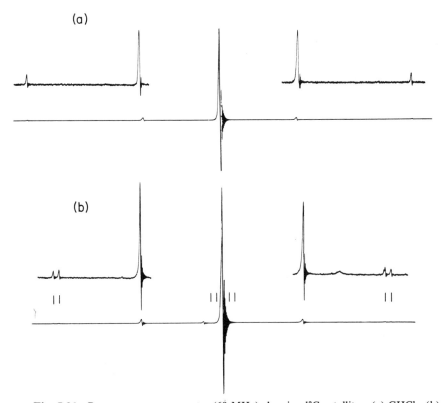

Fig. 7.21 Proton resonance spectra (60 MHz) showing ^{13}C satellites. (a) $CHCl_3$; (b) $CHCl_2CHCl_2$, with the AB portion of an ABX simulation of the spectrum. The sharp lines close to the main resonance in each case are spinning sidebands; the ^{13}C satellites are near the edges of the figure. The satellites shown in the simulated spectrum are obscured by the strong central peak; their positions are estimated and cannot be determined from the observed spectrum.

many other instances certain asymmetric features are not as easily discernible. For example, in a substituted ethane CH_2X—$CPQR$, where X, P, Q, and R are any substituents, we find that the two hydrogen nuclei may have different chemical shifts. To see how such chemical nonequivalence arises, consider the three stable (staggered) conformers IV–VI, which appear as shown when viewed along the C—C bond. At very low tempera-

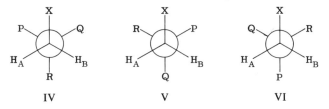

IV   V   VI

ture or under conditions of severe steric hindrance rotation about the C—C single bond might be so restricted that conversion of one conformer to another is very slow. Under such circumstances, as we have seen in Chapter 1 and shall examine in more detail in Chapter 11, the observed spectrum is just the *superposition* of the spectra of the three individual conformers. However when rotation is rapid, as is normally the case with molecules of this sort near room temperature, the observed spectrum represents the *average* of the chemical shifts and coupling constants found in the individual conformers, each weighted according to the fraction of time the molecule spends in that conformation.*

Suppose the three conformations pictured are equally populated, as they would be if completely free rotation occurs about the C—C bond. The chemical shift of H_A in conformation IV is probably influenced by the groups P and R adjacent to it. It might at first appear that there is an equal contribution to the chemical shift of H_B in conformation VI, where groups P and R are adjacent to H_B. However, closer examination of the entire molecule shows that differences exist between the conformations: groups P, X, and Q are neighbors in IV, while R, X, and Q are adjacent in VI. Thus while the "immediate" environments of H_A in IV and H_B in VI are the same, steric or electronic effects from the remainder of the molecule can in principle lead to different chemical shifts for H_A in IV and H_B in VI. A similar analysis applies to all other potentially equivalent pairs. Hence we conclude that the average chemical shift of H_A is not necessarily equal to that of H_B. Whether such a difference is observed depends, of course, on the net result of the magnetic effects involved and on the experimental resolution. The point is that we should always expect such differences and regard equivalence in the observed chemical shifts of H_A and H_B as fortuitous.

The asymmetry responsible for the nonequivalence of the chemical shifts of H_A and H_B need not be due to an immediately adjacent asymmetric carbon atom. For example, the CH_2 protons in

$$C_6H_5-\underset{\underset{O}{\parallel}}{S}-O-CH_2-CH_3$$

have been shown to be nonequivalent and to form the AB part of an ABX_3 system.[123] Many examples are known of chemical shift nonequivalence among protons and other nuclei many bonds away from a site of asymmetry.[124] Figure 7.22 gives a simple example of nonequivalence, in this case of two methyl groups.

It should be emphasized that the presence of some sort of asymmetry

---

* Justification for this statement and precise definitions of "slow" and "rapid" rotation will be given in Chapter 11.

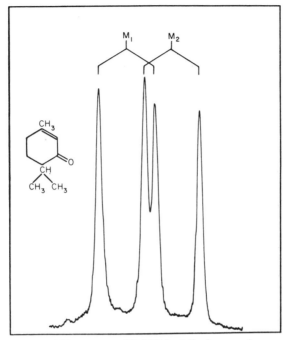

Fig. 7.22 Proton NMR spectrum (60 MHz) of the isopropyl group of piperitone in CDCl₃. Each of the nonequivalent methyl groups, M_1 and M_2, is coupled to the isopropyl proton.

is a necessary condition for chemical nonequivalence of two protons (or two methyl groups, etc.). This does not mean, however, that the molecule must be completely devoid of a plane of symmetry. For example, in the situation we have been considering, suppose that R is the group CH_2X, giving the molecule VII. While this molecule has a plane of symmetry, H_A

and H_B are, as we have seen, nonequivalent. Alternatively, suppose that X is CPQR, giving the molecule VIII. This can exist as a d,l pair or as a meso compound. In the d and l forms H_A and H_B are equivalent but in the meso compound they are nonequivalent.

The relationships existing within molecules possessing some asymmetric characteristics have been treated in detail by several authors. Protons H_A and H_B in a molecule such as IV–VI, (with P, Q, and R different) are said to be *diastereotopic,* whereas if P = Q, they would be *enantiotopic*. In the latter case, as in the case of two enantiomeric molecules, the NMR spectrum normally does not distinguish between the two, and the chemical shifts of H_A and H_B are the same. However, in a *chiral* solvent the interactions between the solvent and the two protons are not necessarily equivalent, and chemical shift differences (usually small) may be found.[125]

Even when nuclei are *chemically* equivalent, *magnetic* nonequivalence may occur. For example in a 1,2-disubstituted ethane, the two protons attached to a given carbon atom are enantiotopic, hence chemically equivalent in an ordinary achiral solvent. Of the three conformations IX–XI, IX and X are mirror images, the average of which must have $\nu_A = \nu_B$ and $\nu_X = \nu_Y$, while XI contains a plane of symmetry, so that the same equalities hold here. On the other hand, H_A and H_B are *not magneti-*

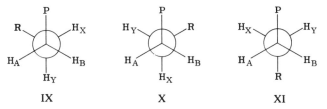

| IX | X | XI |

cally equivalent, since the vicinal coupling constants are not necessarily averaged to the same value. For example, in conformation IX, where H_A and H_X are coupled by a trans coupling constant, substituents P and R are adjacent, while in XI, where H_A and H_Y are trans coupled, P and R are far apart. Thus $J_{AX}(trans) \neq J_{AY}(trans)$, and similar inequalities hold for the other conformations. Hence, this is an AA'XX' system, not an A_2X_2 system. In some individual cases the differences in the average J's may, of course, be so small that deviations from an A_2X_2 pattern are not observed.

Finally, we should point out that our entire discussion of nonequivalence has been predicated on the assumption of equal populations for all conformers. In most cases where asymmetry is present, there are differences in energy, hence in populations, of the conformers. Such differences can significantly enhance the magnitude of the nonequivalence.

7.27 Polymer Configuration

As indicated in Section 6.3, NMR is of great value in the determination of the structure of polymers. We now investigate briefly the application of NMR to the determination of the configuration of a polymer composed of asymmetric monomer units. For example, in a vinyl polymer the CH_3 and CH_2 chemical shifts are strongly dependent on the relative configuration (handedness) of adjacent monomeric units. An *isotactic* sequence is one in which all monomer units have the same configuration (*ddd* or *lll*); a *syndiotactic* sequence is one in which the configurations alternate (e.g., *dld*); a *heterotactic* sequence is one in which a more nearly random configurational arrangement occurs (e.g., *ddl*). We can picture the sequences as shown. If we consider the M—C—X groups adjacent to

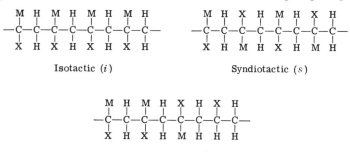

a given methylene group, we see that in the syndiotactic sequence the two methylene protons are in the same environment and are chemically equivalent, whereas in the isotactic sequence they are chemically nonequivalent and should give rise to an AB spectrum. An example is shown in Fig. 7.23.

The foregoing analysis of the expected methylene spectrum depended only on the configuration of two adjacent monomer units, a *dyad*. The environment of a given methyl group, on the other hand, must depend on the relative configurations of both of the neighboring M—C—X groups, hence on a *triad* sequence. The three lines for the methyl resonances in Fig. 7.23 are attributable to the *s*, *i*, and *h* triad sequences. In many instances further effects of groups farther away can be discerned, and longer sequences must be considered. Bovey[83] presents a lucid description of the classification and analysis of such polymer sequences.

7.28 Use of Liquid Crystals as Solvents

Liquid crystals are known to form a partially ordered structure, but with more intermolecular motion than exists in ordinary crystals. Small anisotropic molecules dissolved in liquid crystals experience partial orien-

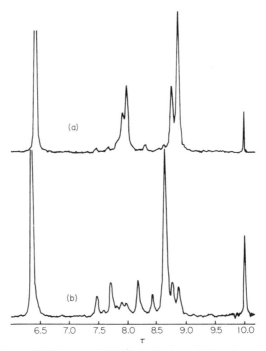

Fig. 7.23   Proton NMR spectra (60 MHz) of polymethyl methacrylate prepared under conditions such that the sample in (a) is primarily syndiotactic, while that in (b) is primarily isotactic. Note that the CH$_2$ resonance at $\tau \approx 8.0$ is predominantly a singlet in (a) but an AB quartet in (b) (Bovey[101]).

tation and thus do not achieve complete cancellation of direct dipole– dipole interactions (see Section 2.6). The spectrum is dependent on both these dipole couplings and ordinary spin couplings, including those that normally would not appear because of magnetic equivalence. In addition, the anisotropy in the chemical shifts is sometimes manifested. The result is that the spectrum becomes extremely complex but can be analyzed to provide information on these anisotropies and dipole interactions unobtainable in other ways. For example, the spectrum of hexafluorobenzene, which is a single line in ordinary solvents, is shown in Fig. 7.24. This spectrum has been completely analyzed to yield the signs and magnitudes of all J's, as well as the values of the dipole couplings. Since the latter interactions vary inversely with the cube of internuclear distance, the results provide a sensitive measure of molecular geometry in solution. Both the theoretical and experimental aspects of this area of research have been well covered in several places.[127]

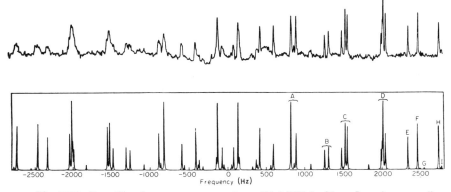

Fig. 7.24 Top: Fluorine resonance spectrum (56.4 MHz) of hexafluorobenzene dissolved in the liquid crystal nematic phase of p,p'-di-n-hexyloxyazoxybenzene at 58°C. Bottom: Computer-simulated spectrum with parameters as follows: Dipole–dipole interactions, D(ortho) $= -1452.67$; D(meta) $= -271.56$; D(para) $= -194.15$; spin–spin couplings, J(ortho) $= -22$; J(meta) $= -4$; J(para) $= +6$ Hz (Snyder and Anderson[126]).

Problems

1. Describe the following spin systems as AB, etc.: $CH_2{=}CHF$; PF_3; cubane; $CH_3CHOHCH_3$; H_2; chlorobenzene; n-propane.

2. Show that for N coupled nuclei there are $\frac{1}{2}N(N-1)$ coupling constants.

3. Verify that ψ_2 in Eq. (7.44) gives the value of E_2 in Eq. (7.43).

4. Which of the following spectra result from AB systems? Spectra are listed as: frequency (relative intensity).
   (a)   100 (1), 108 (2.3), 120 (2.3), 128 (1).
   (b)   100 (1), 104 (2.4), 113 (2.4), 120 (1).
   (c)   100 (1), 110 (4), 114 (4), 124 (1).
   (d)   100 (1), 107 (15), 108 (15), 115 (1).

5. The following quartet appears as part of a rich NMR spectrum: 100 (1), 108 (3), 116 (3), 124 (1). (a) Give *two* possible explanations of its origin. (b) By what experiment could this ambiguity be resolved? (c) Predict the frequencies and relative intensities in each case.

6. Find ν_A and ν_B from each of the following AB spectra: (a) 117, 123, 142, 148; (b) 206, 215, 217, 226.

7. Verify the nonmixing of basis functions of different symmetry for $\Phi_{2'}$ and $\Phi_{3'}$ (Eq. (7.60)) by calculating $\mathscr{H}_{2'3'}$.

8. Use the ABC basis functions in Table 7.2, the rules of Section 7.13, and the theorem on factoring of the secular equation according to F_z to construct the Hamiltonian matrix for the ABC system.

9. Analyze the spectrum in Fig. 7.5 to obtain ν_A, ν_B, and J_{AB}.

10. Derive Eqs. (7.67) from (Eqs. (7.64).

11. Show from Table 7.5 that a change in sign of J_{AB} leads to the same spectrum (both frequencies and intensities).

12. Show from Table 7.5 that the centers of the $(ab)_+$ and $(ab)_-$ quartets are separated by $\frac{1}{2}|J_{AX} + J_{BX}|$.

13. Use the procedure of Table 7.6 to analyze the spectra in Fig. 7.10. Make the calculation both for the "correct" ab subspectra and for the subspectra "incorrectly" assigned.

14. In which of the following molecules is virtual coupling likely to appear?

(a) [structure: Ph—O ring with O—Ph] ; (b) $CH_3CHOHCH_3$; (c) $CH_3CH_2OCH_2CH_3$; (d) $CH_3CH_2CH_2CH_2CH_2OH$

15. Determine the structural formulas of compounds in Spectra 21 and 22, Appendix C. Analyze the ABX portions of the spectra by the procedure of Table 7.6.

16. Use the procedure of Table 7.6 to analyze the ABX patterns in Spectra 23 and 24, Appendix C. Could first-order analysis be used for these spectra? Why?

17. Analyze the AA'XX' Spectra 25 and 26, Appendix C. From the values of the coupling constants deduce the correct isomeric structure of each compound (cf. Fig. 5.3).

18. Determine the structure and all six coupling constants in the molecule giving Spectrum 27, Appendix C.

19. Determine the structures of the molecules giving Spectra 28–30, Appendix C.

20. Give the type of spectrum (e.g., AB_2X_2) expected for each of the following ethane derivatives, where Q, R, and S are substituents that do not spin couple with the protons. Assume that rotation about the C—C bond is rapid and that there is a large chemical shift difference between protons on different carbon atoms. (a) CH_3—CH_2R; (b) CH_2Q—CH_2R; (c) CH_3—$CQRS$; (d) CH_2Q—$CHRS$; (e) CHQ_2—$CHRS$.

21. Repeat Problem 20 for slow rotation about the C—C bond.

22. Show that for vinyl polymers there are exactly six different tetrad sequences.

Chapter 8

Relaxation

In Chapter 2 we found that a perturbed nuclear spin system relaxes to its equilibrium state or steady state by first-order processes characterized by two relaxation times: T_1, the spin–lattice, or longitudinal, relaxation time; and T_2, the spin–spin, or transverse, relaxation time. Thus far in our treatment of NMR we have not made explicit use of relaxation phenomena, but for much of the discussion in the remainder of this book we require a more detailed understanding of the processes by which nuclei relax. There is a great deal of information of chemical value in the study of relaxation processes, and with the widespread use of pulse FT methods (Chapter 10) it is now possible to carry out such investigations in complex molecules.

8.1 Molecular Motions and Processes
for Relaxation in Liquids

For spin–lattice relaxation to occur there must be some means by which the nuclear spin system can transfer its excess energy to the surroundings. Since we are interested primarily in high resolution NMR, we shall restrict ourselves to consideration of relaxation in liquids, where the molecules are in rapid, random motion. As first shown by Bloembergen, Purcell, and Pound (BPP),[128] the fluctuating magnetic or electric fields arising from this motion constitute one important ingredient of the relaxation process. The other is the specific means by which a nucleus can interact with its surroundings.

Turning first to random molecular motion, consider the length of time a "typical" molecule remains in any given position before a collision causes it to change its state of motion. For a small molecule in a non-viscous liquid this period of time may be of the order of 10^{-12} sec; for a polymer it is usually several orders of magnitude longer. In the BPP theory this time is called the *correlation time* τ_c, and is usually defined by the following equation:

$$k(\tau) = k(0)e^{-\tau/\tau_c} \tag{8.1}$$

Here $k(\tau)$ is a *correlation function* —a function that defines the position of the molecule at time τ relative to its position at an arbitrary initial time, $k(0)$. The equation says that the new position, $k(\tau)$, is related to the initial position in an exponential manner; i.e., the two positions are much more likely to be different (uncorrelated) after a long time τ. The exponential time constant is τ_c—a very small value for a molecule with a fast average motion. Since molecular motions are random, τ_c can describe only an average. To proceed further we need to know just what ranges of frequencies of motion are present, and this is most easily done by the process of Fourier analysis. (Fourier methods were referred to in another connection in Section 3.6 and will be discussed further in Chapter 10.) $k(\tau)$ is a function of time, but we can define a related function of frequency by the Fourier transform equation

$$J(\omega) = \int_{-\infty}^{\infty} k(\tau)e^{-i\omega\tau}\, d\tau \tag{8.2}$$

$$= \int_{-\infty}^{\infty} k(0)e^{-\tau/\tau_c}e^{-i\omega\tau}\, d\tau \tag{8.3}$$

$$= A[\tau_c/(1 + \omega^2\tau_c^2)], \tag{8.4}$$

where $J(\omega)$ is called the *spectral density function* and A is a constant, which can readily be calculated.*

Plots of J versus ω for different values of τ_c are shown in Fig. 8.1. The values of τ_c are measured relative to the reciprocal of the nuclear Larmor frequency ω_0. For either a very short or very long τ_c the value of $J(\omega)$ at ω_0 is relatively small. $J(\omega_0)$ reaches its maximum when $\tau_c = 1/\omega_0$; i.e., when the average molecular tumbling frequency is equal to the nuclear precession frequency. Under these circumstances energy transfer between precessing nuclei and randomly tumbling molecules is most efficient, and T_1 is a minimum.

---

* To account for three dimensional motion (either rotational or translational) there are actually three correlation functions k_1, k_2 and k_3, and there is a spectral density function corresponding to each. All of the latter functions are of the form of Eq. (8.4) but differ in the value of A.

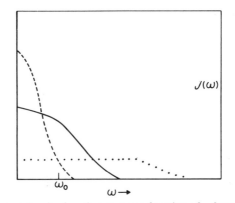

Fig. 8.1   The spectral density function $J(\omega)$ as a function of ω for various values of τ_c.

Our discussion here is entirely in terms of *a* correlation time. Only for a small rigid molecule undergoing completely isotropic motion is this description adequate. In molecules where motion about one axis is preferred, two or three different correlation times are applicable; in molecules where internal rotation can occur several correlation terms must be considered; and in more complex systems (e.g., water in biological cells) a number of translational and rotational correlation times may be pertinent. In such cases a plot of J versus ω consists of a superposition of several curves of the type given in Fig. 8.1, and a study of the frequency dependence of T_1 can be quite informative.

Returning to the simple situation of a single τ_c, we can recast the results of Fig. 8.1 into a plot of T_1 versus τ_c, as shown in Fig. 8.2. No coordinate scale is given, since it depends upon the specific types of interactions to be discussed below. T_1 goes through a minimum at $\tau_c = 1/\omega_0 = 1/2\pi\nu_0$, as we discussed above.

With NMR frequencies in the range of about 1–300 MHz for various nuclei and magnetic field strengths, the minimum in the T_1 curve can come from about 5×10^{-10} to 2×10^{-7} sec. Small molecules (MW < 200) almost always lie on the left of the minimum unless the solvent is very viscous or the observation frequency is extremely low. Small polymers, such as proteins of MW 10,000–25,000 may be in the vicinity of the minimum and could fall to either side depending on the Larmor frequency.

Figure 8.2 also shows T_2 as a function of τ_c. The calculation of the functional form of T_2 proceeds in much the same way as that of T_1. However, since T_2 involves exchange of energy *between* nuclei, rather than between the nuclei and the environment (lattice), the dependence on molecular motion is somewhat different from that of T_1. It can be shown[129]

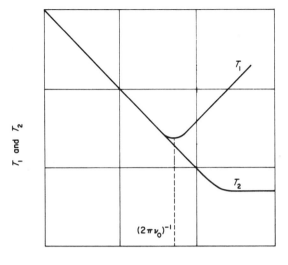

Fig. 8.2 A log–log plot of T_1 and T_2 versus the correlation time τ_c.

that the processes giving rise to spin–spin relaxation (T_2 processes) depend on high frequency (short τ_c) motions in the same way as T_1 processes, but low frequency motions (and as we shall see later, other low frequency processes, such as chemical exchange) significantly shorten T_2. Hence T_2 decreases monotonically with increasing τ_c and ultimately approaches a limiting value that is characteristic of a completely rigid solid lattice.

Regardless of the types of molecular motion, relaxation occurs only if there is some specific *interaction* between the nucleus and its environment that can result in energy exchange. Six types of interaction have been identified, as follows:

1. Nuclear magnetic dipole–dipole interaction
2. Chemical shift anisotropy
3. Spin–rotation interaction
4. Nuclear electric quadrupole interaction
5. Scalar coupling effects
6. Effect of paramagnetic species

All of these interactions, except number 4, involve the *magnetic* moment of the nucleus.

In the following treatment of each of these relaxation mechanisms, we shall focus on the relaxation *rate*, $R_1 \equiv 1/T_1$, since the overall relaxation rate is the sum of the rate produced by each mechanism.

8.2 Nuclear Magnetic Dipole Interactions

Any magnetic nucleus in a molecule supplies an instantaneous magnetic dipole field, which is proportional to the magnetic moment of the nucleus. As the molecule tumbles in solution under the influence of Brownian motion, this field fluctuates in magnitude and direction. Thus there is produced an oscillating magnetic field, with the frequencies of oscillation just like those that we have been discussing. Just as a precessing nuclear moment can interact with a coherent applied rf magnetic field (Section 2.2), so it can interact with the molecular magnetic field component at the Larmor precession frequency. As we have seen, the effectiveness of the fluctuating field in bringing about relaxation depends on the Fourier component of motion at the Larmor frequency, but it depends also on the magnitudes of the nuclear moments and on the distance between the interacting nuclei. A detailed treatment of the relaxation process shows that[129]

$$
R_1 = \frac{2\gamma_i^2\gamma_j^2\hbar^2 S(S + 1)}{15r_{ij}^6} \left[\frac{3\tau_c}{1 + \tau_c^2\,\omega_I^2} + \frac{\tau_c}{1 + \tau_c^2(\omega_I - \omega_S)^2} \right.
$$
$$
\left. + \frac{6\tau_c}{1 + \tau_c^2(\omega_I + \omega_S)^2} \right]. \quad (8.5)
$$

This equation refers to the relaxation of one nucleus i with nuclear spin I by another nucleus j, with nuclear spin S. The relaxation rate R_1 of nucleus i depends on the squares of both its own magnetogyric ratio and that of the other nucleus; it also depends on the inverse sixth power of the distance between the nuclei. Thus nearby nuclei with large γ (such as ^{1}H) will be most effective in relaxation. (The first term of the dependence on τ_c is just that in Eq. (8.4), while the second and third come about from related processes that we shall not discuss in detail.[130,131]

A similar treatment for transverse relaxation yields the results

$$
R_2 = \frac{\gamma_i^2\gamma_j^2\hbar^2 S(S + 1)}{15r_{ij}^6} \left[4\tau_c + \frac{\tau_c}{1 + \tau_c^2(\omega_I - \omega_S)^2} + \frac{3\tau_c}{1 + \tau_c^2\,\omega_I^2} \right.
$$
$$
\left. + \frac{6\tau_c}{1 + \tau_c^2\,\omega_S^2} + \frac{6\tau_c}{1 + \tau_c^2(\omega_I + \omega_S)^2} \right]. \quad (8.6)
$$

For $\tau_c \gg 1/\omega$, the last four terms approach zero, and the expression is dominated by the term in τ_c alone. Thus T_2 becomes shorter with increasing τ_c, but ultimately approaches a limit (not demonstrated in Eq. (8.6)). The long correlation time permits dipole–dipole interactions to become effective in leading to broad lines, that is, short T_2's. The static local magnetic field at a nucleus i due to another nuclear moment j is

$$H(\text{local}) = \pm \frac{\mu}{r_{ij}^3}(3 \cos^2 \theta_{ij} - 1). \tag{8.7}$$

In this expression r_{ij} is the magnitude of a vector joining nuclei i and j, while θ_{ij} is the angle between this vector and the applied field $\mathbf{H}_0$. In a solid, where the θ_{ij} are fixed, the local field due to nearby nuclei can be quite large, about 14 G for a proton at a distance of 1 Å. Since near neighbors have random orientations of spins with respect to $\mathbf{H}_0$, the spread of fields experienced by nuclei in the sample can be of this order of magnitude. The corresponding line width is about 60 kHz, giving T_2 of the order of 10^{-5} sec.

When the molecules in which the magnetic nuclei reside are in motion, we must consider an average value of θ_{ij}. If the motion is random and rapid (short τ_c) $\cos^2 \theta_{ij}$ can be averaged over all space to give $\langle \cos^2 \theta_{ij} \rangle_{\text{av}} = \frac{1}{3}$. This value inserted in Eq. (8.7) gives zero local field from this source. (More accurate calculations show that a very small residual effect remains.) Thus for small τ_c we expect T_2 to approach T_1, a result that can be seen from Eqs. (8.5) and (8.6), which both reduce to

$$R_1 = R_2 = \frac{4\gamma_i^2\gamma_j^2\hbar^2 S(S + 1)}{3r_{ij}^6} \tau_c . \tag{8.8}$$

The region of $\tau_c \ll 1/\omega_0$ is often called the *extreme narrowing* condition.

Nuclei i and j may be in the same molecule, in which case it is a molecular rotation that causes them to move relative to each other and to the direction of $\mathbf{H}_0$, or they may be in different molecules and move when the molecules undergo translational motion. The results are analogous, but Eqs. (8.5)–(8.8)* are interpreted in terms of a rotational correlation time in the first case and a translational correlation time in the second case.

Nuclei of small magnetic moment, such as ^{13}C, are usually relaxed by intramolecular interaction with nearby nuclei of large γ, such as ^1H. Protons, on the other hand, are relaxed usually by other protons, both intra- and intermolecularly, depending on the average distances involved. For example, pure liquid benzene has a proton $T_1 \approx 19$ sec, while a dilute solution of benzene in CS_2 (which has no magnetic nuclei) shows $T_1 \approx 90$ sec. (Viscosity variation, hence a change in $\tau_c(\text{rot})$, may account for a part

---

* Equations (8.5)–(8.8) apply strictly only to two nuclei with different γ in the same molecule. Detailed expressions for other situations (which are quite similar) are given elsewhere.[131]

of this difference.) Natural abundance ^{13}C in benzene has $T_1 \approx 29$ sec, nearly independent of concentration, since the distance to protons in adjacent molecules is large. Further examples of T_1's will be given in Section 8.9.

8.3 Relaxation via Chemical Shift Anisotropy

In Chapter 2 we saw that the magnetic field at the nucleus is given by

$$\mathbf{H}(\text{nucleus}) = \mathbf{H}_0 - \sigma \mathbf{H}_0, \tag{8.9}$$

where σ, the shielding factor, is a tensor. We found that an anisotropy in σ may result in pronounced effects on chemical shifts. An anisotropy in σ may also furnish a mechanism for relaxation, since as the molecule tumbles in solution, the field at the nucleus is continually changing in magnitude. The components of random tumbling motion at the Larmor frequency can then lead to spin–lattice relaxation.

If σ is axially symmetric, theory predicts that in the extreme narrowing limit usually found for small molecules

$$R_1 = \frac{2}{15} \gamma^2 H_0^2 (\sigma^\| - \sigma^\perp)^2 \tau_c, \tag{8.10}$$

where $\sigma^\|$ and $\sigma^\perp$ refer to the components of the shielding tensor parallel and perpendicular to the axis of symmetry. The interesting point in Eq. (8.10) is that R_1 increases quadratically with increasing magnetic field. Experimental evidence for such variation has been reported for several nuclei with large ranges of chemical shifts (e.g., ^{31}P, ^{113}Cd). For ^{13}C, other relaxation mechanisms usually dominate, so the effect of chemical shift anisotropy is usually masked. However, in certain molecules (e.g., $^{13}CS_2$ at low temperature) other mechanisms are very inefficient, so relaxation occurs principally from the chemical shift anisotropy. Since the range of ^1H chemical shifts is small, $\sigma^\| - \sigma^\perp$ is too small to be significant.

8.4 Spin–Rotation Relaxation

The spin–rotation interaction arises from magnetic fields generated at a nucleus by the motion of a *molecular* magnetic moment which arises from the electron distribution in a molecule. Let us consider one particular electron and a given nucleus. As the molecule rotates, this electron rotates about the nucleus at a radius, say, R. The rotational frequency V is given by[129]

$$V = hJ/2\pi I, \tag{8.11}$$

where the molecule is in the Jth rotational state and I is the moment of inertia of the molecule. The current generated by the electron is then

$$i = (e/c)V, \tag{8.12}$$

and the magnetic moment associated with this circulating current is

$$\mu_J = i(\pi R^2) = (eh/2\pi Mc)J \approx \mu_N J, \tag{8.13}$$

and $I = MR^2$, where M is the nuclear mass and μ_N is the nuclear magneton. This magnetic moment generated by the motion of the electron produces a local magnetic field at the resonant nucleus of the order of $\mu_N J/R^3$. For the hydrogen molecule this local field is about 35 G and for HCl the field at the hydrogen nucleus is about 10 G. Since this effect is proportional to the rotational velocity (i.e., inversely proportional to the moment of inertia of the molecule), we may anticipate that in general the smaller the molecule the more important will be the spin–rotation interaction. Furthermore, symmetric molecules with little or no intermolecular interactions (such as hydrogen bonding) will be most affected since they will have rather greater angular velocities. For liquids undergoing isotropic molecular reorientation the relaxation rate can be shown to be

$$R_1 = (2\pi IkT/h^2)C_{\text{eff}}^2\tau_J, \tag{8.14}$$

where C_{eff}^2 is the average component of the spin–rotation tensor, and τ_J is the angular momentum correlation time, which is a measure of the time a molecule spends in any given angular momentum state.

The important distinction between this relaxation mechanism and the others discussed is that τ_J becomes *longer* as the sample temperature increases, whereas τ_c becomes shorter. As the temperature becomes quite high and the sample vaporizes, collisions become more infrequent and the molecule remains in a given angular momentum state for a longer period of time. On the other hand, the higher the temperature the faster a molecule reorients and the shorter becomes τ_c. The result of this is, of course, that for the spin–rotation interaction the relaxation time T_1 becomes longer as the temperature decreases. This behavior is opposite to that observed for the other relaxation mechanisms.

Spin–rotation interaction is known to be the dominant relaxation mechanism for ^{13}C in CS_2 (except at very high fields and low temperatures) and is important for ^{13}C in methyl groups, which reorient rapidly by *internal* rotation, even in large molecules.

In general, one can anticipate that spin–rotation interactions might well be important with nuclei that have a large range of chemical shifts (e.g., ^{19}F, ^{13}C, ^{15}N). The relation between R_1 (spin–rotation) and the chemical shift arises because both the chemical shift and the spin–

rotation tensor components of any given molecule depend on the electron distribution in a molecule. A distribution which results in large chemical shifts will also lead to large spin–rotation interactions.

8.5 Electric Quadrupole Relaxation

We saw in Chapter 2 that nuclei with $I \geq 1$ have an electric quadrupole moment. If such a nucleus is in a molecule where it is surrounded by an asymmetric electric charge distribution, it can be relaxed quite efficiently by electric quadrupole interactions. Molecular tumbling now causes fluctuating electric fields, which induce transitions among the nuclear quadrupole energy levels. The resulting nuclear relaxation is observed in the NMR just as though the relaxation had occurred by a magnetic mechanism.

In the simplified case of rapid molecular tumbling and axial symmetry of the molecular electric field, theory gives[129]

$$R_1 = R_2 = \frac{3}{40} \frac{2I + 3}{I^2(2I - 1)} \left(\frac{e^2Qq}{\hbar}\right)^2 \tau_c, \qquad (8.15)$$

where $(e^2Qq/\hbar)$, called the *quadrupole coupling constant,* is made up of the nuclear quadrupole moment Q, the electric field gradient q, and the fundamental constants e and $\hbar$. The quadrupole coupling constant is zero for a highly symmetric situation (e.g., Cl^- ion, tetrahedral $^{14}NH_4^+$), but can be very large in other cases. For asymmetric ^{14}N bonds, $e^2Qq/\hbar$ is typically a few MHz, leading to T_1 in the range of 10–20 msec; for ^2H the range is generally only 100–200 kHz; and for chlorine and bromine in asymmetric covalent bonds, values over 100 MHz may be found, giving T_1 in the range of microseconds. It is apparent that where quadrupole relaxation exists it is normally the dominant relaxation process.

8.6 Scalar Relaxation

As we observed in Section 8.1, any process that gives rise to a fluctuating magnetic field at a nucleus might cause relaxation. For example, when two nuclei I and S are spin-coupled, the value of J, the scalar spin–spin coupling constant, measures the magnitude of the magnetic field at I arising from the spin orientation of S. As S relaxes, I experiences a magnetic field fluctuation; likewise if J changes, because of bond breaking in chemical exchange processes, I experiences a similar fluctuation. We shall discuss three situations where relaxation might occur by these processes:

1. $R_1^S \gg 2\pi J$ and $R_1^S \gg R_1^I$. This situation occurs, for example, where $I = \frac{1}{2}$ (e.g., ^{1}H or ^{13}C) and is coupled with a moderately large J to a quadrupolar nucleus that relaxes rapidly. Because of the rapid relaxation of S the expected splitting of the resonance lines of I does not occur. Equations for this type of scalar relaxation have been derived:

$$R_1^I = \frac{8\pi^2 J^2}{3} S(S + 1) \frac{T_1^S}{1 + (\omega_I - \omega_S)^2 (T_1^S)^2}, \tag{8.16}$$

$$R_2^I = \frac{4\pi^2 J^2}{3} S(S + 1) \left[T_1^S + \frac{T_1^S}{1 + (\omega_I - \omega_S)^2 (T_1^S)^2} \right]. \tag{8.17}$$

Here S is the spin of the quadrupolar nucleus, ω_I and ω_S are the Larmor frequencies of the two nuclei, and T_1^S is the longitudinal relaxation time of S.

The form of Eq. (8.16) is such that a significant effect will occur only if $\omega_I \approx \omega_S$. This is rare but does occur occasionally (e.g., for $I = {}^{13}$C and $S = {}^{79}$Br, $\omega_C - \omega_{Br} = 0.054$ MHz at 14 kG leading to a significant scalar coupling contribution to R_1 in CH_3Br and similar molecules). In general, R_2^I is much more readily affected; for example, the relaxation of ^{14}N often leads to a shortening of $T_2(^1H)$ by this process, thus accounting for the broad lines often found for protons attached to a nitrogen. However, if T_2^S is *very* short (e.g., ^{35}Cl), there is little effect on T_2^I. From an analysis of T_1 and T_2 data it is often possible to determine the magnitude of a coupling constant that cannot be observed directly; e.g., $J(^1H-^{35}Cl)$ in $CHCl_3$ can be estimated as 6.9 Hz.[129]

2. $R_1^S \ll 2\pi J$ and $R_1^S > R_1^I$. This situation occurs, for example, where $I = {}^{13}$C and $S = {}^1$H. The proton relaxation is not nearly rapid enough to affect the spin–spin splitting, nor does it influence $T_1(^{13}C)$. But it does shorten $T_2(^{13}C)$ substantially; for example, in $^{13}CH_3COOCD_3$, where $J = 130$ Hz, $T_1(^1H) = 12.5$ sec, $T_1(^{13}C) = 19.2$ sec, while $T_2(^{13}C)$ is only 6.1 sec.[132] In this rapidly tumbling molecule, $T_1(^{13}C)$ and $T_2(^{13}C)$ would have been expected to be equal in the absence of the scalar coupling relaxation.

3. J is a function of time.   In the case of a chemical exchange that causes bond breaking, the magnetic field at I fluctuates because J is modulated from its normal value to zero. The resulting effects are clearly analogous to the case 1 or 2, depending on the rate of exchange relative to J. As viewed from nucleus I, it makes little difference whether nucleus S re-

laxes or exchanges, so the same equations apply. We shall discuss exchange in more detail in Chapter 11.

8.7 Relaxation by Paramagnetic Substances

Paramagnetic atoms or molecules (i.e., those with one or more unpaired electrons) can contribute to relaxation of nuclei in two ways: (1) dipolar relaxation by the *electron* magnetic moment; (2) transfer of unpaired electron density to the relaxing atom itself. The basic equations for relaxation by a paramagnetic agent, derived by Solomon and Bloembergen,[133] which are not reproduced here, consist of terms analogous to those in Eqs. (8.5) and (8.6) for the dipolar relaxation and terms similar to those in Eqs. (8.16) and (8.17) since a scalar coupling with the unpaired electrons is involved. A complication arises regarding the correlation times, since several processes can determine the values—the usual rotational correlation time τ_c, the spin–lattice relaxation time of the electron, and the rate of any ion–ligand exchange processes. The scalar portion of the interaction contributes little to T_1 but often dominates T_2.

As we pointed out in Section 4.16, lanthanide shift reagents owe their utility partly to the fact that the electron spin–lattice relaxation time for the lanthanides is very short, so that NMR lines are not exceptionally broad. On the other hand, there are "shiftless" paramagnetic reagents that shorten both T_1 and T_2 to a moderate degree without causing contact or pseudocontact shifts. We shall mention an application of these reagents in Chapter 10.

Finally, we should note that the most common paramagnetic relaxation agent is molecular oxygen. At atmospheric pressure enough oxygen dissolves in most solvents to provide a relaxation rate for protons of about 0.5 sec^{-1} and to make it the dominant mode of relaxation for most protons in small diagmagnetic molecules.

8.8 Some Chemical Applications

With the widespread use of pulse Fourier transform methods in NMR, it has become possible (as we shall see in detail in Chapter 10) to easily measure relaxation times of individual nuclei in complex molecules. During the last five years the study of relaxation, especially of nuclei such as ^{13}C, ^2H, ^{31}P and ^{15}N, has become very popular, and the interpretation of these data has provided a great deal of new information on

CH$_3$ -CH$_2$ -CH$_2$ -CH$_2$ -CH$_2$ -CH$_2$ -CH$_2$ -CH$_2$ -CH$_2$ -CH$_2$ OH

3.1   2.2   1.6   1.1   0.84 0.84 0.84 0.77 0.77 0.65

Fig. 8.3 Values of T_1 for various ^{13}C atoms. Top: *n*-decanol (Doddrell and Aller-hand[135]); bottom: phenol (Levy *et al.*[136]).

molecular structure and on molecular dynamics in the liquid phase. We casn present here only a few illustrations.

Carbon-13 is an attractive nucleus for study because of its widespread occurrence in all organic molecules. Of the six mechanisms of relaxation we discussed, all except quadrupolar relaxation have been shown to be applicable under some circumstances,[134] but in general, dipolar interactions with nearby protons usually prove to be dominant, especially when the carbon is directly bonded to one or more hydrogen atoms. (Even when other mechanisms compete, it is possible to extract that part of the total relaxation rate due to dipolar interactions by means of a measurement of the nuclear Overhauser effect, as we shall see in Chapter 9.)

Two examples of the use of ^{13}C relaxation times in the study of molecular motions are given in Fig. 8.3, where T_1 for each carbon is listed. Both decanol and phenol lie in the category of ''small'' molecules, where rapid molecular tumbling places them in the extreme narrowing condition. As seen from Eq. (8.8), T_1 is inversely proportional to the correlation time. In decanol the increase in T_1 along the chain results from substantial internal, or segmental motion, so that each carbon has its own effective τ_c. Near the polar end of the molecule τ_c represents nearly the value for overall tumbling of the molecule, but farther down the chain the effective τ_c is the resultant of molecular and internal motions. The resultant shorter τ_c leads to a longer T_1.

A similar situation occurs in phenol, where the molecule is rigid but molecular motion is anisotropic. The more rapid motion about the C_1–C_4 axis has no effect on C_4, but reduces τ_c for C_2 and C_3 below that resulting from tumbling of the molecule about the other two axes. In the case of C_2, C_3, and C_4 of phenol, as well as in that of the CH$_2$ groups in decanol, comparisons of T_1's are valid since each carbon atom is bonded to the same number of protons and the bond distances are constant. For comparison of the CH$_3$ data in decanol with the CH$_2$'s we would need to multiply the observed T_1 of the CH$_3$ carbon by $\frac{3}{2}$. Note that in phenol, C_1, with no directly bonded proton, has a much longer T_1.

Information on molecular motions can also be obtained from relaxation of quadrupolar nuclei, such as ^{2}H. As seen from Eq. (8.15), T_1 depends on the correlation time and the quadrupole coupling constant. This latter quantity can be obtained from microwave spectra of gases or by pure nuclear quadrupole resonance in the solid phase. If a value of the quadrupole coupling constant is available for the molecule in question or a sufficiently similar molecule, then τ_c may be obtained directly from the measurement of T_1.

Another useful application of quadrupole relaxation is the use of ^{35}Cl and ^{79}Br as probes of macromolecular structure. As we mentioned in Section 8.6, the ion has a small (ideally zero) quadrupole coupling constant, but when covalently bonded can have a very large e^2Qq. A solution with a large concentration of Cl$^-$ and a small concentration of, say, a protein to which Cl$^-$ can bind at specific sites, will show a T_1(Cl) that is the weighted average of the relatively long T_1 of the free ion and the very short T_1 of the bound form. Thus T_1 can be used as a measure of the fraction of Cl$^-$ bound and hence provide information about competitive binding at the sites.[137]

Proton NMR relaxation studies are useful in many ways. Often the in-

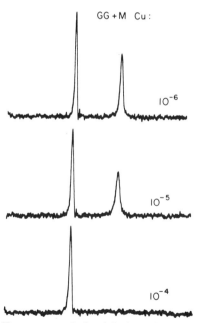

Fig. 8.4 Proton NMR spectrum of glycylglycinate ion, NH$_2$CH$_2$C(O)NHCH$_2$COO$^-$ (0.5 M in D$_2$O) as a function of addition of Cu^{++} (10^{-6} to 10^{-4} M). Reprinted with permission from Li *et al.*, *J. Am. Chem.* **84**, 4650 (1962). Copyright by the American Chemical Society.[138]

terpretation of proton T_1's is difficult because both intra- and intermolecular dipolar interactions must be considered and because paramagnetic impurities have a large effect on proton relaxation times. On the other hand, a great deal of information is available on sites of binding of paramagnetic ions. For example, Fig. 8.4 shows the proton NMR spectra of the glycylglycinate anion in the presence of Cu^{2+} ion. The paramagnetic cupric ion is known to complex with the amino terminal end of the glycylglycinate, so that the CH_2 protons adjacent to the nitrogen are principally relaxed by the copper. The ligand is present in large excess, and the copper ion exchanges rapidly among various ligand molecules. The effect on T_2 is observable because the line width for a $1:1$ ratio of ligand to Cu^{2+} would be several kHz; hence the average that is observed is still fairly large. Spin–spin relaxation by Cu^{2+} and many other paramagnetic ions is often dominated by the scalar coupling interaction, whereas spin–lattice relaxation is normally dipolar. Measurements of T_1 then can sometimes be used to obtain relative distances from a paramagnetic ion to various protons in a ligand. Gd^{3+} has been used in this way to probe nucleotide conformation.[139] In deriving distances one must insure that internal motions do not cause different τ_c's to apply to the nuclei being studied.

8.9 Measurement of Relaxation Times

Both T_1 and T_2 may be measured directly by high resolution NMR techniques if they are in the appropriate ranges. We saw in Section 2.5 that repetitive scans through a resonance after it has initially been saturated may permit a determination of T_1 from the growth of the signal amplitude. This procedure is limited to values of T_1 greater than a few seconds in order that a sufficient number of scans be obtained to permit accurate measurement. In general, the longer T_1, the better this direct method will be. With the rapid scan correlation method (Section 3.6) the time scale can be shortened to T_1's in the range of tens of milliseconds. With regard to T_2, the direct method of measurement from high resolution spectra is limited to relatively short values as determined from line widths by the relation

$$T_2 = 1/\pi\nu_{1/2}. \tag{8.18}$$

For long T_2 the *true* line width is very small, but the *observed* width is then determined by magnetic field inhomogeneity. Since width due to inhomogeneity is likely to be at least 0.1 Hz, only lines that are appreciably broader can be used for this purpose; hence for such direct measurement T_2 must be smaller than about 0.5 sec.

For longer values of T_2 and for a wide range of T_1's, measurements are usually made by some sort of *pulse* technique. A short rf pulse is applied to the sample and a resultant resonance property studied. We shall defer further discussion of these methods to Chapter 10.

Problems

1. Acetonitrile, $CH_3C\equiv N$, has a known geometry, with distances from the nitrile carbon to each hydrogen of 2.14 Å and to the nitrogen of 1.14 Å. Find the relative values of R_1's for (a) dipolar relaxation of the nitrile carbon by the protons, (b) dipolar relaxation of this carbon by the nitrogen (^{14}N), and (c) chemical shift anisotropy relaxation at 68 MHz. Take $\sigma^\| - \sigma^\perp = 300$ ppm. Assume τ_c to be the same for the three processes.

2. Calculate the scalar coupling contribution to R_1 and R_2 of the nitrile carbon in acetonitrile if $T_1(^{14}N) = 1.8$ msec and $J(^{13}C^{-15}N) = -17$ Hz. What is the minimum linewidth of the ^{13}C line for this carbon?

3. Find the ratio R_1/R_2 for dipolar relaxation at frequencies $\nu = 0.8\nu_0; \nu_0; 1.2\nu_0; 1.5\nu_0; 2\nu_0$.

4. For ^{13}C in CH_3OH, T_1 is found to increase with temperature from $-75°C$, pass through a maximum at 35°C, and decrease at higher temperatures. What mechanisms of relaxation are most likely to account for these results?

5. For deuterium in perdeuterated toluene, the T_1's are T_1(ring) $= 0.94$ sec and T_1(methyl) $= 4.9$ sec at 25°C. (a) Assuming isotropic rotation for the ring, find τ_c(ring) and τ_c(methyl) if the quadrupole coupling constants are 193 kHz for the ring deuterons and 165 kHz for the methyl deuterons. (b) Predict the value of $T_1(^{13}C)$ for the methyl carbon due to dipolar relaxation. Assume $r(C-H) = 1.09$ Å.

6. Determine the structure of the molecule giving Spectrum 31, Appendix C.

7. Determine the structure of the molecule giving Spectrum 32, Appendix C. Account for the appearance of the spectrum by a modified first-order analysis.

Chapter 9

Theory and Application of
Double Resonance

Our discussion of NMR thus far has dealt with the situation in which the sample is subjected to only one radio-frequency field, namely, that needed to observe the resonance. We now wish to consider the effects of applying simultaneously two (or more) rf fields at different frequencies, one used to observe resonance and the other(s) to perturb the nuclear spin system. This procedure is generally called nuclear magnetic *double resonance*, abbreviated as NMDR, or *multiple resonance* if two or more perturbing fields are employed. We shall see that if the perturbing field is sufficiently strong and is applied at the resonant frequency of one of a pair of spin-coupled nuclei, the other nucleus behaves in its NMR as though it were no longer coupled. This technique, *spin decoupling,* is a powerful tool in unraveling the spectra of many complex molecules. On the other hand, if the strength of the perturbing field is rather small, additional splittings and/or changes in relative intensities of lines may occur. These branches of double resonance, *spin tickling* and *nuclear Overhauser effects,* are also, as we shall see, of great value in unraveling complex spectra.

9.1 Notation and Terminology

We shall continue to use the notation already employed, in which H_0 denotes the large applied magnetic field, which is taken to be along the z axis, while H_1 denotes the radio-frequency field, which can be thought of as rotating in the xy plane at a frequency ν_1 in the same direction in which

199

the nuclei precess. We now denote by H_2 the second rf field, also rotating in the xy plane in the same direction as H_1, but with frequency ν_2.

If the frequencies ν_1 and ν_2 are both near the Larmor frequency of one type of nucleus (e.g., the proton), so that they differ only by an audio frequency (from a few hertz to a few thousand hertz), we speak of *homonuclear* double resonance (or homonuclear decoupling or tickling). If ν_1 and ν_2 differ by a radio frequency (e.g., 60 MHz for ^{1}H and 24.3 MHz for ^{31}P), then we speak of *heteronuclear* double resonance. In the heteronuclear case, a commonly used notation lists first the nucleus that is observed at frequency ν_1 followed, in brackets, by the nucleus that is irradiated at frequency ν_2; for example, ^{1}H{^{19}F}.

9.2 Experimental Techniques

The apparatus used for double resonance studies is essentially that used in ordinary single resonance experiments with the additional provision for supplying to the transmitter coil of the probe the second radio frequency at the desired power level. The procedures normally used are quite different for homonuclear and heteronuclear double resonance.

For heteronuclear double resonance the second radio frequency is generally supplied by a second stable rf oscillator operating at frequency ν_2. This oscillator must be tunable over the range of resonance frequencies (chemical shifts) expected for the nucleus to be irradiated. Commercial units of moderate stability are available that cover a range of several kilohertz around a crystal-controlled frequency. Probably the most satisfactory source of the second rf field is a frequency synthesizer, which can be set to better than 0.1 Hz over a range of many megahertz and maintained constant to about 0.01 Hz. For many double resonance experiments the stability of two separate crystal-controlled oscillators is insufficient, so the two rf sources must be locked together, with both frequencies ultimately derived from a single crystal.

Since two different rf frequencies are applied to the sample, the transmitter coil in the probe must be tuned to accept both frequencies. Such double tuning is a relatively simple electronic problem. Alternatively two separate coils may be employed.

We have pointed out in previous chapters that an NMR spectrum may be scanned by varying H_0 while holding ν_1 constant (field sweep) or by varying ν_1 while holding H_0 constant (frequency sweep). For single resonance studies, both procedures produce identical spectra, but this is not true for double resonance, as we shall see in Section 9.3. We can easily recognize the difference between the two types of sweep in an A{X} experiment. In a frequency sweep H_0 is constant so that the Larmor frequency of X, ν_X, is also constant. Hence ν_2 may be fixed so that $\nu_2 = \nu_X$ or so that $(\nu_2 - \nu_X)$ is any desired value. In a field sweep, however, ν_X will be swept past ν_2 as the field is swept. In general, the frequency sweep method provides a "cleaner" arrangement, but many instruments (especially external lock systems) are limited to field sweep capabilities only. In some field sweep instruments the value of ν_2 can be swept automatically in synchronism with the field sweep. Such *tracked field sweep* double resonance arrangements are almost indistinguishable in their results from true frequency sweep instruments.

There is a second type of frequency sweep method possible in double resonance studies. This method requires that H_0 be maintained constant and that ν_1 be held constant at a value exactly on one of the A resonance lines. Frequency ν_2 is then swept in the vicinity of ν_X and perturbations in the A spectrum are recorded. This technique has been called INDOR (internuclear double resonance).[140] We shall see some examples of the use of INDOR in Section 9.7. It is apparent that ν_1 can be maintained precisely on one of the spectral lines of A only if a very good field–frequency control is used (usually an internal lock; see Section 3.4).

In principle, *homo*nuclear double resonance can be accomplished by the use of two rf oscillators or frequency synthesizers locked together. With modern pulse Fourier transform spectrometers this is the arrangement that is invariably used. With continuous wave spectrometers, however, the experimental arrangement is simpler and less costly if the methods of audio modulation described in Section 3.3 are employed. For example, in proton resonance at 60 MHz an audio frequency $\nu_1'(\sim 2000 \text{ Hz})$ is often applied to modulate the radio frequency or, equivalently, the magnetic field. The resonance signal is then phase detected at ν_1', so that the NMR is carried out at a frequency of $(60 \text{ MHz} \pm \nu_1')$, depending on whether the upper or lower sideband is used. If an additional audio signal at frequency ν_2' is used for the modulation but not the detection, the nuclei experience the observing frequency $\nu_1 = 60 \text{ MHz} \pm \nu_1'$, as well as the frequency $\nu_2 = 60 \text{ MHz} \pm \nu_2'$. The modulation indices of the frequencies ν_1' and ν_2' can be adjusted to obtain the desired power level in each sideband.

As in the case of heteronuclear double resonance, there are three types of sweep possible. The ordinary frequency sweep, usually obtained by varying ν_1' while holding H_0, ν_2', and the 60-MHz frequency (or whatever the nominal resonance frequency of A is) constant, is the simplest to comprehend and normally gives the most easily interpretable results. The situation is exactly the same as in the heteronuclear case. The other frequency sweep, in which ν_2' is varied while H_0, ν_1', and the 60 MHz are held constant, is also exactly the same as in the heteronuclear case. Field sweep, in which H_0 is varied while all frequencies remain constant, brings ν_X and ν_2 into coincidence only at one point in the scan. In the homonuclear case, it is often convenient to think of a field sweep as though the field were held constant while both ν_1' and ν_2' move across the spectrum at a constant separation $(\nu_1' - \nu_2')$. If two coupled nuclei are separated in chemical shift by $(\nu_1' - \nu_2')$, one of the pair will be irradiated by field H_2 while the second is being observed. We shall see examples of spin decoupling experiments carried out in this way in Section 9.5.

In some experiments we wish to irradiate with H_2, not at a specific frequency, but covering a band of frequencies (often 10^2–10^4 Hz). Such irradiation is normally accomplished by modulating the radio frequency ν_2, which is placed at the center of the frequency range to be covered. Pseudorandom noise, generated digitally, or other methods of modulation can be used, but regardless of the exact procedure the process is usually called *noise-modulated decoupling* (or often, *noise decoupling*).

In some experiments using pulse excitation it is desirable for the field H_2 to be turned on only for a portion of the experiment. This procedure is usually termed *gated decoupling* (or more generally gated double resonance).

9.3 Theory of Double Resonance

The general theory of double resonance is quite complex and is far beyond the scope of this book. We shall concentrate merely on indicating the general approach and discussing some of the most important results.[141]

In our general consideration of complex spectra in Chapter 7 we found, in Section 7.5, that the Hamiltonian could be written as the sum of two parts,

$$\mathcal{H} = \mathcal{H}^{(0)} + \mathcal{H}^{(1)}.$$

$\mathcal{H}^{(0)}$ refers to the interaction of the nuclei with the magnetic field H_0, while $\mathcal{H}^{(1)}$ takes account of spin–spin coupling between the nuclei. The effect of the rf field H_1, used to observe resonance, was introduced as a time-dependent perturbation after the equation had been solved with the above Hamiltonian.

In our consideration of double resonance phenomena we can still introduce H_1 as a time-dependent perturbation, since it is a weak field. However, we cannot validly treat H_2 in a similar manner because the energy of interaction between H_2 and the nuclear spin system is often of the same magnitude as, or even greater than, the energy of interaction (coupling) between nuclei. Hence the effect of H_2 must be included by adding an extra term to the Hamiltonian:

$$\mathcal{H}(\nu_2 t) = \mathcal{H}^{(0)} + \mathcal{H}^{(1)} + \mathcal{H}^{(2)}(\nu_2 t). \tag{9.1}$$

We have noted specifically that since $\mathcal{H}^{(2)}$ is a periodic function of time, the total Hamiltonian will also vary periodically. Fortunately, a general theorem by Larmor (mentioned in Section 2.2) shows that the effect of a magnetic field on a set of spins is equivalent to subjecting the spins to a rotating coordinate system. By using the Larmor theorem and making a simple transformation of coordinates from the fixed laboratory frame of reference to a frame of reference rotating at ν_2, the frequency of the time-dependent portion of $\mathcal{H}$, we can convert $\mathcal{H}$ to a time-independent $\mathcal{H}_R$:

$$\mathcal{H}_R = \mathcal{H}_R^{(0)} + \mathcal{H}^{(1)} + \mathcal{H}_R^{(2)}$$

$$= \sum_i (\nu_i - \nu_2)(I_z)_i + \sum_{i<j}\sum J_{ij} \mathbf{I}_i \cdot \mathbf{I}_j \frac{1}{2\pi} \sum_i \gamma_i H_2 (I_x)_i. \tag{9.2}$$

(H_0 is still taken along the negative z axis, while H_2 is along the positive x axis.) From comparison of Eq. (9.2) with Eqs. (7.17) and (7.18), it is apparent that the effects of transforming to the rotating frame is to cause all chemical shifts to be measured relative to ν_2 (i.e., $(\nu_i - \nu_2)$ appears instead of ν_i in the first term) and to introduce an interaction term between H_2 and the x component of each spin.

The diagonalization of $\mathcal{H}_R$ is a considerably harder task than the diagonalization of $\mathcal{H}$ in the absence of the second rf field. The solution has been carried out in different ways, depending on the relative values of the parameters involved. One particularly important case is that where $|\nu_i - \nu_j| \gg J_{ij}$ for all i and j. This situation often occurs for heteronuclear double resonance; it also occurs in those cases of homonuclear double resonance that would be simply analyzed by first order in the absence of H_2. The solution for this case has been obtained,[142] and some typical results are shown schematically in Fig. 9.1 for an A{X} experiment in an AX system. The "offset parameter" Δ measures the difference between the frequency of the perturbing field and the resonance frequency of X. When $\Delta > 3$, there is no observable effect; but as ν_2 approaches ν_X, changes in line frequency and intensity occur. For $\gamma H_2/2\pi > 2J$ the A doublet collapses to a singlet at ν_A when $\nu_2 = \nu_X$. This is *spin decoupling*.* When

---

* An often quoted qualitative "explanation" for spin decoupling is that the high power in H_2 causes such rapid transitions among the X nuclei that the energy of the A nucleus responds only to the average X energy, which is zero. Hence the A energy levels behave as though the X nucleus were not coupled to A. This rationalization for the observed phenomenon clearly does not account for the effects shown in Fig. 9.1 for weaker perturbing fields. It also is inadequate in explaining completely the observations in strong fields when an A_nX system is considered.

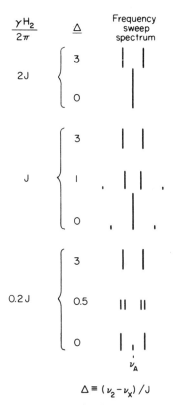

$$\Delta \equiv (\nu_2 - \nu_X)/J$$

Fig. 9.1 Schematic spectra expected for the A resonance of an AX system when a second rf field is applied near the resonance frequency of X. See text for definition of symbols. Based on plots of Freeman and Whiffen.[142]

systems of more than two nuclei are studied, the situation may be considerably more complex. If several magnetically equivalent nuclei are being observed (e.g., A_nX system), the collapse of the A spectrum is incomplete, and even with a large H_2 centered at ν_X a small residual splitting or broadening is observed. On the other hand, when only a single nucleus is being observed while a magnetically equivalent set is being irradiated (e.g., AX_n), complete decoupling can be achieved with $\nu_2 = \nu_X$ and H_2 sufficiently large. In the AX_n case residual splittings can be observed when $\nu_2 \neq \nu_X$, and they are related to J_{AX} by the relation[143]

$$\frac{J_r}{(J_{AX}^2 - J_r^2)^{1/2}} = \frac{\nu_2 - \nu_X}{\gamma H_2/2\pi}. \tag{9.3a}$$

If $\gamma H_2/2\pi \gg \nu_2 - \nu_X$, a slightly simpler relation applies:

$$\frac{J_\mathrm{r}}{J_\mathrm{AX}} = \frac{\nu_2 - \nu_\mathrm{X}}{\gamma H_2/2\pi}. \tag{9.3b}$$

Equation (9.3) is often applied to the interpretation of ^{13}C spectra, as we shall see in Section 9.5.

Figure 9.1 also shows an interesting effect in the frequency sweep spectrum with $\gamma_\mathrm{X} H_2/2\pi = 0.2J$ and $\Delta = 0.5J$; that is, a weak perturbing field applied exactly at the frequency of *one* of the two X lines. This is an example of *spin tickling*. We see that each of the A lines splits into a very closely spaced doublet. We shall discuss tickling in more detail in Section 9.9.

The results in Fig. 9.1 were obtained with the assumption that all J's are negligible relative to $|\nu_\mathrm{A} - \nu_\mathrm{X}|$. In many cases of homonuclear double resonance this condition is not satisfied. These results are still valid approximately for moderate chemical shift differences, except that optimum decoupling occurs not when $\nu_2 = \nu_\mathrm{X}$, but when

$$(\nu_1 - \nu_2) = (\nu_\mathrm{A} - \nu_\mathrm{X}) - \frac{(\gamma H_2/2\pi)^2}{2(\nu_\mathrm{A} - \nu_\mathrm{X})}. \tag{9.4}$$

Equation (9.4) results from the presence of two effects: first, the additional field strength arising from H_2, which increases the Larmor frequency of A; and second, the effect of the nonnegligible J in altering the effective fields seen by the nuclei in the rotating frame of reference. For accurate measurements of chemical shifts by homonuclear spin decoupling (see Section 9.7) the value of H_2 must be determined and the correction of Eq. (9.4) applied.

Our discussion of the theory of NMDR began with the observation that H_2 may be large enough so that it cannot be treated merely as a perturbation. In some cases, however, use is made of a rather weak H_2 so that the effects we have considered can be ignored. In this application H_2

Table 9.1

TYPES OF DOUBLE RESONANCE

$\gamma H_2/2\pi$ (Hz)	Designation	Typical effect
$>2J^a$	Spin decoupling	Collapse of multiplets
$\sim J$	Selective spin decoupling	Partial collapse of multiplets
$\sim\nu_{1/2}{}^b \ll J$	Spin tickling	Splitting of lines
$<\nu_{1/2}$	Nuclear Overhauser effect	Change in area under line

[a] J is the spin–spin coupling constant between the irradiated and observed nuclei.
[b] $\nu_{1/2}$ is the width at half-maximum intensity of the irradiated line.

is used, not to alter energy levels, but to bring about changes in populations of the levels. This technique is called the *nuclear Overhauser effect* (by analogy to a similar effect found by Overhauser[144] for interactions between electrons and nuclei). The theory of the nuclear Overhauser effect (NOE) depends on relaxation processes, rather than perturbations of energy levels. It is discussed in Section 9.4.

We have described several types of double resonance experiments, which depend upon the strength of the perturbing field H_2. Table 9.1 summarizes these branches of NMDR, giving for each type the typical effect on NMR lines. All four kinds of study can be carried out with either homonuclear or heteronuclear double resonance.

9.4 The Nuclear Overhauser Effect

The nuclear Overhauser effect refers to changes in intensities in a spectrum on double resonance that result from dipolar relaxation. (Other NMDR intensity alterations are, by analogy, sometimes called "generalized nuclear Overhauser effects.") We can best understand the basis of the NOE by considering a two spin system, AX, each nucleus with $I = \frac{1}{2}$, and relaxed *solely* by magnetic dipole interaction with the other one. We shall assume no spin–spin coupling, but the presence of first-order coupling does not affect the results. The four energy levels of this system were shown in Fig. 7.1 and are depicted in a slightly different form in Fig. 9.2. Because of the selection rule $\Delta m = \pm 1$, absorption of rf energy causes transitions only among the levels indicated in Fig. 9.2a but dipolar relaxation processes can cause transitions among all the levels, as shown in Fig. 9.2b. By considering in detail the Hamiltonian for dipolar relaxation in a system in the extreme narrowing condition (i.e., short τ_c) Solomon[133] showed that the relaxation transition probabilities $W_0 : W_1 : W_2 = 2 : 3 : 12$. In the absence of rf radiation, relaxation proceeds along these various pathways until the Boltzmann equilibrium is obtained. The intensity of the line at ν_A is then proportional to the population differences $(N_2 - N_1) \approx (N_4 - N_3)$, as shown in Section 2.4. Now suppose the two X transitions are irradiated at ν_X with a value of H_2 large enough to saturate the transitions. Then the population of the levels must be $N_1 = N_3$ and $N_2 = N_4$. Relaxation still operates with the same probabilities, but under these constraints a non-Boltzmann distribution is established, with gain in populations of levels 2 and 4. The result is that the intensity of the A transition changes from a relative value of unity to a value of $1 + \gamma_X/2\gamma_A$. The NOE, given the symbol η, is defined as the difference in intensities; i.e.,

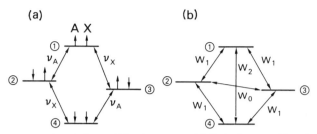

Fig. 9.2 Energy levels for an AX spin system. (a) Paths for transitions induced by absorption of radiation at the resonance frequencies of A and X. (b) Paths for spin–lattice relaxation, resulting in $\Delta F_z = k$ for pathway indicated by W_k.

$$\eta = \gamma_X/2\gamma_A. \tag{9.5}$$

(Often the value given in publications for the NOE is actually $1 + \eta$, so care must be used in defining the values quoted.)

For the homonuclear case, where $\gamma_A = \gamma_X$, the NOE is 50%, while for the important situation where $A = {}^{13}C$ and $X = {}^{1}H$, the NOE is 1.988. Where one γ is negative, the NOE leads to emission; e.g., $A = {}^{15}N$, $X = {}^{1}H$, $\eta = -4.93$.

Our derivation of the NOE results was based on three assumptions: (1) the extreme narrowing condition; (2) a two-spin system; (3) solely dipolar relaxation. When condition (1) is not satisfied, there is a dependence of η on τ_c and ω_0, such that near $\tau_c = 1/\omega_0$ the value of the NOE undergoes a marked decrease. In the limit of $\tau_c \gg 1/\omega_0$ the homonuclear case leads to $\eta = -1$, rather than $+0.5$,[145] while for ${}^{13}C\{{}^{1}H\}$ experiments η drops to 0.15. When condition (2) is not satisfied and there are more than two spins in sets (e.g., A_2X_3), the results are unchanged. However, where there are three or more different kinds of spins (A observed, X irradiated, Y not irradiated) the results can be much more complicated.[146] Such "three-spin effects" have sometimes been overlooked, and erroneous conclusions drawn. When condition (3) is not satisfied, the result in equation 9.5 can easily be corrected for other relaxation mechanisms. Suppose

$$R_1 = R_1^d + R_1^o, \tag{9.6}$$

where R_1^d refers to dipolar relaxation, and R_1^o to all other mechanisms. Then

$$\eta = \tfrac{1}{2} \cdot \frac{\gamma_X}{\gamma_A} \cdot \frac{R_1^d}{R_1}. \tag{9.7}$$

A measured NOE obviously can be used to assess the relative importance of dipolar relaxation. Other examples of the use of the NOE will be given in the applications of NMDR in the following sections.

The NOE can most easily be measured in pulse experiments by the technique of gated decoupling (see Section 9.2). The perturbing field H_2 is turned off for a period of several times T_1 to insure that equilibrium populations occur in all energy levels, and an observation pulse (H_1) is then applied. In the study of ^{13}C and other nuclei where it is desirable to obtain the spectrum under conditions of complete proton decoupling, the decoupler (H_2) is turned on approximately coincident with the pulse so that during data acquisition information on the decoupled spectrum is obtained. This procedure is also used in some instances to suppress the NOE so that the resultant areas under the spectral lines can be used quantitatively as a measure of the number of nuclei present (see Chapter 13).

Occasionally one wishes to obtain a coupled (i.e., undecoupled) ^{13}C spectrum without sacrificing the sensitivity improvement provided by the NOE. In this case the decoupler is gated on during a waiting period between pulses but turned off just before the observation pulse and left off during data acquisition. Most commercial pulse Fourier transform spectrometers provide for both types of gated decoupling.

9.5 Structure Elucidation

Spin decoupling has proved to be an invaluable tool in the elucidation of the structures of many complex molecules. Often a knowledge of which protons in a molecule are spin coupled to each other can provide a piece of information that makes possible an unequivocal structural assignment. A typical example of this use of spin decoupling is given in Fig. 9.3. In this case the structure of an unknown molecule was restricted by other data to two possibilities, I and II. In each case the olefinic proton would be expected to have a chemical shift in the vicinity of 700 Hz (7.0 ppm) from TMS and to be split approximately into a triplet by coupling to the adjacent CH_2 protons and to be further split by the four-bond coupling through the double bond. In structure I, the CH_2 to which the olefinic proton is coupled would be expected to resonate near 2.5 ppm since it is adjacent to a doubly bonded carbon, while in structure II it would appear at much lower field since it is adjacent to both a double-bonded carbon and oxygen. The decoupling experiment shown in Fig. 9.3c demonstrates clearly that the CH_2 resonates near 2.5 ppm and hence rules out structure II.

When coupling constants are large (e.g., $^1J(^{13}C-H) \approx 120$ Hz), a large amount of power is needed for decoupling, and it is usually difficult

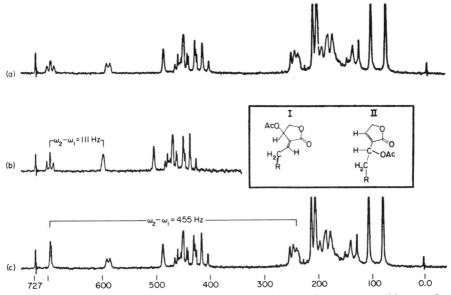

Fig. 9.3 Use of spin decoupling in structure elucidation of an unsaturated lactone, I or II. (a) Single resonance proton NMR spectrum at 100 MHz. (b) Field sweep double resonance spectrum with irradiation field 111 Hz to the high field side of the observing frequency. (c) Field sweep double resonance spectrum with H_2 455 Hz to the high field side of the observing frequency (Shoolery[147]).

to achieve selective decoupling at one chemical shift without affecting others nearby. In ^{13}C NMR selective decoupling is used to great advantage in some instances, but usually some measure of selectivity is obtained in a simple manner by *off-resonance decoupling* of protons, as shown in Fig. 9.4. With $\gamma H_2/2\pi \approx 10J$, ν_2 is set to the low frequency side of all aliphatic protons. The residual splittings (see Eq. (9.3)) now appear with magnitudes inversely proportional to the difference between ν_2 and the proton chemical shift. The multiplicities immediately give information on the nature of the carbon atom (e.g., CH_3, CH_2), and the magnitudes permit assignments of the ^{13}C spectrum relative to the ^{1}H spectrum.

Spin tickling is also applied to problems of structure elucidation. Adjustment of parameters, particularly the strength of H_2, is generally more difficult for tickling studies, so that the less demanding technique of spin decoupling is usually preferred. Tickling has the advantage, however, of causing only slight perturbations, which are less likely to interfere with the recording of the spectrum or with the maintenance of an internal lock signal. In general, the high power needed for complete spin decoupling precludes the irradiation at frequencies much less than 40 Hz from the ob-

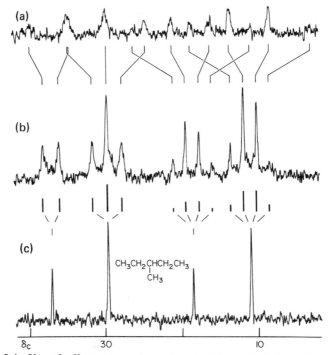

Fig. 9.4 Use of off-resonance decoupling in ^{13}C NMR study of 3-methylpentane (25 MHz). (a) No decoupling; (b) Off-resonance proton decoupling; (c) Complete decoupling by use of noise modulation (Stothers[45]).

serving frequency, but in tickling, closer approach to the observing frequency is possible.

The nuclear Overhauser effect also has great potential in the elucidation of the structures of certain complex molecules. As we saw in Section 9.4, a homonuclear NOE can be as large as 50%. Applications to structural elucidation thus far have been principally in ^{1}H NMR. The experiment consists of a selective irradiation of one proton resonance and the observation of integrated intensity changes in other lines. It is rare that anything approaching a 50% intensification occurs, since dipolar interactions other than the one being studied usually compete. For valid results it is essential that samples be thoroughly deoxygenated, that other paramagnetic impurities be absent, that the solvent contain no magnetic nuclei (or at least none with large magnetic moment) and that the sample concentration be kept low to reduce intermolecular dipolar relaxation. With adequate care instrumental error can be reduced to about 2%. Most ^{1}H{^{1}H} NOE measurements have been made by conventional continuous wave methods, but pulse Fourier transform techniques can be used also. A good discussion of both cw and FT measurements of NOE's has been given recently by Saunders and Easton.[148]

NOE data can be extremely valuable in stereochemical elucidation, since the NOE depends on the rate of dipolar relaxation, which is inversely proportional to the sixth power of the distance between nuclei, according to Eq. (8.8). However, from Eq. (9.7), it is clear that if the *only* relaxation process for the observed nucleus is dipolar relaxation by the irradiated nucleus, the NOE has no dependence on internuclear distances; hence, attempts to relate quantitatively the magnitude of the NOE to distance cannot be applicable in all cases. On the other hand, it is found in practice that the term R_1^0 in Eq. (9.6) is often significant and sufficiently constant to provide an empirical relation between NOE and r^{-6} over an appreciable range.[148] When several different NOE's between various pairs of protons can be measured separately, then it is possible in principle to obtain internuclear distances.[146]

Because of experimental errors and the limits of applicability of the theory, quantitative interpretations of NOE results should be treated cautiously. In many cases, however, the magnitude of a carefully measured NOE can provide definitive information on a structure, particularly where the possibilities are limited.

9.6 Location of "Hidden" Lines

Removal of splittings by spin decoupling often permits the observation of resonances that would otherwise be undetectable. Such decoupling is almost essential for the study of ^{13}C and many other nuclei of low sensitivity and low abundance. Most ^{13}C resonance lines are split into multiplets by spin coupling to protons one, two, or three bonds away, hence are frequently undetectable because of low signal/noise for each component of the multiplet. A $^{13}C\{H\}$ double resonance can collapse many multiplets. The ^{13}C signal resulting from the coalescence of the multiplets is usually further enhanced by the presence of the nuclear Overhauser effect as a by-product of the decoupling, as we have seen. Since several protons with different chemical shifts might be coupled to a given ^{13}C, it is usually desirable to decouple over a range of proton frequencies with noise modulation (Section 9.2).

In the proton resonance spectra of complex molecules one often encounters the situation where a multiplet is hidden under other peaks, so that an accurate measurement of the chemical shift of the proton in question is impossible from the single resonance spectrum. In some cases spin decoupling may be used to collapse the multiplet and permit the observation of the collapsed line above interfering peaks. In other instances it is better to turn the experiment around and to irradiate in the vicinity of the overlapping peaks and to observe optimum decoupling of the other portion of the spin multiplet, which may be in a clear region of the spectrum. The latter procedure is also useful when a resonance is unobservable, not because of overlapping peaks, but because it is inherently broad due to coupling with a nucleus, such as ^{14}N, which relaxes at a moderately rapid rate (see Section 8.6). An alternative to such a homonuclear decoupling is to perform heteronuclear decoupling directly on the rapidly relaxing nucleus.

9.7 Determination of Chemical Shifts

When nuclei A and X are coupled, it is sometimes preferable to determine the spectrum of X, or at least its chemical shift, by a double resonance experiment, rather than by direct study of X. This situation occurs principally when A is a nucleus of high sensitivity, such as ^1H or ^{19}F, and X is a nucleus of low sensitivity, such as ^{29}Si or ^{15}N (the latter usually in isotopically enriched compounds). Orders of magnitude improvement in sensitivity can be achieved by this method. For small values of J a simple decoupling experiment suffices to locate the X chemical shift, while for large J, tickling is usually a more efficient procedure, with the optimum splitting of an A line furnishing the criterion.

These decoupling and tickling experiments suffer from two disadvantages: First, a subjective judgment is required as to the pattern for the "best" decoupling or tickling; second, the technique does not lend itself to time-averaging procedures for further signal enhancement. Both of these shortcomings are avoided in the elegant INDOR technique, where a rather weak perturbing field can be swept in the vicinity expected for the resonance of X, while the magnetic field and observation frequency are held rigidly fixed at one of the peaks in the A spectrum. As ν_2 passes in turn through the X lines, a change in intensity or frequency of the A line may occur, leading to a vertical movement of the recorder pen. Figure 9.5 shows a clear example of an INDOR spectrum, which in this case resembles an inverted X (^{29}Si) single resonance spectrum. The sweep rate of ν_2 must be kept very low to avoid transient effects, and in more complex spin systems the observed pattern may not be simply a replica of the X spectrum.

Homonuclear INDOR is also extremely useful in the determination of chemical shifts of protons in a region of the spectrum that has a multitude of lines. The use of INDOR has been reviewed.[150]

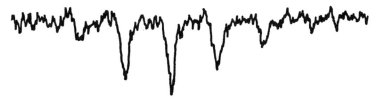

⊢ 50 Hz ⊢

Fig. 9.5 Use of internuclear double resonance (INDOR) to study the ^{29}Si resonance spectrum of SiF$_3$^{29}SiF$_2$SiF$_3$ by monitoring the intensity of one of the ^{19}F lines while weakly irradiating in the vicinity of the ^{29}Si frequency (Johannesen[149]).

In the determination of chemical shifts by double resonance, it is, of course, essential that all frequencies be derived from a single source. In addition, careful consideration must be given to the exact manner in which the measurements are made, since this will affect the comparability of measurements for different substances. If the field H_0 is locked to the basic A frequency by means of an internal lock system employing an internal reference, such as TMS, then the X decoupling frequencies for two or more compounds will be directly representative of the X chemical shifts in those compounds. On the other hand, if the magnetic field is swept, so that the value of H_0 when the double resonance is made is that required for resonance of the A nucleus, then a correction for the chemical shift of A in the different compounds must be made in order to obtain chemical shifts of X. With careful attention to such details, one can make very accurate chemical shift determinations in various nuclei, and in principle they could all ultimately be referred to TMS.

9.8 Relative Signs of Coupling Constants

Double resonance usually provides the best means of determining relative signs of coupling constants. In a weakly coupled (first-order) system it is, in fact, the only simple method of obtaining this information. We can illustrate the reasoning behind the use of double resonance by considering the weakly coupled AMX system. We know that the ordinary spectrum appears to be the same if any one of the signs of the three J's is changed. However, as we saw in Fig. 7.10, a change in sign of one J actually interchanges pairs of lines. Double resonance serves as a probe of these line positions.

From a simple consideration of the origin of the lines in an AMX spectrum we can write the following expressions for the frequencies of the four A, the four M, and the four X lines:

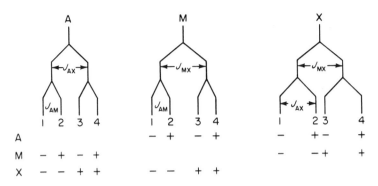

Fig. 9.6 Schematic representation of an AMX spectrum. Spin states $(+\frac{1}{2}$ or $-\frac{1}{2})$ are denoted by + or − signs. The situation depicted is that with all three J's positive.

$$\nu(A_i) = \nu_A + J_{AM}m_M + J_{AX}m_X;$$
$$\nu(M_i) = \nu_M + J_{AM}m_A + J_{MX}m_X; \qquad (9.8)$$
$$\nu(X_i) = \nu_X + J_{AX}m_A + J_{MX}m_M.$$

The small m's can independently assume the values $+\frac{1}{2}$ or $-\frac{1}{2}$ to account for all 12 lines. The index i runs from 1 to 4 and denotes the four A, four M, and four X lines in order of increasing frequency, that is, in the order in which they appear in the spectrum. For example, if both J_{AM} and J_{AX} are positive (and if γ's are positive), line A_1 arises from a "flip" of the A spin, while spins M and X remain oriented with the field (i.e., $m_M = m_X = -\frac{1}{2}$).

Suppose we wish to determine the relative signs of J_{AX} and J_{MX}. The experiment is most easily carried out and the situation most easily explained if J_{AX} and J_{MX} are both somewhat larger than the third coupling constant, J_{AM}. Figure 9.6 has been drawn for this situation, and in the following discussion we shall assume that this is the case. We must focus our attention on the A and M parts of the spectrum. Lines A_1 and A_2 differ only in the value of m_M assigned to them; they have the same value of m_X and may be said to correspond to the same X state. Identical statements may be made about the pairs of lines A_3 and A_4, M_1 and M_2, and M_3 and M_4. If J_{AX} and J_{MX} are both positive, then the low-frequency pair of A lines, A_1 and A_2, arise from the $-\frac{1}{2}$ X state, and the low-frequency pair of M lines, M_1 and M_2, also arise from the $-\frac{1}{2}$ X state. The high-frequency pairs in each case then, of course, arise from the $+\frac{1}{2}$ X state. If both J_{AX} and J_{MX} are negative, the $-\frac{1}{2}$ X state gives rise to the higher-frequency pair in *both* A and M spectra. If, however, J_{AX} and J_{MX} have opposite signs, then the $-\frac{1}{2}$ X state is responsible for the low-frequency pair in one case and the high-frequency pair in the other.

By adjusting the strength of H_2 so that

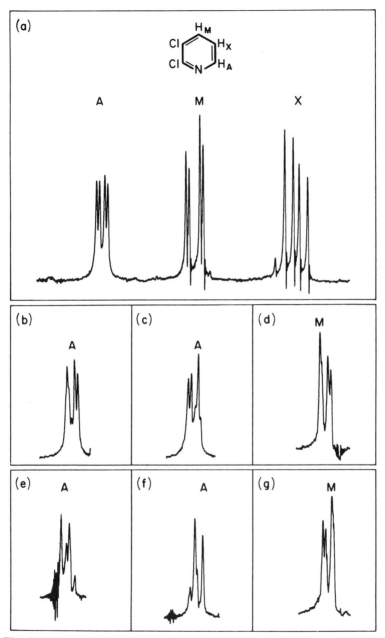

Fig. 9.7 Determination of the relative signs of coupling constants in 2,3-dichloro-pyridine at 100 MHz. (a) Unperturbed spectrum; (b)–(g) selectively decoupled portions of spectrum with decoupling frequency centered as follows: (b) lines M_1 and M_2, (c) M_3 and M_4, (d) A_1 and A_2. (e) X_3 and X_4, (f) X_1 and X_2, (g) A_3 and A_4. See Fig. 9.6 for numbering of lines. (The beat pattern seen in some spectra results from unwanted 60 Hz modulation.)

$$|J_{AM}| < \frac{\gamma H_2}{2\pi} < |J_{AX}|, |J_{MX}|, \tag{9.9}$$

and placing ν_2 at the average frequency of M_3 and M_4, we can obtain *selective decoupling* of only half the molecules—those with a specific X state, either $+\frac{1}{2}$ or $-\frac{1}{2}$. By observing the collapse of either the upper or lower pair of A lines, we can infer that J_{AX} and J_{MX} have the same or opposite signs, respectively. An example of this procedure is shown in Fig. 9.7.

If the value of H_2 can be limited according Eq. (9.9) the decoupling covers only a limited range of frequencies and produces the type of selective decoupling illustrated in Fig. 9.7. If J_{AM} is not appreciably smaller than the other J's, the selective decoupling experiment can often be carried out anyway, but the collapse of a portion of the spectrum will not be so clear.

Spin tickling may also be used for the determination of the relative signs of J's. This technique is particularly valuable with strongly coupled spin systems. Since the sign determination is only one aspect of a general analysis, we shall defer a discussion of this application to the next section.

9.9 Determination of Energy Level Arrangements

The analysis of a complex spectrum to obtain chemical shifts and coupling constants is often a perplexing problem when it involves a strongly coupled system, such as ABC or AA'BB'. The basic problem is that the observed frequencies are not related directly to the chemical shifts and coupling constants, but represent only differences in the energy levels. If a complete and accurate energy level diagram can be constructed from the observed spectrum, the determination of the chemical shifts and coupling constants can be carried out rapidly and reliably by an iterative computer program. Spin tickling and other double resonance techniques are extremely valuable in constructing such an energy level diagram.

In building up an energy level diagram, we shall rely on *connected transitions,* which are spectral transitions with an energy level in common. Two cases may be distinguished: If the common level lies between the other levels involved in the transitions, the transitions are said to constitute a *progressive* pair; if the common level is higher or lower than both of the other levels, the transitions form a *regressive* pair.

We saw in Fig. 9.1 that irradiation with a weak H_2 of one of the X lines in an AX system caused each of the A lines to split into a closely spaced doublet. A more careful analysis of the energy levels involved in an AX or

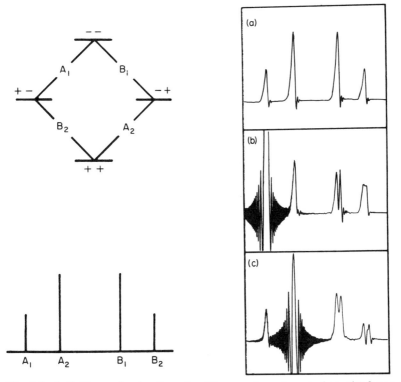

Fig. 9.8 Left: Energy level pattern, transitions, and spectrum (schematic) for an AB system. Right: Proton NMR spectrum (frequency sweep) for 2-bromo-5-chlorothiophene (a) unperturbed, (b) with ν_2 centered at line A_1, and (c) with ν_2 at line A_2. The beat pattern results from the passage of ν_1 through ν_2 during the scan (Freeman and Anderson[151]).

an AB system (Fig. 9.8) shows that transitions A_1 and A_2 have no energy level in common, but that A_1 and B_1 form a regressive pair, while A_1 and B_2 constitute a progressive pair. It has been shown that a weak irradiation of one of a pair of connected transitions causes the other to split, the two components of the split peak being extremely sharp for a regressive pair of transitions but rather broad for a progressive pair.[151] Transitions not connected to the irradiated one are unaffected by the tickling irradiation. These points are clearly demonstrated for an AB system in Fig. 9.8.

The generalizations given in the previous paragraph apply to any spin system, not just AB. Spin tickling may thus be used to identify progressive and regressive connected transitions and hence aid in the construction of the energy level diagram. An example of the repeated use of tickling in an ABC system is given in Fig. 9.9. The four tickling experi-

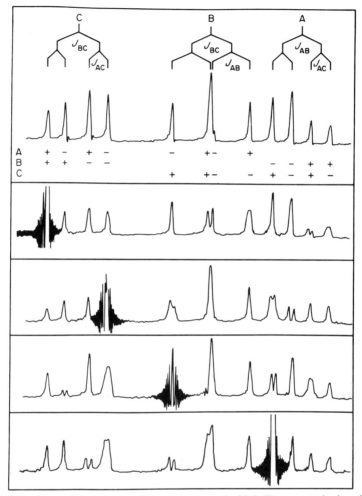

Fig. 9.9 Proton NMR spectra (frequency sweep) of 2,3-dibromopropionic acid. Top: Single resonance spectrum. Others: ν_2 centered on line giving beat pattern (Freeman and Anderson[151]).

ments illustrated are sufficient to map out an array of connected transitions. From the arrangement of energy levels the relative signs of the coupling constants can be determined, and from the observed line frequencies energy values can be assigned to each state.

Spin tickling has been applied to more complex systems, as well. The analysis of an AA′BB′ spectrum, which is a formidable task by using an iterative computer program and only the ordinary spectrum, can be accomplished readily by systematic repeated tickling experiments.[120] The signs of all J's are determined unambiguously by this procedure, whereas a computer-aided analysis often does not permit the establishment of some signs.

9.10 "High Resolution" Spectra in Solids

We saw in Sections 2.6 and 8.2 that the broad NMR lines characteristic of solids are due to internuclear magnetic dipolar interactions that are very effective in essentially rigid materials. However, double resonance methods can be used to "decouple" such dipolar couplings in much the same way that the effect of scalar couplings can be eliminated by ordinary decoupling (Section 9.3). Since dipolar couplings (D_{ij}) are orders of magnitude larger than typical scalar couplings (J_{ij}), considerably larger values of H_2 are needed. With powerful amplifiers and suitable probe design to insure the efficient utilization of the decoupling power, it is now possible to achieve line widths in solids of less than 100 Hz—not very narrow by standards of high resolution in liquids, but much less than the 10 kHz that might be expected.

With such experiments it is possible to observe not only chemical shifts of different nuclei but also the *anisotropy* in the chemical shift—a feature that is not affected by the dipolar decoupling but is averaged out when molecules tumble rapidly in liquids (see Sections 4.8 and 8.3). An example, the ^{13}C spectrum of linear polyethylene, is shown in Fig. 9.10. The three principal values of the ^{13}C chemical shift tensor are indicated. The line shape here is typical of that found in solids; it arises from the weighted average of the resonances from randomly oriented molecules. For polyethylene the total range of chemical shift tensor components is 40 ppm, a typical value for aliphatic carbons. Aromatic carbons have anisotropies in the range of 200 ppm, while typical carbon anisotropies in carbonyl groups are nearly as large, about 180 ppm.[153]

Dipolar decoupling experiments, as we have described them, are effective for observing NMR spectra of nuclei such as ^{13}C or ^{15}N, which are present at only low abundance and do not interact significantly with each other. The dipolar interactions that lead to line broadening are then due almost entirely to protons; hence ^1H decoupling can be effective. (In a somewhat more sophisticated version of the experiment, *cross polarization,* this decoupling provides a bonus in that the low ^{13}C sensitivity is increased by the large proton polarization.[154]) For study of ^1H or similar abundant, high magnetic moment nuclei, a different technique is required, because it is now *homonuclear* dipolar interactions that dominate the line

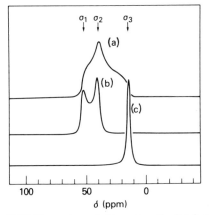

Fig. 9.10 Carbon-13 NMR spectrum (with proton dipolar decoupling) of polyethylene (solid). (a) Polycrystalline form, with the three principal components of the chemical shift tensor indicated. (b) Polyethylene fiber oriented with the fiber axis perpendicular to H_0. (c) Fiber axis oriented parallel to H_0 (VanderHart[152]).

broadening. By using specially designed rf pulse sequences, such studies can be carried out, but we shall not discuss them here.[155]

Problems

1. The proton NMR spectrum of a complex molecule shows clearly the A doublet of an AB spectrum, but the B portion is obscured by other spectral lines. It is desired to use double resonance to determine the chemical shift ν_B. (a) If $J_{AB} = 10$ Hz, what is the minimum value of H_2 needed to insure complete decoupling of the A doublet? (b) If this amount of power is used and if optimum decoupling is obtained when ν_1 is at 173 Hz (the frequency of the collapsed A doublet) and $\nu_2 = 243$ Hz, what is ν_B? (All frequencies are measured with respect to TMS as zero.)

2. For the methyl carbon in toluene $T_1(^{13}C)$ has been measured as 17.4 sec, with an NOE $\eta_{CH} = 0.63$. With

$$1/T_1 = R_1 = R_1^d + R_1^o,$$

   find R_1^d and R_1^o.

3. Find η for $^{29}Si\{^1H\}$ in a molecule tumbling fast enough to be in the extreme narrowing range.

4. Find the structures of molecules giving Spectra 33–35, Appendix C.

5. A frequency sweep selective decoupling experiment is to be performed to determine the relative signs of J_{AM} and J_{AX} in a three-proton AMX system. If J_{MX} is the smallest of the three coupling constants in absolute value, sketch the spectrum and indicate where the decoupling field should be applied. Predict the results for J_{AM} and J_{AX} of the same sign and of opposite signs.

6. How would your answer to problem 5 have changed, if nucleus M were ^{15}N ($\gamma < 0$), while the other nuclei remained protons?

Chapter 10

Pulse Fourier Transform Methods

The use of a short, intense rf pulse to excite nuclei and to permit the observation of nuclear resonance was suggested in one of the first papers on NMR.[7] The technique has long been of value in measuring relaxation times in systems with only one NMR line, but its use for complex molecules was virtually precluded by the experimenter's inability to disentangle signals from various chemically shifted nuclei. During the mid-1960's, however, two events brought pulse methods to the attention of chemists. First, Ernst and Anderson[25] showed, in their now classic paper, that ordinary "slow-passage" spectra could be obtained with a great saving in instrument time by use of pulse NMR observations and Fourier transform (FT) mathematics. Second, during the same period practical minicomputers were developed for on-line laboratory service, and advances in computer software drastically reduced the time needed for FT computations.

At present virtually all of the more sophisticated NMR spectrometers use pulse FT methods, and as costs for much of the apparatus decline, these methods are being extended to less expensive instruments. In this chapter, we shall explore many of the ways pulse FT techniques differ from ordinary continuous wave methods.

10.1 RF Pulses and the Free Induction Decay

As we saw in Chapter 2, the aim of an NMR experiment is to perturb the macroscopic magnetization $\mathbf{M}$ from its equilibrium position along the z axis and thus to generate a component of magnetization in the xy plane that induces an electrical signal in the receiver coil along the y axis. In the

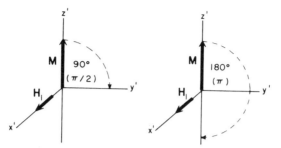

Fig. 10.1 Precession of M about H_1 in the rotating frame. Flip angles of 90° and 180° ($\pi/2$ and π rad, respectively) are illustrated.

pulse method this perturbation is achieved by applying a very intense rf field for a short time. The process is pictured quite simply in Fig. 10.1 if we use the rotating frame of reference (Section 2.10). In this frame H_1, the rf field, can be taken to be along the x' axis (i.e., the x axis of the rotating frame). Just as described in Section 2.2 H_1 interacts with M, and according to the Larmor equation M precesses in the $y'z'$ plane at a rate

$$\omega = \gamma H_1 \quad \text{rad/sec.} \tag{10.1}$$

If the rf field is allowed to remain on for only a short time t_p, then at the end of that time M will have gone through an angle

$$\theta = \gamma H_1 t_p \quad \text{rad}$$
$$= \frac{360}{2\pi} \gamma H_1 t_p \quad \text{deg.} \tag{10.2}$$

γ is a property of the nucleus being studied, but both H_1 and t_p are under the experimenter's control. Usually, for reasons that we shall see later, it is desirable to use H_1 as large as available in the spectrometer (typically 10–100 G) and to adjust t_p (the *pulse width*) to obtain the desired "flip angle" of M. Thus we can apply a 30° pulse, a 90° pulse, or any value we wish. As we shall see, 90° and 180° pulses have special importance. Usually the time required for a 90° pulse is in the range 1–100 μsec.

Consider a sample that has only one NMR line (e.g., H_2O), subjected to a 90° pulse. The entire magnetization then lies along the y' axis, as shown in Fig. 10.2, where it induces a signal in the receiver coil. The magnitude of M_{xy} determines the strength of the observed signal (called a *free induction signal,* since the nuclei precess "freely" without applied rf). As transverse relaxation occurs, the signal decays. In a perfectly homogeneous field the time constant of the decay would be T_2; but, in fact, the free induction signal decays in a time T_2^* that often is determined primarily by field inhomogeneity, since nuclei in different parts of the field precess

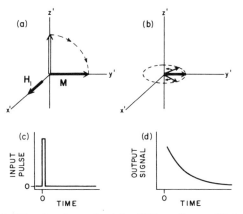

Fig. 10.2 (a) A 90° pulse along x' rotates **M** from the equilibrium position to the y' axis. (b) **M** decreases as the magnetic moments dephase. (c) The input signal, a 90° pulse, corresponding to (a). (d) Exponential free induction decay, corresponding to (b).

at slightly different frequencies, hence quickly get out of phase with each other. Thus the signal decays with a characteristic time T_2^*, defined as follows:

$$\frac{1}{T_2^*} \equiv \frac{1}{T_2} + \gamma \, \Delta H_0, \tag{10.3}$$

where ΔH_0 represents magnetic field inhomogeneity across the sample volume. Figure 10.2d shows the pure exponential decay that results from an rf pulse applied exactly at the resonance frequency of a single type of nucleus. This decay directly measured the decrease in M_{xy}.

Suppose now that the rf is slightly different from the Larmor frequency of the nuclei. If we again consider the frame rotating at the radio frequency, then immediately after the 90° pulse **M** lies along the y' axis.† However, **M** now rotates relative to the rotating frame, and the detector displays not only the exponentially decaying value of M_{xy} but also the interference effects as M_{xy} and the reference frequency to the phase sensitive detector alternately come in and out of phase with each other. A typical response, as shown in Fig. 10.3a, is analogous to the ringing seen in an ordinary high resolution NMR spectrum.

We can extend the discussion of the last paragraph by considering a system that contains several nuclei of the same species (e.g., protons) that differ in Larmor frequency because of chemical shifts and/or spin–spin coupling. Each line in the ordinary spectrum corresponds to a different

---

† One might inquire just how the magnetization can be tipped into the xy plane by an **H₁** not at the resonance frequency. We answer this question in Section 10.3.

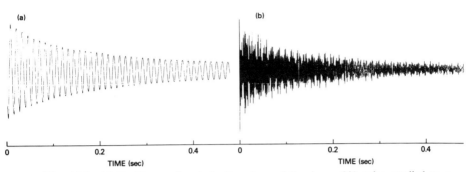

Fig. 10.3 (a) Oscillatory free induction decay following a 90° pulse applied to a sample with a single resonance line (^1H in $CHCl_3$). (b) Interference effects in the FID when several resonance frequencies are present (^1H spectrum of 1,7-dimethylcytosine in DMSO). Note that the spectrum (shown in Fig. 10.4) has a large peak due to water in the solvent. The FID, too, shows this frequency as the dominant regular oscillation.

frequency in the free induction signal, and each gives rise to a pattern like that of Fig. 10.3a. The interference between all these signals then generates a response of the sort depicted in Fig. 10.3b, and we turn to Fourier transform procedures to extract those component frequencies.

Measurement of the free induction decay (FID) is the basic way in which the magnitude and other characteristics of **M** are determined. The FID following a 90° pulse provides the spectral information needed in Fourier transform NMR, and the FID resulting from sequences of two or more pulses is used in the determination of relaxation times.

10.2 Fourier Transformation of the FID

It is well known in physics and engineering that any signal (electrical or mechanical) that is varying with time contains certain characteristic frequencies; it is the province of Fourier analysis to determine the frequencies and their amplitudes. We cannot discuss this branch of applied mathematics in detail, but we can define the essential relations. The Fourier transform $F(\omega)$ of a function of time $f(t)$ is defined as

$$F(\omega) = \int_{-\infty}^{\infty} f(t)e^{-i\omega t}\, dt. \tag{10.4}$$

It is known that if $f(t)$ is the free induction decay following a pulse, then $F(\omega)$ represents the ideal slow passage spectrum.[156] So in principle instead of measuring the spectrum by conventional cw methods (which may take many minutes), we can record the FID (usually in a second or so) and from Eq. (10.4) obtain the same information. In practice there are a number of important points to be kept in mind.

First, we wish to measure the FID over a short period of time, not between the infinite limits given in Eq. (10.4). Second, we must obtain a finite number of digital samples of the signal in order to carry out the calculation in a digital computer. And third, there may be instrumental artifacts that must be accounted for and corrected in the calculation.

Sampling rate and "foldover." In general, the FID contains frequencies covering a range of hundreds or thousands of Hz—whatever the expected chemical shift range for the nucleus being studied. Because the higher frequencies oscillate rapidly in time, it is necessary to sample more often to acquire a true representation of the high frequency information. In fact, the sampling theorem says that a minimum of two samples (data points) must be acquired during the period of a sine wave for a faithful representation, so if f_{max} represents the highest frequency present, we must sample at a rate no less than $2f_{max}$ data points/sec.[157] NMR frequencies are in the MHz range but fortunately, because of the way the FID is detected, the electrical signal being digitized contains only *differences* in frequency between the actual nuclear precession frequency and the frequency at which H_1 (the pulse) is applied. So if the pulse is applied at one extreme of the chemical shift range, f_{max} represents the other extreme—rarely more than 1–25 kHz.

What happens if frequencies are present higher than our anticipated f_{max} when we digitize data at $2f_{max}$ points/sec? Some reflection on the properties of sine waves indicates that a frequency $(f_{max} + \Delta f)$ cannot then be represented uniquely, but rather is indistinguishable from $f_{max} - \Delta f$. The result is that the spectrum we obtain after Fourier transformation has all lines at frequencies higher than f_{max} *folded over* into the range $0 - f_{max}$. An example is given in Fig. 10.4. (Usually phase distortions occur in the folded peaks; this aids in identification.) In most FT spectrometers f_{max} is selected by the operator as the "spectral width," and the computer automatically samples at $2f_{max}$ points/sec.

Acquisition time. Two factors govern how long a FID should be sampled: resolution and signal/noise. The Fourier transform of a Lorentzian NMR line is an exponential decaying to infinite time. Any truncation of the decay will in principle lead to a line distortion; specifically truncation at T sec will broaden the line to approximately $1/T$ Hz. Of course, each component frequency in the FID has a time constant T_2^*, corresponding to a line width of $1/\pi T_2^* \approx 1/3T_2^*$. Hence it is seldom of any advantage to acquire data beyond $3T_2^*$, since at that point truncation adds

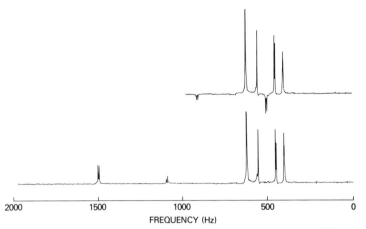

Fig. 10.4 "Foldover" of spectral lines as a result of sampling the FID at too low a rate. Top: Spectrum (220 MHz) of 1,7-dimethylcytosine in DMSO-d_6. Spectral width, 2 kHz; filter, 2 kHz. Bottom: Spectral width, 1 kHz; filter, 2 kHz. Note peaks folded back in frequency and out of phase.

little to the linewidth in the Fourier transformed spectrum. From the standpoint of signal/noise (S/N), the FID decreases in amplitude with time, while the noise remains constant, so again truncation of the FID is indicated. As we shall see in Section 10.4, optimum S/N is usually achieved by truncation at $T \approx 1.5 T_2^*$, at the expense of some resolution.

Discrete Fourier Transform. For computer calculation the analog of Eq. (10.3) is the discrete Fourier transform, a finite sum, of the form

$$F_\nu = \sum_{t=0}^{N-1} f_t \, e^{-2\pi i \nu t/N} \qquad (10.5)$$

The algorithm used in the computer works most efficiently with 2^n points, so N (the total number of data points) is invariably chosen to be 4096, 8192, etc. (usually referred to as 4K, 8K, etc.) Clearly N must be greater than or equal to the product of the acquisition time and $2f_{max}$. Often the total computer memory available sets an upper limit on N, so either the resolution (reciprocal of the acquisition time) or spectral width (f_{max}) must be correspondingly restricted. In other instances, when the memory is larger than $2f_{max} T$, N can be increased by adding a string of zeroes to the end of the FID. This process of *zero-filling* results in better definition of the line shape in the Fourier transformed spectrum.

10.3 Instrumental Requirements

Although we are not concerned here with details of spectrometer design, there are a few important points worth mentioning on instrumental aspects.

RF power. We pointed out that in pulse experiments H_1 is normally in the range $10-100$ G, rather than in the range of $10^{-4}-10^{-5}$ G used in cw studies. The large value of H_1 is needed in order to excite nuclei over the whole chemical shift range. We saw in Section 2.10 that $\mathbf{M}$ always responds to $\mathbf{H}_{eff}$, as defined in Eq. (2.52). When $\mathbf{H}_1$ is applied at a particular frequency, each nucleus i over the chemical shift range experiences a slightly different $\mathbf{H}_{eff}$:

$$\mathbf{H}_{eff}(i) = \mathbf{H}_1 + \frac{1}{\gamma}(\omega_{rf} - \omega_i). \tag{10.6}$$

Since we would like each $\mathbf{M}_i$ to precess through the same angle in the $y'z'$ plane, we want to make $\mathbf{H}_{eff}$ coincide in magnitude and direction as closely with $\mathbf{H}_1$ as possible; i.e., we want

$$\mathbf{H}_{eff} \approx \mathbf{H}_1. \tag{10.7}$$

For this relation to be satisfied,

$$|\mathbf{H}_1| \gg |\omega_{rf} - \omega_i|/\gamma \tag{10.8}$$

for each nucleus. If Δ represents the range of frequencies to be expected (in our usual practical units of Hz), then

$$|H_1| \gg 2\pi\Delta/\gamma. \tag{10.9}$$

The value of H_1 is often expressed by stating the width of a 90° pulse, t_{90}, which can be obtained by substituting the relation 10.9 into Eq. (10.2):

$$90 = \frac{360}{2\pi}\gamma H_1 t_{90} \gg \frac{360}{2\pi}\gamma\left(\frac{2\pi\Delta}{\gamma}\right)t_{90},$$

$$t_{90} \ll \frac{1}{4\Delta}. \tag{10.10}$$

For wide frequency ranges (e.g., ^{13}C at high field) the inequalities 10.9 and 10.10 are not strictly obeyed in most commercial spectrometers. The result is that lines far from the pulse frequency suffer both phase and amplitude distortions. The former are easily corrected in the computer, as we shall see in Section 10.4, but the latter are not. Usually if the condition is met that

$$t_{90} \leq \frac{1}{4\Delta}, \tag{10.11}$$

which is somewhat weaker than the inequality 10.10, only 2–3% amplitude distortion occurs.

Low-pass filter. Electrical filters are needed in all NMR spectrometers to limit the frequency range of the noise that is detected. Random noise covers an unlimited frequency range, and as we saw, high frequency information (noise, as well as signal) is folded back by the digitization process into the spectral range of interest. A low pass filter, applied before digitization, excludes noise outside this range and markedly improves signal/noise.

Phase-sensitive detection. NMR spectrometers normally use a phase-sensitive detector, in which the NMR signal is referenced in phase to the oscillator supplying the rf power. A signal precisely at the reference frequency is detected as a dc voltage, while each additional frequency appears as an ac voltage at the difference frequency ($\omega_i - \omega_{rf}$). There is no way that the detector can distinguish whether this difference is positive or negative, so all negative frequency differences are folded over into the positive domain. Thus additional noise always appears in the spectrum, and if the pulse frequency is set within the spectral range, rather than at one extreme, NMR lines appear at spurious frequencies.

These problems are overcome by the use of *quadrature phase detection* (QPD), in which two phase-sensitive detectors are referenced at precisely a 90° phase difference between them. With this arrangement positive and negative frequency differences are disentangled. QPD has come into use in many NMR spectrometers since it improves S/N by $\sqrt{2}$ (by preventing foldover of noise) and it permits a much more efficient use of rf power (by placement of the pulse frequency at the center of the spectrum).

10.4 Data Processing in the Computer

In addition to carrying out the Fourier transformation itself, the computer is often used to control many aspects of the experiment (e.g., to determine pulse widths and spacings), and it performs several other essential computations. Of course, if S/N is to be improved by time averaging, the computer must coherently add successive FID's. After acquiring the data the computer may be used to weight the data in the time domain in some desired manner. The most common process of this sort is the appli-

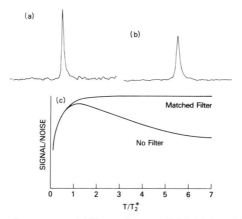

Fig. 10.5   Use of an exponential filter to improve S/N. (a) Acquisition time, $T = 4T_2^*$, no exponential filter. (b) $T = 4T_2^*$, matched filter. (c) Relative S/N as a function of acquisition time with no exponential filter and with a matched filter.

cation of an exponential weighting function, or *exponential filter*. In multiplying the FID by a decreasing exponential, the points in the FID at longer time are discriminated against, and since these points have poorer S/N in the time domain, the overall peak S/N in the transformed spectrum can be improved, a process often called *sensitivity enhancement*. It is known[25] that for improving S/N the optimum value of the exponential time constant is T_2^*, the time constant of the FID itself. The resulting transformed spectrum has linewidths twice as great, but this is sometimes a small price to pay for the improved S/N. Figure 10.5a shows an example of S/N improvement with the use of an optimum (or "matched") exponential filter. Figure 10.5b demonstrates quantitatively the effect of the matched filter in relation to the acquisition time of the FID. As we saw in Section 10.2, truncation of the FID before about $3T_2^*$ results in a broadening of the spectral line and resultant decrease in peak height (since the area is constant). On the other hand, longer data acquisition results in increased noise from the FID with little additional signal. The role of the matched filter is to suppress this unwanted noise.

It is clear that a negative time constant can be used in the exponential filter to extend the FID to longer time, hence to improve the resolution in the transformed spectrum at the expense of poorer S/N. This procedure is used, but in practice is generally limited to rather modest resolution enhancement.

Another major type of processing required is phase correction to insure that all lines in the transformed spectrum have the desired absorption mode character. There are inherent in the pulse experiment several sources of phase distortion that vary with frequency, so that across a transformed spectrum there will be varying degrees of dispersion mode character introduced: For example, during the pulse itself a time of about $1-100$ μsec is required for the magnetizations corresponding to each line $\mathbf{M}_i$ to precess from the z' axis in the rotating frame to the $x'y'$ plane. Ideally all $\mathbf{M}_i$ are aligned along the y' axis after the pulse (see Fig. 10.2), but during the pulse period the $\mathbf{M}_i$ will have moved in the rotating frame through an angle proportional to $(\omega_i - \omega_{\mathrm{rf}})$. For a difference of 1 kHz and $t_{\mathrm{p}} = 20$ μsec, this amounts to about 7°; for 10 kHz it would be 70°. Other sources of phase distortion occur, which are of similar magnitude (e.g., limited value of H_1, as discussed previously).

Fortunately most phase distortions are linear in frequency and can easily be corrected. The Fourier transformation of N data points provides $N/2$ "real" points and $N/2$ "imaginary" points. These two transformed spectra contain essentially the same information but are 90° out of phase with each other. Computer algorithms exist that add frequency-dependent mixtures of these two spectra to produce a distortion-free final spectrum.

10.5 Sensitivity Enhancement by Time Averaging

The most common use of pulse FT methods is to permit rapid acquisition of a single response so that many repetitions may be time averaged to improve S/N. The principle of coherent averaging of n scans to improve S/N by $\sqrt{n}$, as discussed in Section 3.5, is equally valid for FID's, and the Fourier transformation and other data processing may then be applied to the time-averaged FID. Enormous savings in time are possible, since a spectrum of, say, 1 Hz resolution can be obtained with an acquisition time for the FID of about 1 sec. By ordinary cw NMR a scan rate of about 1 Hz/sec would be required to produce 1 Hz resolution, so that a spectral width of 1 kHz would require 1000 sec. Thus the FT method saves a factor of 1000 in time. With wider spectral widths even greater savings are possible. (The computer processing time is usually negligible compared with the time for a number of repetitive runs.)

However, there is one aspect that can limit the gain in sensitivity for pulse FT. Tht is the time required for spin–lattice relaxation to restore the magnetization to equilibrium along the z axis. If $T_1 \gg T_2^*$, little restoration will have occurred at the end of the data acquisition (e.g., $3T_2^*$), and a repetition of a 90° pulse will result in only a very weak FID. Under these circumstances one must either wait longer between pulses or use a smaller value of the "flip angle" than 90°, so that a substantial z component of $\mathbf{M}$ remains. For a series of 90° pulses the optimum pulse spacing is 1.27 T_1.

Actually, slightly better results are obtained by pulsing again immediately after each data acquisition but using a smaller flip angle. Of course, a knowledge of the T_1's to be expected is needed in order to apply these relations, but often an adequate estimate can be made from previous studies of similar molecules.

A recent paper discusses the practical conditions for optimizing the signal/noise by correct choice of the various parameters—flip angle, pulse repetition time, and exponential filter time constant.[158]

10.6 Measurement of Relaxation Times

Pulse methods are usually the most convenient (and sometimes the only) means of measuring relaxation times. The procedure is to apply a sequence of two or more pulses, the first to perturb the nuclear magnetization from its equilibrium state, the other(s) to cause further motion of **M** and to permit a measurement of a FID. Fourier transformation of the FID leads to a "partially relaxed spectrum."

Measurement of T_1. Probably the most accurate means for measuring T_1 is the *inversion-recovery*, or $180°,\tau,90°$, method. A $180°$ pulse first inverts **M** along the negative z' axis, as shown in Fig. 10.6. During a waiting time τ longitudinal relaxation occurs, causing M_z to go from the value of $-M_0$ through zero to its equilibrium value of M_0. If at a time τ after the $180°$ pulse, a $90°$ pulse is applied, also along the x' axis, **M** is rotated to the y' axis. A free induction signal results, the initial height of which is proportional to the magnitude of **M**, hence to the value of M_z, at the time t. If the system is now allowed to return to equilibrium by waiting at least five times T_1 † and the $180°,\tau,90°$ sequence repeated for a different value of τ, the decay rate of M_z can be established, as indicated in Fig. 10.6c.

Quantitatively, the decay of M_z is given by the Bloch Eqs. (2.50) with $M_x = M_y = 0$:

$$dM_z/dt = -(M_z - M_0)/T_1. \tag{10.12}$$

Integration of Eq. (10.12) with $M_z = -M_0$ at $t = 0$ gives

$$M_z = M_0(1 - 2e^{-t/T_1}). \tag{10.13}$$

This is the function plotted in Fig. 10.6c. In practice, Eq. (10.13) is often recast into the form

† At $5T_1$, $M_z = 0.993M_0$.

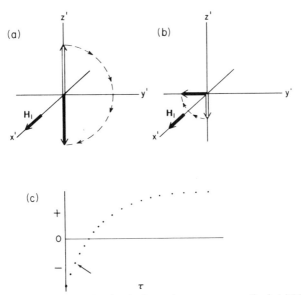

Fig. 10.6 Determination of T_1 by the inversion-recovery method. (a) **M** is inverted by a 180° pulse at time 0. (b) After a time interval τ, a 90° pulse rotates the remaining **M** to the y' (or $-y'$) axis. (c) The initial amplitude of the FID (for a single-line spectrum) or the height of a given line in the Fourier transformed spectrum is plotted as a function of τ. Note that each point results from a separate 180°,τ,90° sequence.

$$\ln(A_\infty - A_\tau) = \ln 2A_\infty - \tau/T_1, \qquad (10.14)$$

where A_τ is the height of a spectral line in the Fourier transformed FID following the 90° pulse at time τ, and A_∞ is the limiting value of A_τ for a very long interval between the 180° and 90° pulses. The value of T_1 is determined from the slope of a plot of $\ln(A_\infty - A_\tau)$ versus τ.

It can be seen from Eq. (10.13) that $A_\tau = 0$ for $\tau = T_1 \ln 2 = 0.69T_1$. Thus T_1 can be found from the pulse spacing τ that results in no free induction signal following the 90° pulse. While useful for a rough measurement, this procedure (the *null method*) is inadequate for accurate determination of T_1, since it is critically dependent on a very homogeneous H_1 and a precise setting of the 180° pulse width.

An example of the spectra obtained in a T_1 determination is shown in Fig. 10.7. This method of display is common since it permits easy visual observation of the differences in relaxation rates of the various lines.

Other, somewhat similar, methods can be used to measure T_1. For example, the *saturation-recovery method* uses a 90°,τ,90° pulse sequence. The first pulse brings **M** to the $x'y'$ plane and thus reduces $M_{z'}$ to zero, and its recovery is monitored in exactly the same way as the 180°,τ,90°

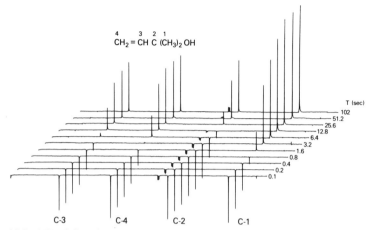

Fig. 10.7 ''Partially relaxed'' FT spectra resulting from several $180°,\tau,90°$ sequences.

method. The $90°,\tau,90°$ method, as well as other methods related to it, are very useful and may be more efficient in measuring long T_1's. However, they are generally more prone to errors.

Measurement of T_2. Unless $T_2 \ll (2/\gamma\,\Delta H_0)$, the contribution of inhomogeneity in $\mathbf{H_0}$ to the free induction decay precludes the use of the FID decay time T_2^* as a measure of T_2. An ingenious method for overcoming the inhomogeneity problem was first proposed by Hahn,[159] who called it the *spin-echo* method. The method consists of the application of a $90°,\tau,180°$ sequence and the observation at a time 2τ of a free induction ''echo.'' The rationale of the method is shown in Fig. 10.8, which depicts the behavior of the magnetization in the rotating frame. In (a) $\mathbf{M}$ is shown being tipped through $90°$ by application of $\mathbf{H_1}$ along the positive x' axis. The total magnetization $\mathbf{M}$ can be thought of as the vector sum of individual macroscopic magnetizations $\mathbf{m}_i$ arising from nuclei in different parts of the sample and hence experiencing slightly different values of the applied field, which is never perfectly homogeneous. There is thus a range of precession frequencies centered about ν_0, which we have taken as the rotation frequency of the rotating frame. In (b) then, the $\mathbf{m}_i$ begin to fan out, as some nuclei precess faster and some slower than the frame. At a time τ after the $90°$ pulse, a $180°$ pulse is applied, also along the positive x' axis, as shown in (c). The effect of this pulse is to rotate each $\mathbf{m}_i$ by $180°$ about the x' axis. Thus those $\mathbf{m}_i$ that are moving faster than the frame (shown in (b) moving toward the observer or clockwise looking down the z' axis) naturally continue to move faster, but in (d) their motion is now away from the observer. At time 2τ all $\mathbf{m}_i$ come into phase along the negative y'

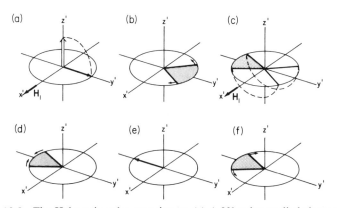

Fig. 10.8 The Hahn spin-echo experiment. (a) A 90° pulse applied along x' at time 0 causes **M** to tip to the positive y' axis. (b) The macroscopic magnetizations, $\mathbf{m}_i$, of nuclei in different parts of the sample dephase as a result of the inhomogeneity in $\mathbf{H}_0$. Those nuclei precessing faster than the average (the rotation rate of the frame) appear in the rotating frame to move toward the observer. Looking down from the positive z' axis they appear to move clockwise, while those slower than average move counterclockwise. (c) A 180° pulse along x' at time τ causes all $\mathbf{m}_i$ to rotate 180° about the x' axis. (d) The faster nuclei, still moving clockwise in the rotating frame, now go away from the observer, while the slower nuclei move counterclockwise toward the observer. (e) At time 2τ the $\mathbf{m}_i$ rephase along the $-y'$ axis. (f) At time $>2\tau$ the $\mathbf{m}_i$ again dephase.

axis, as shown in (c). The continuing movement of the $\mathbf{m}_i$ causes them again to lose phase coherence in (f).

The rephasing of the $\mathbf{m}_i$ causes a free induction signal to build to a maximum at 2τ, but the signal is, of course, negative relative to the initial free induction decay since rephasing of the $\mathbf{m}_i$ occurs along the *negative* y' axis. If transverse relaxation did not occur, the echo amplitude might be just as large as the initial value of the free induction following the 90° pulse. However, each $\mathbf{m}_i$ decreases in magnitude during the time 2τ because of the natural processes responsible for transverse relaxation in the time T_2. Thus the echo amplitude depends on T_2, and for a sample with only a single resonance frequency, T_2 may in principle be determined from a plot of peak echo amplitude as a function of τ. In the more general case, since the echo is just two FID's back-to-back, Fourier transformation of half the echo results in a spectrum with line heights dependent on their individual T_2. As in the measurement of T_1 by the 180°,τ,90° method, it is necessary to carry out a separate pulse sequence for each value of τ and to wait between pulse sequences an adequate time (at least five times T_1) for restoration of equilibrium.

The spin-echo technique, as we have described it, is limited in its range of applicability because of the effect of molecular diffusion. The precise refocusing of all m_i is dependent upon each nucleus remaining in a constant magnetic field during the time of the experiment (2τ). If diffusion causes nuclei to move from one part of an inhomogeneous field to another, the echo amplitude is reduced. The effect of diffusion is dependent on the spatial magnetic field gradients (G), the diffusion coefficient (D), and the time during which diffusion can occur. The amplitude of the echo for a pulse separation τ is

$$A(\text{echo at } 2\tau) \propto \exp[-(2\tau/T_2) - \tfrac{2}{3}\gamma^2 G^2 D\tau^3]. \qquad (10.15)$$

Equation (10.15) shows that the echo amplitude does not decay in a single exponential fashion. Because of the τ^3 dependence, the effect of diffusion is particularly pronounced for large values of τ and thus affects most the measurement of long T_2's. Diffusion coefficients are readily measured by the spin-echo technique.

For measurements of T_2 an improvement on the simple spin-echo method (due to Carr and Purcell)[159] is used in which 180° pulses are applied at τ, 3τ, 5τ, etc., to form echoes at 2τ, 4τ, 6τ, etc. Hundreds or thousands of such echoes may be formed at sufficiently close time intervals to preclude appreciable diffusion. The heights of the echoes then more accurately reflect T_2's. In practice, small mis-settings of pulse width cause cumulative errors, which can be overcome by a shift in rf phase, as suggested by Meiboom and Gill.[159] These methods have been discussed in a simple manner by Farrar and Becker.[157]

Measurement of $T_{1\rho}$. One additional relaxation time that we have not yet mentioned is called T_1 *in the rotating frame* and symbolized $T_{1\rho}$. It is of increasing utility in chemical applications and can be measured by pulse FT methods.

Normally, $T_{1\rho}$ is measured by a so-called *spin-lock* pulse sequence, which consists of a 90° pulse applied with H_1 along the x' axis as usual, followed immediately by a long "pulse" (possibly up to seconds in length) applied along the y' axis. The 90° pulse rotates **M** to the y' axis (as in Fig. 10.2), but its decay by transverse relaxation is now prevented by the magnetic field it encounters along y'. Since $\mathbf{H}_1$ is much larger than inhomogeneities in $\mathbf{H}_0$, $\mathbf{H}_{\text{eff}} \approx \mathbf{H}_1$, and with $\mathbf{H}_{\text{eff}}$ and **M** co-linear, no precession takes place. Thus **M** is "locked" along the y' axis. **M** does indeed relax, but the relaxation is now somewhat akin to a longitudinal relaxation since it is *along H_1*; hence the name "T_1 in the rotating frame." In the limit, as H_1 becomes smaller, $T_{1\rho}$ must approach T_2, and in liquids the two are often indistinguishable.

Measurement of $T_{1\rho}$ is easily accomplished by applying a spin-lock sequence and turning off $\mathbf{H}_1$ at a time τ, as indicated in Fig. 10.9. The resulting FID can be transformed to give a "partially relaxed spectrum,"

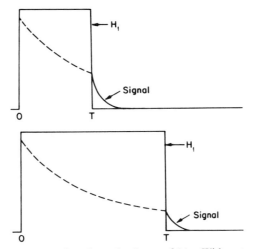

Fig. 10.9 Measurement of $T_{1\rho}$ from the decay of $M_{y'}$. With a strong H_1 applied, $M_{y'}$ decreases for a time T, then H_1 is turned off and the free induction decay obtained and Fourier transformed. The process is repeated for other values of T, and from the set of partially relaxed spectra the value of $T_{1\rho}$ for each spectral line can be obtained.

and the process repeated (later after $5T_1$) with a different value of τ. The measurement of $T_{1\rho}$ by FT methods is often easier than the use of a Carr–Purcell sequence to measure T_2, with much the same information obtained. We shall mention another application of $T_{1\rho}$ in Chapter 11.

10.7 Two-Dimensional FT-NMR

Two-dimensional Fourier transform NMR has been developed during the last few years to help unravel some of the complexities in ordinary NMR spectra, including those obtained by conventional pulse FT methods. The technique is simple: data are acquired as a function of time (called t_2, but having no relation to the relaxation time T_2), just as in ordinary FT-NMR. However, prior to data acquisition a perturbation of some sort is applied for a time t_1. Fourier transformation of the FID as a function of t_2 for fixed t_1 gives a spectrum similar to that obtained by ordinary FT-NMR; but further Fourier transformation with t_1 as the independent variable now gives a "2-dimensional" spectrum that can be plotted as a function of two independent frequencies, ν_1 and ν_2. Often, information of interpretative value can be extracted from this 2-D plot, the exact nature depending on the specific perturbation that was applied.

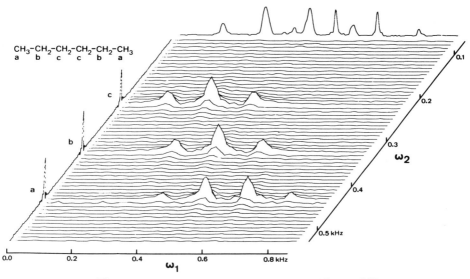

Fig. 10.10   2-D ^{13}C spectrum of *n*-hexane (Müller *et al.*[160]).

One illustration of a 2-D spectrum is given in Fig. 10.10, which was obtained from a ^{13}C NMR experiment, as follows: With the decoupler turned off a 90° pulse rotates the magnetization to the *xy* plane, where it precesses under the influence of the Hamiltonian that includes all ^{13}C—H couplings. After a time t_1, the decoupler is turned on to completely decouple all protons, and the FID is obtained. After equilibrium is reestablished, the experiment is repeated for other times t_1, and the resultant FID's subjected to a two-dimensional Fourier transformation. The result is a set of spectra, as shown in Fig. 10.10. The projection along the ω_1 axis is the spectrum that would have been obtained by ordinary FT methods with no decoupling, while the projection along ω_2 is the same as a completely proton-decoupled spectrum. However, the 2-D plot contains additional information, as the multiplets (in this case a quartet and two triplets) are disentangled and can be readily associated with the single lines along ω_2. The information in this example is similar to that obtained from an off-resonance decoupling experiment (see Section 9.5), but in more complex cases the 2-D approach usually gives a more clearly defined result.

A virtually unlimited number of other 2-D experiments can be carried out. For example, in addition to the 90° pulse at time zero, a 180° pulse can be applied at time $t_1/2$, so that an echo forms at t_1. The nature of the information obtained in the 2-D spectrum depends on whether the 180°

pulse is applied to the observed nuclei or to some or all of the nuclei to which they are coupled. Two-dimensional spectra obtained with appropriate pulse sequences can be used to investigate connections between various transitions, to study double quantum transitions and to examine various spectral features in solids. Two-dimensional capability is rapidly becoming an almost essential feature of FT-NMR spectrometers.

Problems

1. What value of H_1 is required for a 90° pulse width of 20 μsec for ^{13}C?

2. Derive an equation analogous to Eq. (10.13) to describe the recovery of M_z from a 90° pulse.

3. It was pointed out in Section 10.5 that a series of 90° pulses should be repeated at $1.27T_1$ to obtain optimum S/N in a given total experimental time. Using the equation derived in the preceding problem to find the extent of recovery of M_z in the time $1.27T_1$, derive an expression for the relative S/N accumulated in $1000T_1$ sec. Compare this result with values calculated for accumulations with 90° pulses repeated at $5T_1$; at $0.5T_1$. (Recall from Section 10.5 that S/N increases with $\sqrt{n}$.)

Chapter 11

Exchange Processes: Dynamic NMR

Nuclear magnetic resonance is widely used to study dynamic processes. In this chapter we shall explore the reasons why NMR spectra are sensitive to such processes, and we shall see how NMR spectra can provide valuable information on the rates and mechanisms of many reactions.

11.1 Spectra of Exchanging Systems

Before taking up the quantitative relations between NMR spectra and rate processes, it may be helpful to develop a semiquantitative treatment based on nuclear precession. While not mathematically entirely sound, this treatment provides some insight into the processes that occur.

Suppose a given nucleus can exchange between two sites, A and B,*

* The two sites might be, for example, an alcohol, ROH, and a phenol, PhOH, in a mixture of these two substances, where the OH proton exchanges between them; or the sites might be the axial and equatorial positions for a proton in a molecule such as cyclohexane, which interconverts between two conformations. In the former case there is a breaking of bonds and a "chemical reaction" type of exchange:

$$ROH_a + PhOH_b \rightleftharpoons ROH_b + PhOH_a$$

In the latter case the exchange is "positional"; e.g.,

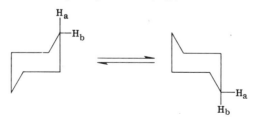

The same treatment applies to both situations, so far as chemical shifts are concerned. (However, see Section 11.3 for a distinction with regard to spin coupling.)

and that in these sites it has resonance (Larmor) frequencies ν_A and ν_B, respectively. We shall arbitrarily take $\nu_A > \nu_B$. To simplify the discussion, we shall assume that the nucleus has an equal probability of being in the two sites; hence the lifetime of the nucleus in state A, τ_A, must equal that in state B:

$$\tau_A = \tau_B = 2\tau. \tag{11.1}$$

(The factor of 2 is used for consistency with the expressions given in the following section.)

Consider a coordinate system that rotates about H_0, in the same direction in which the nuclei process, at a frequency

$$\nu_0 = \tfrac{1}{2}(\nu_A + \nu_B). \tag{11.2}$$

In the rotating coordinate system a nucleus at site A precesses at $(\nu_A - \nu_0)$, while a nucleus at site B precesses at $(\nu_B - \nu_0)$; that is, it appears in this rotating frame to be precessing in a direction opposite that of the nucleus in site A. We can now distinguish the following four cases regarding exchange rates.

1. *Very slow exchange.* The lifetime at each site, 2τ, is long, so that a given nucleus enters site A and precesses many times at frequency $(\nu_A - \nu_0)$ before leaving site A and entering site B. The result is that interaction with the rf field occurs, and in the fixed laboratory frame of reference a resonance line appears at ν_A. An identical situation occurs for the nucleus at site B. Thus the spectrum consists of two sharp lines at ν_A and ν_B, just as it would in the absence of exchange. (See Fig. 11.1a.)

2. *Moderately slow exchange.* The lifetime 2τ is now somewhat smaller than the value in the preceding paragraph. By the familiar Heisenberg uncertainty principle,

$$\Delta E \cdot \Delta t \sim h, \tag{11.3}$$

where ΔE is the uncertainty in energy corresponding to an uncertainty in time of measurement Δt. In our case $\Delta t \approx \tau$ so that the uncertainty in energy is reflected in an increased line width*

$$\nu_{1/2} \approx \frac{\Delta E}{h} \approx \frac{h}{\Delta t \cdot h} \approx \frac{1}{\tau}. \tag{11.4}$$

Thus the resonance lines at ν_A and ν_B are broadened, as indicated in Fig. 11.1b, c.

---

* The argument is similar to that given in Section 2.6 for the relation between T_1 and line width.

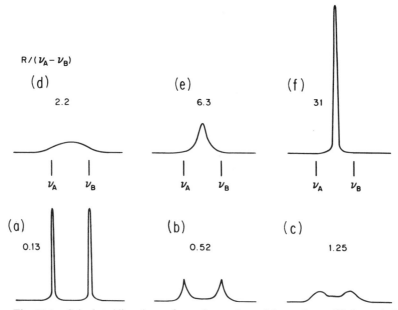

Fig. 11.1 Calculated line shapes for various values of the exchange lifetime relative to the difference in frequency of the two equally populated sites. Exchange rate $R \equiv 1/2\tau$, where $2\tau = \tau_A = \tau_B$. From tables based on an equation similar to Eq. (11.6).

3. *Very fast exchange.* A nucleus enters site A, where in the rotating frame it begins to precess at $(\nu_A - \nu_0)$. But before it can complete even a small portion of a single precession, its lifetime in site A expires, and it enters site B. It now begins to precess in the opposite direction in the rotating frame, but again undergoes essentially no precession before it must again leave site B and reenter A. The result is that in the rotating frame the nucleus remains stationary, and thus in the laboratory frame it appears to be precessing at the frequency with which the coordinate system rotates, ν_0. Hence, as shown in Fig. 11.1f, a sharp resonance line appears at ν_0, the average of the two Larmor frequencies, even though no nucleus actually precesses at that frequency.

4. *Intermediate exchange rate.* Between cases 3 and 4 there is a range of lifetimes that lead to an intermediate type spectrum, a broad line spanning the frequency range $(\nu_A - \nu_B)$, as indicated in Fig. 11.1d.

An example of exchange, the hindered internal rotation in *N,N*-dimethylformamide, was given in Fig. 1.6.

11.2 Theory of Chemical Exchange

The most widely used theoretical treatment of exchange phenomena in NMR[161] is based on extensions of the Bloch equations (see Section 2.8) to include exchange terms. We shall not reproduce the derivations of the necessary equations, which are well summarized elsewhere,[162] but shall indicate some of the more useful results of this treatment.

For the present we shall continue the assumptions we made in the preceding section: that there are only two equally populated sites (i.e., that Eq. (11.1) holds). In addition, we assume that in the absence of exchange the lines would have negligible width; that is, that

$$\frac{1}{T_{2(A)}} \approx \frac{1}{T_{2(B)}} \approx 0. \tag{11.5}$$

Under these conditions the shape of the resonance line(s) is

$$g(\nu) = K \frac{\tau(\nu_A - \nu_B)^2}{[\frac{1}{2}(\nu_A + \nu_B) - \nu]^2 + 4\pi^2\tau^2(\nu_A - \nu)^2(\nu_B - \nu)^2} \tag{11.6}$$

where K is a normalizing constant. Figure 11.1 gives plots of Eq. (11.6) for several values of τ. It is apparent that the shape of the curve depends only on the ratio $R/(\nu_A - \nu_B)$, where $R = 1/2\tau$ is the exchange rate. Thus our terminology of Section 11.1, "slow" and "fast," refers to the number of exchanges per second relative to $(\nu_A - \nu_B)$, measured in hertz.

It can be seen from Fig. 11.1 that for slow exchange the two peaks observed are separated by less than $(\nu_A - \nu_B)$. By finding the maxima of Eq. (11.6) it can be shown (see problem 3 at the end of this chapter) that

$$\frac{\text{separation of peaks}}{(\nu_A - \nu_B)} = \left[1 - \frac{1}{2\pi^2\tau^2(\nu_A - \nu_B)^2}\right]^{1/2}. \tag{11.7}$$

Hence τ may be easily determined.

Equation (11.7) shows that the peaks draw together as τ decreases and coalesce (separation of peaks = 0) for

$$\tau = \frac{1}{\sqrt{2}\,\pi(\nu_A - \nu_B)}. \tag{11.8}$$

Comparisons between exchange rates in different systems are often stated in terms of the conditions (temperature, pH, etc.) for coalescence of the peaks.

Other procedures for finding τ from Eq. (11.6) are based on the ratio of the height of the peaks to that of the minimum between the peaks, or on a computer-aided fitting of the entire curve, the so-called *total line shape* (TLS) analysis. The TLS method is by far the most successful of these approaches, but even so there are a number of pitfalls. For example, if measurements are made over a range of temperatures to bring about appreciable variation in a particular reaction rate, it is usually necessary to assume that $(\nu_A - \nu_B)$, which is measured under conditions of slow exchange at low temperature, either remains constant over the temperature range employed or varies in some predetermined manner (often a linear extrapolation). Such an assumption is frequently unwarranted. Likewise, the effect of finite T_2^0 (T_2 in the absence of exchange) is normally assumed to be constant over the temperature range. In principle, the TLS method can be applied iteratively, with the computer using a least squares criterion to provide the best fit to each observed spectrum from independent variation of τ, $(\nu_A - \nu_B)$, and possibly T_2^0. In practice, the fit is insensitive to some of these parameters over at least part of the range, so that in all but the most favorable cases large errors may occur.[163] In general, the greatest accuracy is obtained in the vicinity of coalescence, with poorer results at both slower and faster rates. The propagation of errors frequently leads to inaccuracy in the value found for the entropy of activation.

We began this section with the assumptions that there are only two equally populated sites and that the resonance lines in the absence of exchange have essentially zero width. These assumptions are unnecessary, but their elimination leads to considerably more complex mathematical expressions. In the two cases of slow exchange and of very rapid exchange, however, simple equations result.

For slow exchange, where the lines are broadened but do not overlap appreciably, the observed width of line j (assumed to be Lorentzian in shape) is

$$(\nu_{1/2})_j = (\nu_{1/2}^0)_j + \frac{1}{\pi \tau_j} \tag{11.9}$$

where $1/\tau_j$ is the probability per unit time of a nucleus at site j moving to a site where it has a different Larmor frequency, and $(\nu_{1/2}^0)_j$ is the width of line j in the absence of exchange. From the relation between $\nu_{1/2}$ and T_2 (Eq. (2.43)), Eq. (11.9) may be recast as

$$\frac{1}{(T_2)_j} \text{ (observed)} = \frac{1}{(T_2)_j} \text{ (no exchange)} + \frac{1}{\tau_j}. \tag{11.10}$$

For very fast exchange, the general result is that the frequency of the single line observed is the average of the Larmor frequencies for the different sites, weighted according to the probability that a nucleus is at each site:

$$\nu \text{ (observed)} = p_A \nu_A + p_B \nu_B + p_C \nu_C + \cdots \qquad (11.11)$$

where the p_i sum to unity. Since τ does not enter Eq. (11.11), we cannot actually determine the exchange lifetimes, but merely establish that the exchange rate is greater than about 50 $(\nu_A - \nu_B)$.

11.3 Collapse of Spin Multiplets

In Sections 11.1 and 11.2 we spoke of exchange of a nucleus between sites where it has different Larmor frequencies. The difference in frequencies might arise from differences in chemical shift or from the presence of spin coupling. For example, in CH_3OH the methyl and hydroxyl protons are spin coupled with $J \approx 5$ Hz. If we assume that δ/J is large enough so that first-order analysis is applicable, then the CH_3 resonance consists of a doublet, as shown in Fig. 11.2. One line of the doublet arises from those molecules (approximately 50%) that have the hydroxyl proton spin oriented *with* H_0, while the other line arises from the molecules with the OH proton spin oriented *against* H_0. If the OH proton exchanges between molecules, the methyl resonance may be affected. Suppose a given molecule of methanol contains an OH proton whose spin is oriented with the field. If this proton is lost and replaced by another proton, either from another CH_3OH molecule or from somewhere else (e.g., an H^+ or H_2O impurity), the CH_3 Larmor frequency will be unchanged if the replacing proton is also spin oriented with the field, but will change by J Hz if the new proton is spin oriented against the field. In the usual *random* process, then, half of all exchanges will result in a change in the CH_3 Larmor frequency, while the other half will not. With this statistical factor taken into account, Eq. (11.6)–(11.10) apply, provided $\delta \gg J$. Where the components of a spin doublet are involved, $(\nu_A - \nu_B)$ in Eqs. (11.6)–(11.8) is, of course, replaced by J.

Figure 11.2 shows the collapse of the spin multiplets of both the CH_3 and OH resonances in CH_3OH with increase in temperature. For the OH there are four states rather than two, so that the equations we have employed must be modified.

Proton exchanges in OH or NH groups are often catalyzed by H^+ or OH^- and so are highly pH dependent. Sometimes exchange at an intermediate rate broadens the resonance line so much that it may pass unobserved.

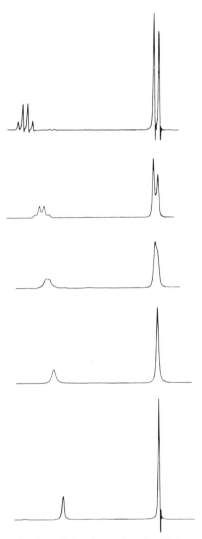

Fig. 11.2 Collapse of spin splitting in methanol with increase in temperature. Both the CH_3 and OH proton resonances are shown. Temperatures (from top): $-54°$, $-20°$, $-10°$, $0°$, $15°$.

The collapse of a spin multiplet to a single line with rapid proton exchange results from the random spin orientations of the exchanging protons. If a *given* proton exchanges rapidly between two or more sites but is never actually replaced by a proton with different spin orientation, the situation is different. For example, the pair of tautomers (I) and (II) result

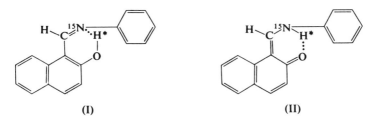

(I)   (II)

from rapid exchange of the proton H* between the oxygen and nitrogen atoms. If only (II) existed, the resonance of H* would be split into a doublet of ~90 Hz by coupling to ^{15}N. Actually the spectrum of this substance in $CDCl_3$ shows H* to give rise to a 35-Hz doublet at 25° and a 52-Hz doublet at −50°, resulting from the rapid tautomerism. In this case

$$J(\text{observed}) = p_I J_I + p_{II} J_{II}$$

where p_I and p_{II} are the fractions of the two tautomers present.

A very common example of this averaging of J's by rapid exchange occurs in molecules where there is rapid rotation about single bonds, such as CH_3—CH_2— (see Table 5.2).

11.4 More Complete Theories of Exchange

The theory discussed in Section 11.2 has been widely used and is adequate for many exchanging systems. However, when nonfirst-order spin coupling is involved in any of the exchange sites, this theory cannot provide a framework for quantitative treatment of the results. In such cases more general developments, based upon the density matrix formalism of quantum mechanics, are required. Such theory is beyond the scope of this book but is discussed elsewhere.[164] An example of one type of system that has been treated by such procedures is given in Fig. 11.3, where calculated spectra are given for two protons that form a coupled AB system. Proton B exchanges with another proton, X, where X is present at much higher concentration than A and B. The qualitative result is the same as would be expected from the simpler theory of Sections 11.2 and 11.3: first, a broadening of the spin "doublets" and ultimately a collapse of the A "doublet." The B lines broaden and eventually, with rapid exchange, are absorbed into the X signal. Thus, while the density matrix treatment must be employed for the calculation of exchange rates, the simpler theories can be useful even in more complex systems in providing a qualitative interpretation.

Not only can the rates of suitable reactions or exchange processes be

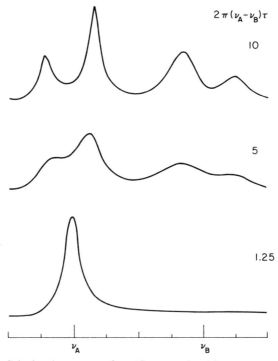

Fig. 11.3 Calculated spectrum of an AB system in which B exchanges with another nucleus, X, present at much higher concentration than A or B. $J_{AB}/(\nu_A - \nu_b) = 0.4$; $J_{AX} = J_{BX} = 0$. Adapted from Kaplan.[165]

determined from the analysis of line shapes, but in many instances the mechanisms of exchange can also be deduced, since the precise line shapes obtained at intermediate exchange rates often depend on the nature of the process. Many examples of the use of NMR line shape analysis are given in the compendium of dynamic NMR studies by Jackman and Cotton.[164]

11.5 Double Resonance and Pulse Techniques

Thus far we have been concerned with the effect of exchange on the ordinary NMR spectrum. Both double resonance techniques (see Chapter 9) and pulse methods (see Chapter 10) provide powerful additional tools for investigating rate processes.

Double resonance is advantageous principally in permitting measurements of rates of slower exchange reactions than can be handled by ordinary single resonance methods. From Eq. (11.10) it is seen that slow exchange rates are normally determined by the change effected in T_2. To bring about an observable change in line width, τ must be less than T_2. As usual, however, it is not the real T_2 of the system that should be used in Eq. (11.10), but the apparent T_2^*, which is often determined by magnetic field inhomogeneity. The double resonance method, however, permits a determination of τ when $\tau < T_1$. Since T_1 is often much greater than T_2^*, *larger* τ's can thus be determined.

The double resonance procedure for an exchange between sites A and B involves measurement of the change in the signal of the A line immediately after the B resonance is saturated by imposition of a second rf field. The A resonance decays as its Boltzmann distribution is disrupted by exchange of nuclei between sites A and B. From the exponential decay rate of the A resonance, τ can be calculated.[166]

In the presence of a perturbing rf field the line shape of an exchanging system is altered in a manner that can sometimes be analyzed to give a sensitive measurement of exchange rate, especially in the slow exchange region.

The spin-echo method is quite useful in extending the range of exchange rates to faster reactions (i.e., *smaller* τ's); it also obviates the necessity for certain assumptions made in the treatment of ordinary NMR spectra. We saw in Section 10.6 that in the Carr–Purcell experiment, 180° pulses at times τ_p, $3\tau_p$, $5\tau_p$, and so on, are followed by echoes at $2\tau_p$, $4\tau_p$, $6\tau_p$, and so on. If τ_p is made short, so that $2\tau_p \ll 2\tau$, the exchange lifetime, then the spin-echo experiment measures the actual T_2, without any contribution from exchange. On the other hand, if $2\tau_p \gtrsim 2\tau$, exchange affects the results, and from a computer-aided treatment of the data, exchange rates can be found over a wide range of exchange rates, including those that are too fast for accurate measurement by TLS analysis.[164] The data treatment used requires that the only nuclei being observed are those undergoing exchange, so the method is limited to simple molecules or those subject to suitable isotopic substitution.

The spin-lock method used to measure $T_{1\rho}$ (see Section 10.6) can also be used to measure exchange rates that are faster than those accessible to TLS methods. Rates as high as 10^6 sec^{-1} have been measured.[167] It is possible for this method to be used with Fourier transform techniques to study more complex molecules.

11.6 CIDNP

It was discovered in 1967 that the NMR spectra of the products of certain free-radical-mediated reactions display very large, but transient, intensity enhancements. It was thought initially that this intensity enhancement results from scalar or dipolar interactions between the nuclear magnetic moments and the magnetic moments of the unpaired electrons in the free radical intermediate. The process was believed to be analogous to the intensity enhancement that results from a nuclear–electron double resonance experiment, termed both the Overhauser effect and "dynamic nuclear polarization." Hence the phenomenon occurring in the free radical reactions was called *chemically induced dynamic nuclear polarization* (*CIDNP*). Further studies disclosed that the mechanism is somewhat different, but the misnomer CIDNP has persisted.

A typical example of CIDNP is shown in Fig. 11.4. The spectra are of the methine quartet in N-methyl-N-(α-phenethyl)aniline, which is the reaction product. Note that 0.3 min after initiation of the reaction the quartet appears with intensities far from the equilibrium values; in fact, two of the four lines show emission, rather than absorption. The original explanation of CIDNP could not account for such effects, nor could it account quantitatively for some large intensity enhancements. The now generally accepted theory of CIDNP is based on the "radical pair" mecha-

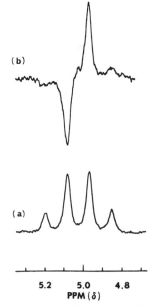

(b)

(a)

5.2   5.0   4.8
PPM (δ)

Fig. 11.4 Proton NMR spectra (60 MHz): (a) Reference spectrum of N-methyl-N-(α-phenethyl)aniline. (b) Spectral scan of 1 min duration started 0.3 min after initiating a reaction between N,N-dimethylbenzylamine and fluorobenzene (in the presence of n-butyllithium and tetramethylenediamine) to form N-methyl-N-(α-phenethyl)aniline. Reprinted with permission from A. R. Lepley, *J. Am. Chem. Soc.* **91**, 1237 (1969).[168] Copyright by the American Chemical Society.

nism, which focuses attention on the fate of pairs of radicals as they interact physically and eventually react chemically to form additional radicals or products. While the radicals are close enough to interact, the overall state of their electron spins may be characterized as either singlet (spin orientations antiparallel) or triplet (spin orientations parallel). The relative energies of the two states are dependent on the well known quantum mechanical "exchange" interaction. As the radical fragments separate, the strength of the exchange interaction decreases until it is of the same magnitude as the magnetic hyperfine interaction between the electron spin and the nuclear spins within the radical. (Other effects of electron spin interactions with nuclear spins were discussed in Sections 4.13 and 8.7.) The nuclear interactions can thus induce changes in electron spin orientations, which cause a radical pair initially in a singlet state to develop triplet character, and *vice versa*. Such interactions can occur, of course, not only between the radicals initially formed by a thermally or photochemically induced bond scission, but also between any pair of radicals that encounter each other by diffusive processes. It is expected that radical pairs in singlet states will react far more rapidly than those in triplet states because of the general lowering of energy in the attractive singlet molecular state. Thus by this rather indirect process, the nuclear spin states in the reactants (and in the radicals initially formed) influence the populations in the spin states of the products.

The details of the radical pair mechanism are complicated, since free radical reactions themselves are complex and proceed by several paths, so that many possible reaction sequences need to be considered. The theory is simplest when the reaction occurs in the high magnetic field of the spectrometer, rather than in a weak field. Since the populations in the various spin states of the product molecules depend on the spin states of the reactants, this mechanism accounts, not only for the large net increase in signal observed in many cases, but also for "multiplet effects" of the sort illustrated in Fig. 11.4. Ultimately spin–lattice relaxation processes in the product molecules restore a Boltzmann distribution and equilibrium values of NMR intensities.

CIDNP studies have been carried out with many reactions, both for clarification of the phenomenon itself and for elucidation of details of the individual reaction mechanisms. Among thermally initiated reactions, decomposition of peroxides and reactions of alkyllithium compounds with organic halides have been especially popular. A variety of photochemical decompositions and molecular rearrangements have also been investigated. Several excellent reviews[169] provide many examples of reactions and of the application of the radical pair theory in a qualitative form to the interpretation of the results.

Problems

1. Find the structures of the molecules giving Spectra 36–38, Appendix C.

2. Estimate the exchange lifetime τ for N,N-dimethylformamide at 108°C and 118°C from the spectra in Fig. 1.6.

3. Derive Eq. (11.7) from (11.6) by setting the first derivative of Eq. (11.6) equal to zero. Note that one of the three possible solutions for ν is not a maximum but the minimum at $\nu = \frac{1}{2}(\nu_A + \nu_B)$.

Chapter 12

Solvent Effects and
Hydrogen Bonding

Thus far our discussions of the factors governing NMR parameters, such as chemical shifts and coupling constants, have emphasized *intra*molecular contributions. Since almost all NMR measurements are made in the liquid phase, it is clear that *inter*molecular interactions might also be important. In this chapter we shall explore some of the theoretically definable effects arising from the solvent medium. We shall see that such effects are particularly important in *proton* NMR, where intramolecular shieldings are small. We shall devote special attention to the strong specific molecular interaction of hydrogen bonding.

12.1 Medium Effects on Chemical Shifts

In treating the intramolecular contributions to the chemical shift in Chapter 4 we found it helpful to classify the various effects involved. Similarly, following Buckingham *et al.*,[170] we can express the total effect of the solvent medium on nuclear shielding as the sum of five terms:

$$\sigma(\text{solvent}) = \sigma_B + \sigma_W + \sigma_A + \sigma_E + \sigma_H, \qquad (12.1)$$

where σ_B is the contribution of the bulk magnetic susceptibility of the medium. As we saw in Chapters 3 and 4, this effect can be allowed for theoretically when dealing with an external reference and is zero for an internal reference. Since in general we use internal references, we shall ignore this effect. It should be noted, however, that by choosing an internal reference we make the implicit assumption that the reference compound itself is not subject to solvent interactions. For a *complete* analysis of solvent effects, an *external* reference must be used and σ_B calculated as discussed in Section 4.4. In general, solvent effects are small for a reference such as TMS, but can be pronounced in some cases, especially where aromatic molecules are present in high concentration.[171]

The term σ_W arises from the effect of the weak van der Waals forces between solute and solvent molecules. Such effects can distort and change the symmetry of the electronic environment of a given nucleus. Theory predicts that σ_W should be negative, and experimental tests of Eq. (12.1) under conditions where σ_W should be one of the dominant terms indicate that a change in proton shielding of the order of 0.1–0.2 ppm might be expected from the effect of van der Waals forces. Generally, large polarizable halogen atoms in the solvent lead to increased negative values of σ_W. For nuclei other than hydrogen this term might be much larger, but is still probably small relative to the ranges of chemical shifts found for such nuclei.[172]

The term σ_A refers to the magnetic anisotropy in the solvent molecules and arises from the nonzero orientational averaging of solvent with respect to solute. The magnetic anisotropy itself was discussed in Sections 4.8 and 4.9, where we saw that aromatic rings and groups such as C=C, C=O, C≡C, and C≡N cause especially large effects. While it is difficult to isolate this effect from others, studies show that aromatic solvents usually lead to positive σ_A's of about 0.5 ppm, while solvents containing triple bonds have negative σ_A's of about 0.2–0.4 ppm.

The large positive σ_A normally found for aromatic solvents has been widely exploited. In complex molecules it is often found that solvent effects are selective. For example, in Fig. 1.5 we saw that the chemical shifts of different methyl groups in a steroid may respond quite differently to change of solvent. Systematic studies of the variation in chemical shifts of protons in well known locations in large molecules have shown considerable regularity in solvent effects.[173] For example, we can denote a solvent shift

$$\Delta_{C_6H_6}^{CDCl_3} = \delta_{CDCl_3} - \delta_{C_6H_6} \qquad (12.2)$$

where the δ's are defined according to Eq. (4.7) with respect to TMS. Thus a positive value of Δ indicates an upfield shift in the proton resonance on going from $CDCl_3$ to benzene as solvent. It has been found that in molecules containing a C=O group, Δ is negative for protons located on the oxygen side of a plane through the carbon atom of the carbonyl group perpendicular to the C=O axis, and positive on the other side of this plane[173] (see Fig. 12.1). This result can be interpreted in terms of a "complex" between the solute and aromatic solvent such that the π electrons of the aromatic ring are near the slightly positively charged carbon atom of the C=O group, while at the same time remaining as far as pos-

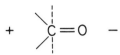

Fig. 12.1 Effect of solvent change on chemical shifts of protons in a molecule containing a C=O group.[173] Dashed line indicates a plane perpendicular to the plane of the C=O group; + or − refers to the sign of $\Delta_{C_6H_6}^{CDCl_3}$.

sible from the negative oxygen, as indicated in Fig. 12.1. Similar generalizations may be made for polar aromatic solvents interacting with polar solutes. In general, the solute–solvent complex is best thought of as a transitory species that biases the otherwise random distribution of solvent molecules around the solute, rather than a distinct separate species. These predicted solvent effects are quite useful in structure elucidation, but like many empirical generalizations, must be used with considerable caution. Steric effects, for example, can sometimes modify substantially the expected orientation of the interacting solute–solvent pair. In spectra with a number of lines lying close together it is often helpful to study the spectrum as a function of changing proportions in a mixed solvent, so that each line can readily be tracked.

The term σ_E arises from the effect of an electric field on the nuclear shielding. Buckingham *et al.*[170] have shown that when a polar molecule or a molecule containing polar groups is dissolved in a dielectric medium, it induces a "reaction field," the effect of which is usually to reduce the shielding around a proton in the solute. Thus σ_E is ordinarily negative, but could be positive for certain molecular geometries. Equations for determining the magnitude of σ_E in terms of molecular properties such as dipole moment and polarizability have been derived but will not be reproduced here. Experimental studies of the electric field effect indicate that σ_E might be as large as 1 ppm for polar molecules in solvents of high dielectric constant.

The term σ_H refers to *specific* solute–solvent interactions, the most important of which is hydrogen bonding. We shall discuss hydrogen bonding in Section 12.4. When such interactions are present, σ_H usually is the dominant term in Eq. (12.1). There is often a tendency in such cases to ignore the other effects but, as we have seen, they can be quite significant even if not dominant.

12.2 Solvent Effects on Coupling Constants

While solvent effects on chemical shifts can be quite large, the effect of solvents on coupling constants is usually small. Careful measurements of coupling constants show, however, that the effects in some cases are not negligible. For example, geminal H—H coupling constants in a number of compounds, such as styrene oxide, formaldehyde, and α-chloroacrylonitrile, vary with solvent by 1–2 Hz. In addition, some one-bond ^{13}C—H and ^{15}N—H coupling constants vary with solvent; in these cases the variation seems to be related to hydrogen bonding and will be discussed in Section 12.4.

Observed vicinal coupling constants in some ethane derivatives are also found to vary with solvent. In such cases, however, the observed J is the weighted average of the J's for the various conformers, so that the observed changes in J reflect mainly the effect of the solvent in altering the proportions of the conformers.

12.3 Solvent Effects on Relaxation and Exchange Rates

We saw in Chapter 8 that T_1 and T_2 depend on the rate of molecular tumbling, which, in turn, is a function of the viscosity of the solution. For small molecules and small polymers, T_1 and T_2 are to the left of the minimum in Fig. 8.1, so that an increase in viscosity reduces the relaxation times and thus broadens the resonance lines. Such effects are found in materials of high viscosity, such as glycerol or ethylene glycol, the line widths of which are markedly temperature dependent. Of the commonly used solvents for proton NMR listed in Table 3.1, dimethyl sulfoxide is the most viscous; at room temperature lines of compounds dissolved in this solvent are generally noticeably broader than in other solvents.

If the solute contains exchangeable protons, its spectrum may be quite dependent on the nature of the solvent insofar as exchange lifetimes are concerned (see Chapter 11). If, in addition, the solvent contains exchangeable protons, there may be a rapid exchange between solute and solvent so that the solvent resonance line due to its exchangeable protons in effect includes also the exchangeable protons of the solute.

12.4 Hydrogen Bonding

Hydrogen bonding is recognized as a relatively strong, specific interaction between molecules or between suitable portions of the same molecule. Proton chemical shifts are extremely sensitive to hydrogen bonding. In almost all cases, formation of a hydrogen bond causes the resonance of the bonded proton to move downfield (i.e., to larger δ values)

Table 12.1

Effect of Hydrogen Bonding on Proton Chemical Shifts[a]

Compound	δ(bonded)[b]	δ(bonded) $-$ δ(free)
$C_2H_5OH\cdots O-C_2H_5$ with H below	5.3	4.6
$C_6H_5OH\cdots O=P(OC_2H_5)_3$	8.7	4.3
$Cl_3CH\cdots O=C(CH_3)_2$	8.0	0.7
$Cl_3CH\cdots$ benzene ring	6.4	-0.9
$C_6H_5C\underset{O\cdots H-O}{\overset{O-H\cdots O}{\diagdown}}CC_6H_5$	14.0	7.8
$C_6H_5NH_2\cdots N$ pyridine	5.3	2.0
$C_6H_5NH_2\cdots O=S(CH_3)_2$	5.0	1.7
$C_6H_5SH\cdots O=C-N(CH_3)_2$ with H below	3.8	0.5
$CH_3C=CH\ CCH_3$ with $O-H\cdots O$ below	15.5	—

[a] Data compiled from work in the author's laboratory and from various literature sources. These data are illustrative but should not be regarded as accurate because of variations caused by differences in solvent, concentration, and temperature.

[b] In parts per million with respect to TMS (internal).

by as much as 10 ppm. In general there is a rough correlation between the NMR shift and the strength of the bond as measured by the enthalpy of its formation.[174] Table 12.1 shows typical changes in proton resonance frequency on hydrogen bonding. Intramolecularly bonded enols and phenols display resonances at especially low field.

Proton NMR has proved to be one of the most sensitive methods of studying hydrogen bonding, both qualitatively and quantitatively. Since hydrogen bonds are usually made and broken very rapidly relative to the chemical shift difference between bonded and nonbonded forms (expressed in hertz), separate lines are not observed for different species; the frequency ν of the single observed line is given by Eq. (11.11). In suitable systems the variation of ν with concentration and temperature can be analyzed to give equilibrium constants and other thermodynamic properties for hydrogen bond formation.[175]

The reason for the pronounced downfield shift of a bonded proton has not been completely explained. However, it seems clear that it is not merely a result of decreased electron density around the proton, for overall the proton is probably in a region of higher electron density in the hydrogen bond $X—H \cdots Y$ than in $X—H$ alone. Nevertheless, the electron distribution in the $X—H$ covalent bond is evidently altered by the electric field from Y in such a way that the proton is deshielded. In addition, the proton may experience a neighbor anisotropy effect (see Section 4.8) from Y. For most acceptors this effect results in a small upfield shift, and for aromatic molecules where the proton is bonded approximately to the center of the π-electron cloud, the ring current effect (Section 4.9) leads to a large upfield shift, which predominates over any deshielding due to other factors. A hydrogen bond to π electrons of an aromatic ring is the only type of hydrogen bond that results in an observed upfield, rather than downfield, shift.

The effects of hydrogen bonding on the chemical shifts of X and Y have not been studied carefully. It appears that changes in their chemical shifts may amount to a few parts per million, but in contrast to proton chemical shifts, such effects are small relative to the ranges covered for typical acceptor atoms, such as N and O.

A few studies of the effect of hydrogen bonding on coupling constants have been directed to $X—H$ one bond couplings. For both $^{15}N—H$ in aniline and $^{13}C—H$ in $CHCl_3$, ^{1}J increases by about 5% on formation of a hydrogen bond.

Problems

1. Find the structures of the molecules giving Spectra 39 and 40, Appendix C.

2. What effect on T_1 (^{13}C) in the hydrogen bonding molecule phenol is expected as the concentration of phenol is increased in the relatively inert solvent CCl_4?

Chapter 13

Use of NMR in Quantitative Analysis

We saw in Chapter 6 that quantitative measurements of the relative areas of the lines in an NMR spectrum can be of great aid in problems of structure elucidation. In many instances such intensity measurements can also be valuable for the quantitative analysis of mixtures of compounds of known constitution. In this chapter we shall indicate some of the advantages and disadvantages of NMR as a quantitative analytical method and describe a few of the systems that have been analyzed in this way.

13.1 Advantages of NMR in Quantitative Analysis

Probably the principal advantage of NMR as a quantitative analytical method relative to most other spectroscopic techniques is the absence in NMR of quantities analogous to the absorption coefficients or extinction coefficients found in other types of spectra. As we saw in Section 2.4, the intensity of an NMR signal for a given nuclear isotope is proportional to the number of nuclei contributing to the signal but is independent of the chemical nature of a given isotopic nucleus. Thus in principle a proton NMR analysis for, say, benzene can be based on a standard signal from naphthalene, acetone, or any other convenient proton-containing molecule. This is a very distinct advantage, for it often means in practice that the compound being analyzed in a mixture need not be available in pure form for use as a standard.

Another advantage of NMR stems from the fact that resonance lines are usually narrow relative to chemical shift differences. As the values of obtainable magnetic fields are increased, the likelihood of appreciable overlap of the proton NMR spectra of different components of a mixture is reduced. For nuclei other than hydrogen, the range of chemical shifts is much larger (see Section 4.5), so that overlap is even less likely. Of course, molecules that contain *only* nuclei that are very similar chemically (e.g., benzene and naphthalene) will generally have spectra that overlap significantly.

13.2 Drawbacks and Problems in the Use of NMR in Quantitative Analysis

There are a number of disadvantages to the use of NMR as a quantitative analytical method, but as we shall see, most of these can be overcome by taking suitable precautions.

Most quantitative spectroscopic methods depend on measurements of the easily determined peak *heights* of spectral lines or bands. While peak heights can be used in NMR in some special cases (see Section 13.3), this procedure is generally avoided since peak heights are highly dependent on instrumental conditions and nuclear relaxation times. As we have seen, it is the *area* under a line that serves as a measure of the number of nuclei responsible for it. The electronic integrators supplied with most commercial cw instruments are accurate to about 1% or 2% of full scale measurement. This figure can be misleading, however, in two ways. First, poor signal/noise ratio leads to greatly reduced accuracy, since amplifier drift and fluctuations in the size of the integral increase. Second, if one component is present at much lower concentration than another, the percentage error in measurement of the quantity of this minor component could be quite high. On the other hand, the precision of the NMR method, like all analytical methods subject to random errors, can be improved by averaging the results of several independent determinations.

The presence of more than one isotope of a given element can cause difficulties in analyses. This problem arises most frequently in organic molecules, where the 1.1% natural abundance of ^{13}C leads to ^{13}C satellites around the main peaks in proton NMR spectra (see Section 7.24). A ^{13}C satellite of a strong line could cause considerable error if it happens to overlap a weak line or group of lines. Provided the analyst is aware of this possibility, the presence of ^{13}C satellites poses no fundamental barrier to accurate analyses since the natural abundance of ^{13}C is known and the area under the satellite peaks can be precisely related to that under the main peaks.* In fact, in analyzing for components present in vastly dif-

---

* If there are n equivalent carbon atoms in the molecule, a simple statistical treatment shows that to a very high degree of approximation $(1.1 \times n)\%$ of the molecules contain one ^{13}C atom and contribute to the ^{13}C satellites.

ferent concentrations, it is sometimes possible to compare the area under the main lines of one compound with that under the ^{13}C satellite lines of the other.

Related to the problem of ^{13}C satellites is the occurrence of spinning sidebands (see Section 3.2). In principle the area of the spinning sidebands should be added to that of the main peak to obtain the true area representative of the quantity of material responsible for the resonance, but since spinning modulates all resonances equally, they can be neglected everywhere. Usually by careful adjustment of magnetic field homogeneity, use of high-precision sample tubes, and optimization of spinning rate, it is possible to reduce spinning sidebands to a negligible level.

The problem of selective saturation of different NMR lines was discussed in Section 3.7. We saw that this potential source of error can be eliminated by careful choice of the magnitude H_1 and of sweep speed. Sweep speeds of 10 Hz/sec with H_1 of 0.1–0.2 mG have been found to be optimum for quantitative analysis of protons in typical organic samples.[176]

With pulse Fourier transform methods, the phenomenon of saturation does not occur in quite the same way. As we saw in Chapter 10, the magnitude of a given spectral line is reduced if the nuclei responsible for it have not completely relaxed when the pulse is applied. Thus if pulses are repeated in a shorter interval than about $5T_1(\max)$, where $T_1(\max)$ is the longest relaxation time for the nuclei in question, the relative areas will be distorted. In ^{13}C NMR, for example, relaxation times of quaternary carbon atoms are often 10–50 times as long as those of protonated carbons. Figure 13.1a illustrates the gross distortions in areas that can result from ignoring relaxation effects. The quaternary carbons give signals that are too small by factors of 4–5. Two solutions to the problem are possible: (1) allow an adequate recovery period between pulses; (2) add a small amount of paramagnetic substance to enhance relaxation rates (see Section 8.8). The former is preferable when feasible, but often the resultant delay times of hundreds of seconds between successive pulses cannot be tolerated. Materials such as chromium acetylacetonate [Cr(AcAc)$_3$] can be very effective, as illustrated in Fig. 13.1b.

Another related source of error in ^{13}C experiments is the differences in nuclear Overhauser enhancement (NOE) that result from the application of proton decoupling. The NOE can be suppressed, as we saw in Section 9.4, by gating the decoupler off during the delay period. The addition of a paramagnetic relaxation agent also provides an effective alternative relaxation path to dipolar relaxation, and thus suppresses the NOE. In Fig. 13.1b gated decoupling was used, along with the addition of Cr(AcAc)$_3$. The importance of the various factors is discussed and very well illustrated in a recent review.[177]

We noted in Section 3.5 that NMR is inherently less sensitive than

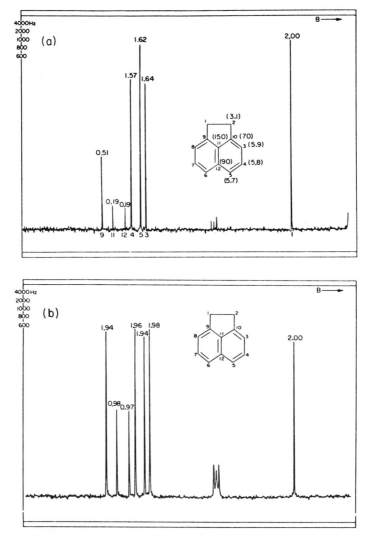

Fig. 13.1 Carbon-13 spectrum of acenaphthene. (a) Run under typical conditions: 640 pulses, 20° flip angle, 1.0 sec pulse repetition time, noise-modulated decoupling applied throughout the experiment. The assignments of the lines to individual carbon nuclei are indicated below each line, while the relative areas (relative to that of carbon 1 set equal to 2.00) are shown above each line. The measured values of T_1 for each carbon are given in parentheses beside the structural formula. (b) Cr(AcAc)$_3$ added to make 0.1 M solution, and experiment performed with the conditions used in (a) except that the decoupler was gated off during a 2 sec delay period after each 1 sec data acquisition period. The areas are within 3% of the theoretical values (Shoolery[177]).

many other spectroscopic methods. For this reason NMR is only rarely used for analyzing trace components, but the rapid relaxation of paramagnetic ions does permit this type of analysis in special circumstances, as we shall discover in Section 13.3.

Particular care must be taken in proton NMR analyses if exchangeable protons are present in one or more of the components. Usually it is poor practice to base the analytical method on an exchangeable proton, since small amounts of water or other substances containing active protons might be present as an impurity. By exchanging rapidly with the proton in question, such impurities can alter the area of its resonance line (see Chapter 11). Even if the exchangeable proton is not used in the analysis, its presence could cause some difficulty if it exchanges at an intermediate rate and gives a very broad line that overlaps or underlies regions that are to be used in the analysis. If such interference occurs, conditions of solvent, temperature, or added acid can usually be adjusted to change the frequency and width of the interfering line.

13.3 Some Analytical Uses of NMR

Probably the simplest use of NMR as a quantitative analytical method is the analysis of isotopic composition. For example, the residual hydrogen in D_2O has been determined by NMR.[178] If instrumental parameters, especially field homogeneity, are maintained constant, peak heights can be used. The procedure involves adding known amounts of H_2O to the sample of D_2O and thus constructing a calibration curve of signal amplitude versus amount of H_2O (or HDO) originally in the D_2O sample. Clearly, this technique can be adapted to other substances provided allowance is made for the effects of spin-spin coupling where the protons are not exchangeable. For example, if proton NMR is used to determine benzene-d_5 in a sample of benzene-d_6, the line of C_6D_5H will be split or broadened relative to that of C_6H_6 because of H—D coupling. The difficulty can be overcome by use of areas, rather than peak heights.

Another early use of NMR in quantitative analysis involves the determination of moisture in solid carbohydrate or proteinaceous material.[179] For this purpose broad line NMR equipment suffices, and the distinction between water and the solid substrate is made on the basis of the vastly different line widths. A similar technique is used commercially to analyze for oil in corn kernels[180] and in seeds. The nondestructive nature of NMR is critical in these applications, since the best seeds found by analysis are planted in order to improve subsequent crops. Pulse methods, usually without Fourier transformation, are often used instead of cw broad line techniques.

Proton NMR has been shown to be useful and fairly accurate in the analysis of mixtures of dinitrotoluenes.[181] It was found that the methyl peaks afforded the best quantitative measure of the amounts of the various isomers and that peak heights, rather than areas, provided the most accurate determinations. In this system NMR apparently overestimates the amounts of minor components (< 10 mole%) by about 0.5% absolute, while it underestimates by about 2% a component present at > 50 mole%.

Proton NMR has been used to determine the amount of "active hydrogen" in a sample by addition of D_2O and measurement of the intensity of the water line.[182] Careful experimental technique is required in determining the amount of HDO in the D_2O added and in some cases correcting for slow exchange of protons.

Number average molecular weights of glycol polyesters and polyalkylene glycols have been found by proton NMR analysis of the relative intensities of the CH_2 groups in the chain and of terminal CH_2 groups.[183] Pyridine was used as the solvent to enhance chemical shift differences, and a trace of HCl was added to promote OH exchange and thus narrow the terminal CH_2 line.

NMR has been employed in the analysis of mixtures of drugs,[184] in one case with the assistance of a computer program to solve the simultaneous equations involved in the analysis.[185]

When rapid proton exchange occurs, the *frequency* of the observed line, rather than line areas, may sometimes be used as a means of quantitative analysis. An example of this technique in the measurement of equilibrium constants was mentioned in Section 12.4. Mixtures of materials with rapidly exchanging protons can be analyzed similarly.[186] Line widths can also be used in special analytical methods. We saw in Section 8.8 that a paramagnetic ion, such as Cu^{2+}, causes broadening of some of the lines of a ligand to which the metal ion is attached. With rapid metal ion exchange, the large paramagnetic broadening effect acts as a "chemical amplifier" to permit the determination of very small amounts of paramagnetic ion by measurement of line widths, rather than line areas.[186] This is one of the few quantitative analytical applications of NMR that permits determination of a substance in "trace" amounts.

Quantitative analysis of multicomponent mixtures by NMR has increased significantly during the last few years, as pulse FT methods and advances in electronics have provided better signal/noise and greater spectrometer stability. The availability of the data in digital form in the computer portion of the instrument permits integration to be carried out by precise digital methods, so that integration errors *per se* are unimportant. Spectrometer noise, phasing misadjustments and spectral line overlap provide the principal instrumental limitations on precision. Several books and articles have been devoted entirely or largely to quantitative NMR measurements.[177,187,188]

Chapter 14

Contemporary Developments in NMR

It seems appropriate to conclude our discussion of NMR by presenting a few brief examples of areas in which NMR is beginning to make a pronounced impact. The single greatest use of NMR continues to be in the structure elucidation of organic compounds. As we have seen, the combination of ^{1}H and ^{13}C NMR provides a powerful tool, which is widely exploited by organic chemists. Improvements in instruments and techniques, especially the introduction of Fourier transform methods, have reduced the amounts of sample needed for NMR and permitted its use in a wider array of organic structure problems. But an even more important consequence of the revolutionary developments in NMR technology is the application of NMR to types of problems in physics, chemistry and, especially, biology that were for 30 years outside the range of NMR approaches.

14.1 Solids

We saw in Chapters 2 and 8 that internuclear magnetic dipolar interactions lead to line widths in solids that are orders of magnitude greater than those in liquids. Likewise, nuclei with quadrupole moments experience electrical interactions with their surroundings that cause extremely broad lines when there are insufficient molecular motions to average out the interactions. For many years NMR studies in solids (and in solid-like, relatively rigid polymers) were very limited in scope and were generally carried out by solid-state physicists, since chemists were unable to extract information on the chemical shifts of different nuclei. That whole

266

picture has now changed, as new methods have been developed to obtain narrow, high-resolution-like lines in solids.

Central to this development has been the use of high power rf pulses (long a standard method for studying broad lines in solids), coupled with the use of Fourier transform methods to extract the frequency spectrum. But there have also been many new conceptual developments in understanding the nature of the interactions and relaxation processes in solids. This whole area is currently a frontier of NMR research, and we cannot begin to describe here the many facets to this work. We can only mention some of the approaches that now seem most promising.

In Section 9.10 we pointed out that double resonance methods can be used to decouple nuclei that are interacting by dipolar interactions, in much the same way that the effect of scalar couplings can be removed by ordinary spin decoupling. This method is particularly effective for nuclei of spin $\frac{1}{2}$ which are present at low abundance (e.g., ^{13}C, ^{15}N) and are relaxed by dipolar interactions with nearby protons. As illustrated in Section 9.10, even a dipolar-decoupled spectrum can be quite broad, since the chemical shift anisotropy leads to different resonance frequencies for molecules with different orientations relative to $\mathbf{H}_0$. Like many magnetic and electrical interactions, the magnitude varies as $(3 \cos^2 \theta - 1)$. We saw in Section 8.2 that rapid, random molecular motion averages such an interaction to zero since the average value of $\cos^2 \theta$ is $\frac{1}{3}$. Another way of obtaining zero for this quantity is to have $\theta = 54.7°$. This observation suggests a means of removing broadening due to chemical shift anisotropy: spin the solid sample about an axis inclined at 54.7° (the "magic angle") to $\mathbf{H}_0$. *Magic angle spinning* is now widely employed in the NMR study of solids. In its simplest modification, spinning occurs at a rate fast compared with the width (in Hz) of the dipolar decoupled line, but with appropriately timed sampling of the FID it is possible to achieve the same effects with slower spinning.

Quadrupolar nuclei interact with surrounding electric fields in such a way that the frequency of the resonance line depends on the angle between the symmetry axes of the nuclear moment and the molecular field gradient, thus giving rise to a broad line for a randomly disposed solid. However, the frequency of the two quantum transition (see Section 7.24) for a nucleus with $I = 1$ (e.g., ^2H) is independent of this orientation. Elegant methods have been developed to observe these normally forbidden transitions to obtain sharp lines in randomly arranged molecular solids.[189]

For nuclei, such as protons, whose line widths are dominated by dipolar interactions with spins of like species, a somewhat different approach is used. By rapidly applying a pulse cycle of the sort $90°_x$, $90°_{-x}$,

90_y°, 90_{-y}°, the nuclear magnetization can be made to spend one-third of its time along each of the three Cartesian axes. The result is the same as though the magnetization had spent all its time along the average direction—54.7° (the magic angle) from the z axis. A careful analysis of this situation shows that the homonuclear dipolar interactions average to zero. While technically difficult to achieve, pulse cycles of this sort, together with more complex variants, have indeed provided substantial narrowing of proton NMR lines in solids.[155]

The study of high resolution NMR spectra in solids is clearly one of the most important areas of current NMR research. This work is not limited to conventional solids, single crystals, and polycrystalline materials, but is also applicable to many types of immobile polymers. Of particular interest is the investigation of biopolymers in membranes. Recent monographs present the theory of high resolution NMR in solids and give examples of the methods.[190]

14.2 Multinuclear NMR

Until rather recently most NMR spectrometers were designed to study a single type of nucleus (e.g., ^{1}H) or a very small number of nuclei. With advances in electronics, it has become feasible to build NMR spectrometers and probes that are able to be used for a very wide range of nuclear resonance frequencies. Increasingly NMR spectrometers with high field superconducting magnets are being used for high sensitivity studies of many nuclei, such as ^{27}Al, ^{113}Cd, and ^{207}Pb, to mention only a few examples. With multinuclear NMR becoming commonplace, we can anticipate much greater contributions of NMR to inorganic chemistry, as well as to problems in organometallic and metalloenzyme studies.

14.3 Biochemical Studies

Since the late 1950's NMR has been applied to structural elucidations of molecules of biochemical interest, and, as indicated in Section 6.3, NMR has proved to be valuable in the study of conformational changes in many biopolymers. Until recently, however, most NMR studies of biochemically significant molecules have been carried out under conditions far from those occurring *in vivo*. With the availability of high sensitivity, high field spectrometers, it is now possible to study NMR spectra in living systems and to investigate metabolic processes. For the most part, ^{1}H NMR is of little use in such studies, since the spectra arising from the many different chemical components are far too complex to interpret. Studies with ^{31}P have proved quite valuable, since many phosphates play crucial roles in cellular processes (e.g., adenosine triphosphate, creatine phosphate, glucose phosphate). In addition, the chemical shift of inorganic phosphate is highly dependent on pH, hence can be used as a sensitive measure of *intra*cellular pH. Carbon-13 studies are also proving to be extremely useful, but in most cases enriched ^{13}C compounds must be used to follow metabolic processes, since accumulation of sufficient signal/noise for natural abundance materials would require too long a time when chemical reactions are occurring.[191]

NMR studies of this sort have been carried out with a variety of materials—simple unicellular organisms, fragments of tissues, and whole mammalian organs. For example, whole, perfused rat hearts have been examined by ^{31}P NMR under conditions of simulated heart attacks.[192]

14.4 NMR Imaging

One interesting and potentially extremely valuable application of NMR in biology and even in medical diagnosis is the generation of two- or three-dimensional images within living subjects (including humans). The principle is simple: Suppose, for the present, that we consider only a single substance, such as water, that is distributed in a nonuniform manner within a macroscopic object which is contained entirely within an NMR probe. If a magnetic field gradient is applied in one direction across the sample, the Larmor equation, $\omega_0 = \gamma H_0$, shows that the resonance frequency of the nuclei depends on the field strength, hence on the position of the molecule in the sample. A measurement of signal as a function of frequency then provides a profile of the amount of water across one direction in the sample. By altering the direction of the magnetic field gradient other profiles can be obtained, and by methods of image reconstruction a two- or even three-dimensional image of the distribution of water in the macroscopic sample can be obtained.[193]

Clear images of the distribution of water in biological objects, including living animals, have been obtained with such image reconstruction methods. However, the technique as we have described it, is inefficient, since a number of separate, time-consuming scans are required to generate a single two-dimensional image. Several other approaches have been developed, most of which use pulse Fourier transform methods.[194] Some approaches require the use of rapidly switched field gradients,[195] and one employs gradients along both H_0 and H_1.[196]

The potentialities of NMR imaging, as a complementary approach to x-ray and ultrasonic imaging, in biological research and medical diagnosis are enormous. While the initial applications have concentrated on water, which is abundant in biological organisms and provides a strong NMR signal, nuclei other than ^1H can be studied, and with appropriate techniques high resolution spectra that differentiate chemically shifted nuclei can be observed. Properties other than resonance frequency (e.g., relaxation times) can also be measured. Current work in this field is directed toward the study of the flow of blood and other body fluids, the differences in NMR properties between normal and diseased tissues, the abnormal distribution of water within the body, and the investigation of biochemical processes (see Section 14.3) within particular organs or parts of organs.

References

1. W. Pauli, Jr., *Naturwissenschaften* **12,** 741 (1924).
2. D. M. Dennison, *Proc. R. Soc. London Ser. A* **115,** 483 (1927).
3. O. Stern, *Z. Phys.* **7,** 249 (1921); W. Gerlach and O. Stern, *Ann. Phys. Leipzig* **74,** 673 (1924).
4. I. Estermann and O. Stern, *Z. Phys.* **85,** 17 (1933).
5. I. I. Rabi, S. Millman, P. Kusch, and J. R. Zacharias, *Phys. Rev.* **55,** 526 (1939); J. M. B. Kellogg, I. I. Rabi, N. F. Ramsey, Jr., and J. R. Zacharias, *ibid.* **56,** 728 (1939).
6. E. M. Purcell, H. C. Torrey, and R. V. Pound, *Phys. Rev.* **69,** 37 (1946).
7. F. Bloch, W. W. Hansen, and M. Packard, *Phys. Rev.* **69,** 127 (1946).
8. W. D. Knight, *Phys. Rev.* **76,** 1259 (1949); W. C. Dickinson, *ibid.* **77,** 736 (1950); G. Lindström, *ibid.* **78,** 817 (1950); W. G. Proctor and F. C. Yu, *ibid.* **77,** 717 (1950).
9. J. T. Arnold, S. S. Dharmatti, and M. E. Packard, *J. Chem. Phys.* **19,** 507 (1951).
10. R. J. Highet, private communication.
11. G. Slomp and F. MacKellar, *J. Am. Chem. Soc.* **82,** 999 (1960).
12. F. A. Bovey, *Chem. Eng. News* **43,** 98 (August 30, 1965).
13. See, for example, H. Goldstein, "Classical Mechanics." Addison-Wesley, Reading, Massachusetts, 1950.
14. For details see, for example, P. L. Corio, "Structure of High Resolution NMR Spectra," pp. 16–19. Academic Press, New York, 1966.
15. See Corio,[14] pp. 60–62.
16. See, for example, J. A. Pople, W. G. Schneider, and H. J. Bernstein, "High Resolution Nuclear Magnetic Resonance," pp. 31–43. McGraw-Hill, New York, 1959.
17. See Pople *et al.,*[16] p. 23.
18. C. P. Slichter, "Principles of Magnetic Resonance," pp. 10–16. Harper, New York, 1963.
19. Slichter,[18] pp. 16–22.
20. F. Bloch, *Phys. Rev.* **94,** 496 (1954); see also G. A. Williams and H. S. Gutowsky, *ibid.* **104,** 278 (1956); and J. I. Kaplan, *J. Chem. Phys.* **27,** 1426 (1957).
21. For comprehensive treatments of modulation in NMR, see (a) W. A. Anderson, *Rev. Sci. Instrum.* **33,** 1160 (1962); (b) O. Haworth and R. E. Richards, *Prog. NMR Spectrosc.* **1,** 1 (1966).
22. R. R. Ernst, *Rev. Sci. Instrum.* **36,** 1689 (1965).
23. R. E. Lundin, R. H. Elskin, R. A. Flath, N. Henderson, T. R. Mon, and R. Teranishi, *Anal. Chem.* **38,** 291 (1966).
24. R. R. Ernst, *Adv. Magn. Resonance* **2,** 1 (1966).

25. R. R. Ernst and W. A. Anderson, *Rev. Sci. Instrum.* **37,** 93 (1966).
26. R. R. Ernst, *J. Magn. Resonance* **3,** 10 (1970); R. Kaiser, *ibid.* **3,** 28 (1970).
27. J. Dadok and R. F. Sprecher, *J. Magn. Resonance* **13,** 243 (1974); R. K. Gupta, J. A. Ferretti, and E. D. Becker, *ibid.* **13,** 275 (1974).
28. See, for example, Pople *et al.*,[16] p. 80.
29. D. J. Frost and G. E. Hall, *Mol. Phys.* **15,** 129 (1968).
30. J. K. Becconsall, G. D. Daves, Jr., and W. R. Anderson, Jr., *J. Am. Chem. Soc.* **92,** 430 (1970).
31. (a) N. C. Li, R. L. Scruggs, and E. D. Becker, *J. Am. Chem. Soc.* **84,** 4650 (1962); (b) D. C. Douglass and A. Fratiello, *J. Chem. Phys.* **39,** 3161 (1963); (c) D. F. Evans, *J. Chem. Soc.* 2003 (1959); (d) H. A. Lauers and G. P. Van der Kelen, *Bull. Soc. Chim. Belges* **75,** 238 (1966); (e) R. F. Spanier, T. Vladimiroff, and E. R. Malinowski, *J. Chem. Phys.* **45,** 4355 (1966).
32. J. R. Zimmerman and M. R. Foster, *J. Phys. Chem.* **61,** 282 (1957); C. A. Reilly, H. M. McConnell, and R. G. Meisenheimer, *Phys. Rev.* **98,** 264A (1955).
33. C. A. Reilly, *J. Chem. Phys.* **25,** 604 (1956).
34. A. L. Van Geet, *Anal. Chem.* **40,** 2227 (1968); *ibid.* **42,** 679 (1970). See also D. S. Raiford, C. L. Fisk, and E. D. Becker, *Anal. Chem.* **51,** 2050 (1979).
35. For example, the "Flath-Lundin Filter Assembly," obtainable from Hamilton Syringe Company, Whittier, California.
36. H. M. Fales and A. V. Robertson, *Tetrahedron Lett.* 111 (1962).
37. J. N. Shoolery, "Microsample Techniques in ^{1}H and ^{13}C NMR Spectroscopy." Varian Associates, Palo Alto, California, 1977.
38. J. N. Shoolery, private communication.
39. F. H. A. Rummens, *Org. Magn. Resonance* **2,** 209 (1970).
40. ASTM Standard E-386-76, American Society for Testing and Materials, Philadelphia, Pennsylvania, 1976.
41. IUPAC Recommendations for the Presentation of NMR Data for Publication in Chemical Journals, *Pure Appl. Chem.* **29,** 627 (1972); **45,** 217 (1976).
42. E. Mohacsi, *J. Chem. Ed.* **41,** 38 (1964).
43. J. W. Emsley, J. Feeney, and L. H. Sutcliffe, "High Resolution NMR Spectroscopy," pp. 1115–1136. Macmillan (Pergamon), New York, 1965.
44. C. F. Hammer, Notes for Georgetown University NMR Workshop (1965).
45. J. B. Stothers, "^{13}C NMR Spectroscopy." Academic Press, New York, 1972.
46. G. C. Levy and G. L. Nelson, "^{13}C NMR for Organic Chemists." Wiley (Interscience), New York, 1972.
47. E. W. Randall and D. G. Gillies, *Prog. NMR Spectrosc.* **6,** 119 (1971).
48. G. A. Webb and M. Witanowski, "Nitrogen NMR." Plenum Press, New York, 1973.
49. H. A. Christ, *Helv. Phys. Acta* **33,** 572 (1960).
50. Emsley *et al.*,[43] p. 1046.
51. E. G. Brame, Jr., *Anal. Chem.* **34,** 591 (1962).
52. L. Cavalli, Fluorine chemical shifts and F-F coupling constants, *in* "Nuclear Magnetic Resonance Spectroscopy of Nuclei Other than Protons" (T. Axenrod and G. A. Webb, eds.). Wiley (Interscience), New York, 1974.
53. H. Finegold, *Ann. N.Y. Acad. Sci.* **70,** 875 (1958).
54. R. A. Y. Jones and A. R. Katritzky, *Angew. Chem., Int. Ed.* **1,** 32 (1962).
55. Emsley *et al.*,[43] pp. 1054–1067.
56. W. E. Lamb, Jr., *Phys. Rev.* **60,** 817 (1941).
57. N. F. Ramsey, *Phys. Rev.* **78,** 699 (1950).

58. For details, see Emsley et al.,[43] pp. 68–76.

59. See Emsley et al.,[43] pp. 779–782.

60. H. Spiesecke and W. G. Schneider, *J. Chem. Phys.* **35**, 722 (1961).

61. H. Spiesecke and W. G. Schneider, *J. Chem. Phys.* **35**, 731 (1961). More extensive tabulations have also been given by S. Castellano, R. Kostelnik, and C. Sun, *Tetrahedron Lett.*, 4635, 5205 (1967).

62. P. Diehl, *Helv. Chim. Acta* **44**, 829 (1961); J. S. Martin and B. P. Dailey, *J. Chem. Phys*, **39**, 1722 (1963); J. J. R. Reed, *Anal. Chem.* **39**, 1586 (1967).

63. See, for example, Emsley et al.,[43] pp. 130 ff.

64. G. J. Karabatsos, G. C. Sonnichsen, N. Hsi, and D. J. Fenoglio, *J. Am. Chem. Soc.* **89**, 5067 (1967).

65. J. W. ApSimon, W. G. Craig, P. V. Demarco, D. W. Mathieson, L. Saunders, and W. B. Whalley, *Tetrahedron* **23**, 2339 (1967).

66. L. M. Jackman, "Applications of NMR Spectroscopy in Organic Chemistry," p. 55. Macmillan (Pergamon), New York, 1959.

67. J. A. Pople, *J. Chem. Phys.* **24**, 1111 (1956).

68. C. E. Johnson, Jr., and F. A. Bovey, *J. Chem. Phys.* **29**, 1012 (1958).

69. This tabulation is reproduced in Emsley et al.,[43] pp. 595–604.

70. J. N. Shoolery, "Technical Information Bulletin," Varian Associates, Palo Alto, California, 1959. The table is reproduced with some additions in Jackman,[66] p. 59, and in Jackman and Sternhell,[71] pp. 181–183.

71. L. M. Jackman and S. Sternhell, "Applications of NMR Spectroscopy in Organic Chemistry," 2nd ed. Pergamon, Oxford, 1969.

72. H. Primas, R. Arndt, and R. Ernst, "Advances in Molecular Spectroscopy," pp. 1246–1252. Pergamon, Oxford, 1962. This table is reproduced in Emsley et al.,[43] pp. 838–841.

73. D. M. Grant and E. G. Paul, *J. Am. Chem. Soc.* **86**, 2984 (1964).

74. L. P. Lindeman and J. Q. Adams, *Anal. Chem.* **43**, 1245 (1971).

75. For a summary, see Levy and Nelson,[46] pp. 38–81.

76. H. Batiz-Hernandez and R. A. Bernheim, *Prog. NMR Spectrosc.* **3**, 63 (1967).

77. E. D. Becker, R. B. Bradley, and T. Axenrod, *J. Magn. Resonance* **4**, 136 (1971).

78. G. N. La Mar, W. D. Horrocks, Jr., and R. H. Holm, "NMR of Paramagnetic Molecules." Academic Press, New York, 1973.

79. For a discussion of many aspects of shift reagents, see R. E. Sievers (ed.), "Nuclear Magnetic Resonance Shift Reagents." Academic Press, New York, 1973.

80. N. F. Ramsey and E. M. Purcell, *Phys. Rev.* **85**, 143 (1952).

81. A. A. Bothner-By, *Adv. Magn. Resonance* **1**, 195 (1965).

82. N. F. Chamberlain, "The Practice of NMR Spectroscopy." Plenum Press, New York, 1974.

83. F. A. Bovey, "NMR Spectroscopy." Academic Press, New York, 1969.

84. S. Castellano, C. Sun, and R. Kostelnik, *J. Chem. Phys.* **46**, 327 (1967).

85. J. B. Lambert, B. W. Roberts, G. Binsch, and J. D. Roberts, *in* "Nuclear Magnetic Resonance in Chemistry," B. Pesce (ed.). Academic Press, New York, 1965.

86. G. Mavel, *Prog. NMR Spectrosc.* **1**, 251 (1966).

87. E. G. Finer and R. K. Harris, *Prog. NMR Spectrosc.* **6**, 61 (1970).

88. M. Karplus, *J. Chem. Phys.* **30**, 11 (1959); *J. Am. Chem. Soc.* **85**, 2870 (1963).

89. R. J. Highet, personal communication.

90. E. W. Garbisch, Jr., *J. Am. Chem. Soc.* **86**, 5561 (1964).

91. See, for example, L. D. Hall and R. B. Malcolm, *Can. J. Chem.* **50**, 2102, 2902 (1972).

92. A. A. Bothner-By and R. H. Cox, *J. Phys. Chem.* **73**, 1830 (1969); R. Wasylishen, *in* "NMR Spectroscopy of Nuclei Other than Protons," T. Axenrod and G. A. Webb (ed.). Wiley (Interscience), New York, 1974.
93. W. B. Jennings, D. R. Boyd, C. G. Watson, E. D. Becker, R. B. Bradley, and D. M. Jerina, *J. Am. Chem. Soc.* **94**, 8501 (1972).
94. J. A. Pople and A. A. Bothner-By, *J. Chem. Phys.* **42**, 1339 (1965).
95. "Varian High Resolution NMR Spectra Catalog," Vols. 1 and 2. Varian Associates, Palo Alto, California, 1963.
96. "Sadtler Standard NMR Spectra." Sadtler Research Laboratories, Philadelphia, Pennsylvania, 1967.
97. R. H. Bible, Jr., "Interpretation of NMR Spectra, An Empirical Approach." Plenum Press, New York, 1965; "Guide to the NMR Empirical Method." Plenum Press, New York, 1967.
98. D. W. Mathieson, "NMR for Organic Chemists." Academic Press, New York, 1967.
99. D. W. Mathieson, "Interpretation of Organic Spectra." Academic Press, New York, 1966.
100. J. C. Maire and B. Waegell, "Structures, Mechanisms and Spectroscopy." Gordon and Breach, New York, 1971.
101. F. A. Bovey, "High Resolution NMR of Macromolecules." Academic Press, New York, 1971.
102. H. F. Epstein, A. N. Schechter, and J. S. Cohen, *Proc. Nat. Acad. Sci. U. S.* **68**, 2042 (1971).
103. A. F. Casy, "PMR Spectroscopy in Medical and Biological Chemistry." Academic Press, New York, 1971; M. B. Hayes, J. S. Cohen, and M. L. McNeel, *Magn. Resonance Rev.* **3**, 1 (1974); F. A. Bovey, *J. Polym. Sci. Macromol. Rev.* **9**, 1, (1974); T. L. James, "NMR in Biochemistry." Academic Press, New York, 1975.
104. P. L. Corio[14] provides a rigorous and detailed treatment.
105. J. W. Emsley, *et al.*,[43] Chapter 8.
106. J. A. Pople *et al.*,[16] Chapter 6.
107. See, for example, H. Eyring, J. Walter, and G. E. Kimball, "Quantum Chemistry," p. 37. Wiley, New York, 1944.
108. See Corio,[14] p. 172.
109. S. Castellano and A. A. Bothner-By, *J. Chem. Phys.* **41**, 3863 (1964).
110. J. D. Swalen and C. A. Reilly, *J. Chem. Phys.* **37**, 21 (1962).
111. R. B. Johannesen, J. A. Ferretti, and R. K. Harris, *J. Magn. Resonance* **3**, 84 (1970).
112. J. D. Swalen, *Prog. NMR Spectrosc.* **1**, 205 (1966).
113. C. W. Haigh, *Ann. Rep. NMR Spectrosc.* **4**, 311 (1971).
114. P. Diehl, H. Kellerhals, and E. Lustig, *NMR Basic Principles Prog.* **6**, 1 (1972).
115. See Diehl *et al.*,[114] p. 89, for addresses of several sources.
116. S. Castellano and J. S. Waugh, *J. Chem. Phys.* **34**, 295 (1961); J. S. Waugh and S. Castellano, *ibid.* **35**, 1900 (1961).
117. P. Diehl, R. K. Harris, and R. G. Jones, *Prog. NMR Spectrosc.* **3**, 1 (1967).
118. P. Diehl, personal communication.
119. G. W. Flynn and J. D. Baldeschwieler, *J. Chem. Phys.* **38**, 226 (1963).
120. E. Lustig, E. P. Ragelis, N. Duy, and J. A. Ferretti, *J. Am. Chem. Soc.* **89**, 3953 (1967).
121. For summaries of systems treated in the literature, see Emsley *et al.*,[43] pp. 661–663, and annual updates in *Nuclear Magnetic Resonance: Specialist Periodical Reports*.
122. W. B. Moniz and E. Lustig, *J. Chem. Phys.* **46**, 366 (1967).
123. J. S. Waugh and F. A. Cotton, *J. Phys. Chem.* **65**, 562 (1961).

124. See, for example, M. Schwarz, R. B. Bradley, H. M. McIntyre, and E. D. Becker, *Org. Magn. Resonance* **6**, 625 (1974).

125. Excellent discussions of chemical nonequivalence in rotational isomers, with a number of examples, are given by Bovey,[83] Chapter 6; W. B. Jennings, *Chem. Rev.* **75**, 307 (1975); and A. Ault, *J. Chem. Ed.* **51**, 729 (1974).

126. L. C. Snyder and E. W. Anderson, *J. Chem. Phys.* **42**, 3336 (1965).

127. P. Diehl and C. L. Khetrapal, *NMR Basic Principles Prog.* **1**, 1 (1969); L. Lunazzi, *Detn. Org. Structures Phys. Methods* **6**, 335 (1976); C. L. Khetrapal and A. C. Kunwar, *Adv. Magn. Resonance* **9**, 301 (1977).

128. N. Bloembergen, E. M. Purcell, and R. V. Pound, *Phys. Rev.* **73**, 679 (1948).

129. See, for example, T. C. Farrar and E. D. Becker, "Pulse and Fourier Transform NMR," Chapter 4. Academic Press, New York, 1971.

130. E. R. Andrew, "Nuclear Magnetic Resonance." Cambridge Univ. Press, London and New York, 1956.

131. A. Abragam, "The Principles of Nuclear Magnetism," Chapter 8. Oxford Univ. Press, London and New York, 1961.

132. R. R. Shoup and D. L. VanderHart, *J. Am. Chem. Soc.* **93**, 2053 (1971).

133. I. Solomon, *Phys. Rev.* **99**, 559 (1955); N. Bloembergen, *J. Chem. Phys.* **27**, 572 (1957).

134. T. C. Farrar, S. J. Druck, R. R. Shoup, and E. D. Becker, *J. Am. Chem. Soc.* **94**, 699 (1972).

135. D. Doddrell and A. Allerhand, *J. Am. Chem. Soc.* **93**, 1558 (1971).

136. G. C. Levy, J. D. Cargioli, and F. A. L. Anet, *J. Am. Chem. Soc.* **95**, 1527 (1973).

137. T. R. Stengle and J. D. Baldeschwieler, *Proc. Natl. Acad. Sci. U. S.* **55**, 1020 (1966); *J. Am. Chem. Soc.* **89**, 3045 (1967).

138. N. C. Li, R. L. Scruggs, and E. D. Becker, *J. Am. Chem. Soc.* **84**, 4650 (1962).

139. C. D. Barry, J. A. Glasel, A. C. T. North, R. J. P. Williams, and A. V. Xavier, *Biochem. Biophys. Res. Commun.* **47**, 166 (1972).

140. E. B. Baker, *J. Chem. Phys.* **37**, 911 (1962).

141. For a review of NMDR, see R. A. Hoffman and S. Forsen, *Prog. NMR Spectrosc.* **1**, 15 (1966).

142. W. A. Anderson and R. Freeman, *J. Chem. Phys.* **37**, 85 (1962); R. Freeman and D. H. Whiffen, *Proc. Phys. Soc. London* **79**, 794 (1962).

143. R. R. Ernst, *J. Chem. Phys.* **45**, 3845 (1966); K. G. R. Pachler, *J. Magn. Resonance* **7**, 442 (1972).

144. A. W. Overhauser, *Phys. Rev.* **92**, 411 (1953).

145. D. Doddrell, V. Glushko, and A. Allerhand, *J. Chem. Phys.* **56**, 3683 (1972); P. Balaram, A. A. Bothner-By, and J. Dadok, *J. Am. Chem. Soc.* **94**, 4015 (1972).

146. J. H. Noggle and R. E. Schirmer, "The Nuclear Overhauser Effect." Academic Press, New York, 1971.

147. J. N. Shoolery, *Discuss. Faraday Soc.* **34**, 104 (1962).

148. J. K. Saunders and J. W. Easton, *Determination Org. Structures Phys. Methods* **6**, 271 (1976).

149. R. Johannesen, *J. Chem. Phys.* **47**, 955 (1967).

150. V. J. Kowaleski, *Prog. NMR Spectrosc.* **5**, 1 (1969); F. W. van Deursen, *Org. Magn. Resonance* **3**, 221 (1971).

151. R. Freeman and W. A. Anderson, *J. Chem. Phys.* **37**, 2053 (1962).

152. D. L. VanderHart, private communication.

153. A. Pines, M. G. Gibby, and J. S. Waugh, *Chem. Phys. Lett.* **15**, 373 (1972).

154. A. Pines, M. G. Gibby, and J. S. Waugh, *J. Chem. Phys.* **59**, 569 (1973).

155. J. S. Waugh, L. M. Huber, and U. Haeberlen, *Phys. Rev. Lett.* **20**, 180 (1968).
156. I. J. Lowe and R. E. Norberg, *Phys. Rev.* **107**, 46 (1957).
157. For detailed discussion of this point and other aspects of FT NMR, see T. C. Farrar and E. D. Becker, "Pulse and Fourier Transform NMR." Academic Press, New York, 1971; and D. Shaw, "Fourier Transform NMR Spectroscopy." Elsevier, Amsterdam, 1976.
158. E. D. Becker, J. A. Ferretti, and P. N. Gambhir, *Anal. Chem.* **51**, 1413 (1979).
159. (a) E. L. Hahn, *Phys. Rev.* **80**, 580 (1950); (b) H. Y. Carr and E. M. Purcell, *ibid.* **94**, 630 (1954); (c) S. Meiboom and D. Gill, *Rev. Sci. Instrum.* **29**, 688 (1958).
160. L. Müller, A. Kumar, and R. R. Ernst, *J. Chem. Phys.* **63**, 5490 (1975).
161. H. S. Gutowsky, D. W. McCall, and C. P. Slichter, *J. Chem. Phys.* **21**, 279 (1953); H. M. McConnell, *ibid.* **28**, 430 (1958).
162. See reviews by C. S. Johnson, Jr., *Adv. Magn. Resonance* **1**, 33 (1965); and L. W. Reeves, *Adv. Phys. Org. Chem.* **3**, 187 (1965).
163. R. R. Shoup, E. D. Becker, and M. L. McNeel, *J. Phys. Chem.* **76**, 71 (1972).
164. L. M. Jackman and F. A. Cotton (ed.), "Dynamic NMR Spectroscopy." Academic Press, New York, 1975.
165. J. I. Kaplan, *J. Chem. Phys.* **28**, 278 (1958).
166. R. A. Hoffman and S. Forsen, *Prog. NMR Spectrosc.* **1**, 15 (1966).
167. C. Deverell, R. E. Morgan, and J. H. Strange, *Mol. Phys.* **18**, 553 (1970). See also, Farrar and Becker,[157] pp. 104–106.
168. A. R. Lepley, *J. Am. Chem. Soc.* **91**, 1237 (1969).
169. A. R. Lepley and G. L. Closs, "Chemically Induced Magnetic Polarization." Wiley (Interscience), New York, 1973; H. R. Ward, *Accounts Chem. Res.* **5**, 18 (1972); R. G. Lawler, *ibid.* **5**, 23 (1972).
170. A. D. Buckingham, T. Schaefer, and W. G. Schneider, *J. Chem. Phys.* **32**, 1227 (1960).
171. See, for example, E. D. Becker, *J. Phys. Chem.* **63**, 1379 (1959).
172. See, for example, Stothers,[45] pp. 493–502.
173. See, for example, J. H. Bowie, D. W. Cameron, P. E. Schutz, D. H. Williams, and N. S. Bhacca, *Tetrahedron* **22**, 1771 (1966).
174. D. P. Eyman and R. S. Drago, *J. Am. Chem. Soc.* **88**, 1617 (1966).
175. See, for example, G. C. Pimentel and A. L. McClellan, "The Hydrogen Bond." Freeman, San Francisco, California, 1960.
176. J. L. Jungnickel and J. W. Forbes, *Anal. Chem.* **35**, 938 (1963).
177. J. N. Shoolery, *Prog. NMR Spectrosc.* **11**, 79 (1977).
178. Varian Associates, *NMR at Work* No. 57 (1958).
179. See, for example, R. A. Pittman and V. W. Tripp, *Appl. Spectrosc.* **25**, 235 (1971).
180. T. F. Conway and F. R. Earle, *J. Am. Oil Chem. Soc.* **40**, 265 (1963).
181. A. Mathias and D. Taylor, *Anal. Chim. Acta* **35**, 376 (1966).
182. P. J. Paulsen and W. D. Cooke, *Anal. Chem.* **36**, 1721 (1964).
183. T. F. Page, Jr., and W. E. Bresler, *Anal. Chem.* **36**, 1981 (1964).
184. D. P. Hollis, *Anal. Chem.* **35**, 1682 (1963).
185. G. D. Haines, Dissertation, Georgetown University, Washington, D.C., 1966.
186. R. J. Day and C. N. Reilley, *Anal. Chem.* **38**, 1323 (1966).
187. F. Kasler, "Quantitative Analysis by NMR Spectroscopy." Academic Press, New York, 1973.
188. D. E. Leyden and R. H. Cox, "Analytical Applications of NMR." Wiley (Interscience), New York, 1977.
189. A. Pines, D. J. Ruben, S. Vega, and M. Mehring, *Phys. Rev. Lett.* **36**, 110 (1976).

190. M. Mehring, "High Resolution NMR in Solids." Springer-Verlag, Berlin, 1976; U. Haeberlen, "High Resolution NMR in Solids: Selective Averaging." *Adv. Magn. Resonance, Supplement 1.* Academic Press, New York, 1976.

191. See, for example, R. G. Shulman, T. R. Brown, K. Ugurbil, S. Ogawa, S. M. Cohen, and J. A. Hollander, *Science* **205,** 160 (1979); C. T. Burt, S. M. Cohen, and M. Barany, *Ann. Rev. Biophys. Bioeng.* **8,** 1 (1979).

192. D. P. Hollis, R. L. Nunnally, G. J. Taylor IV, M. L. Weisfeldt, and W. E. Jacobus, *J. Magn. Resonance* **29,** 319 (1978).

193. P. C. Lauterbur, *Nature* **242,** 190 (1973).

194. A. Kumar, D. Welti, and R. R. Ernst, *J. Magn. Resonance* **18,** 69 (1975).

195. P. Mansfield and I. L. Pykett, *J. Magn. Resonance* **29,** 355 (1978).

196. D. I. Hoult, *J. Magn. Resonance* **33,** 183 (1979).

Appendix A

General NMR References

A. Comprehensive Treatments

1. J. A. Pople, W. G. Schneider, and H. J. Bernstein, "High Resolution NMR." McGraw-Hill, New York, 1959.
2. J. W. Emsley, J. Feeney, and L. H. Sutcliffe, "High Resolution NMR," Vols. 1 and 2. Macmillan (Pergamon), New York, 1966.
3. F. A. Bovey, "NMR Spectroscopy." Academic Press, New York, 1968.
4. A. Abragam, "The Principles of Nuclear Magnetism." Oxford Univ. Press, London and New York, 1961.
5. C. P. Slichter, "Principles of Magnetic Resonance." 2nd ed. Springer-Verlag, Berlin, 1978.

B. Introductory Books

1. R. J. Abraham and P. Loftus, "Proton and Carbon-13 NMR Spectroscopy." Heyden, London, 1978.
2. E. J. Haws, R. R. Hill, and D. J. Mowthorpe, "The Interpretation of Proton Magnetic Resonance Spectra." Heyden, London, 1973.
3. R. H. Bible, "Introduction to NMR Spectroscopy." Plenum Press, New York, 1965.
4. J. D. Roberts, "Nuclear Magnetic Resonance." McGraw-Hill, New York, 1959.
5. J. D. Roberts, "Introduction to Spin–Spin Splitting in High Resolution NMR." Benjamin, New York, 1961.
6. A. Carrington and A. D. McLachlen, "Introduction to Magnetic Resonance." Harper, New York, 1966.
7. E. F. Mooney, "An Introduction to F-19 NMR Spectroscopy." Heyden, London, 1970.

C. NMR Techniques

1. T. C. Farrar and E. D. Becker, "Pulse and Fourier Transform NMR." Academic Press, New York, 1971.
2. D. Shaw, "Fourier Transform NMR Spectroscopy." Elsevier, Amsterdam, 1976.

3. K. Müllen and P. S. Pregosin, "Fourier Transform NMR Techniques: A Practical Approach." Academic Press, New York, 1976.
4. J. H. Noggle and R. E. Schirmer, "The Nuclear Overhauser Effect." Academic Press, New York, 1971.

D. Organic Applications

1. L. M. Jackman and S. Sternhell, "Applications of NMR Spectroscopy in Organic Chemistry," 2nd ed. Pergamon, Oxford, 1969.
2. F. W. Wehrli and T. Wirthlin, "Interpretation of Carbon-13 NMR Spectra." Heyden, London, 1976.
3. G. C. Levy and G. L. Nelson, "^{13}C NMR for Organic Chemists." Wiley (Interscience), New York, 1972.
4. J. B. Stothers, "^{13}C NMR Spectroscopy." Academic Press, New York, 1972.
5. G. C. Levy and R. L. Lichter, "^{15}N NMR Spectroscopy." Wiley (Interscience), New York, 1979.
6. J. C. Randall, "Polymer Sequence Determination: Carbon-13 NMR Method." Academic Press, New York, 1977.
7. D. W. Mathieson (ed.), "NMR for Organic Chemists." Academic Press, New York, 1967.

E. Biochemical Applications

1. T. L. James, "NMR in Biochemistry." Academic Press, New York, 1975.
2. R. A. Dwek, I. D. Campbell, R. E. Richards, and R. J. P. Williams (eds.), "NMR in Biology." Academic Press, New York, 1977.
3. P. F. Knowles, D. Marsh, and H. W. E. Rattle, "Magnetic Resonance of Biomolecules." Wiley (Interscience), New York, 1976.
4. A. F. Casy, "PMR Spectroscopy in Medical and Biological Chemistry." Academic Press, New York, 1971.

F. Serial Publications

1. J. S. Waugh (ed.), *Advances in Magnetic Resonance.* Academic Press, New York.
2. J. W. Emsley, J. Feeney, and L. H. Sutcliffe (eds.), *Progress in NMR Spectroscopy.* Pergamon, Oxford.
3. P. Diehl, E. Fluck, and R. Kosfeld (eds.), *NMR: Basic Principles and Progress.* Springer-Verlag, Berlin and New York.
4. E. F. Mooney (ed.), *Annual Reports on NMR Spectroscopy.* Academic Press, New York.
5. R. J. Abraham (senior reporter), *Specialist Periodical Reports: NMR.* The Chemical Society, London.

6. C. P. Poole, Jr. (ed.), *Magnetic Resonance Reviews*. Gordon and Breach, New York.
7. L. J. Berliner and J. Reuben, *Biological Magnetic Resonance*. Plenum Press, New York.
8. J. H. Bradbury (ed.), *Bulletin on Magnetic Resonance*. Franklin Institute Press, Philadelphia.

G. Compilations of Data

In addition to compilations in the foregoing list, see the tabulations in Sections 4.12 and 5.8

Nuclear Spins, Magnetic Moments, and Resonance Frequencies[a]

Z	Element	A	Spin I	NMR frequency (MHz for a 10-kG field)	Natural abundance (%)	Relative sensitivity at constant field	Magnetic moment μ (multiples of the nuclear magneton $eh/4\pi Mc$)	Electric quadrupole moment Q (multiples of barns (10^{-24} cm^2))
0	n	1*	$\frac{1}{2}$	29.167	—	0.322	-1.91315	—
1	·H	1	$\frac{1}{2}$	42.5759	99.985	1.00	2.79268	—
1	·H	2	1	6.53566	1.5×10^{-2}	9.65×10^{-3}	0.857387	2.73×10^{-3}
1	·H	3*	$\frac{1}{2}$	45.4129	—	1.21	2.97877	—
2	·He	3	$\frac{1}{2}$	32.433	1.3×10^{-4}	0.442	-2.1274	—
3	·Li	6	1	6.2653	7.42	8.50×10^{-3}	0.82192	6.9×10^{-4}
3	·Li	7	$\frac{3}{2}$	16.546	92.58	0.293	3.2560	-3×10^{-2}
3	·Li	8*	2	6.300	—	2.59×10^{-2}	1.653	—
4	·Be	9	$\frac{3}{2}$	5.9834	100	1.39×10^{-2}	-1.1774	5.2×10^{-2}
5	·B	10	3	4.5754	19.58	1.99×10^{-2}	1.8007	7.4×10^{-2}
5	·B	11	$\frac{3}{2}$	13.660	80.42	0.165	2.6880	3.55×10^{-2}
6	·C	13	$\frac{1}{2}$	10.7054	1.108	1.59×10^{-2}	0.702199	—
7	·N	13	$\frac{1}{2}$*	4.91	—	1.53×10^{-3}	$(-)0.322$	—
7	·N	14	1	3.0756	99.63	1.01×10^{-3}	0.40347	7.1×10^{-2}
7	·N	15	$\frac{1}{2}$	4.3142	0.37	1.04×10^{-3}	-0.28298	—
8	O	15*	$\frac{1}{2}$	11.0	—	1.70×10^{-2}	0.719	—
8	·O	17	$\frac{5}{2}$	5.772	3.7×10^{-2}	2.91×10^{-2}	-1.8930	-2.6×10^{-2}

Z	Element	A	Spin I	NMR frequency (MHz for a 10-kG field)	Natural abundance (%)	Relative sensitivity at constant field	Magnetic moment μ (multiples of the nuclear magneton $eh/4\pi Mc$)	Electric quadrupole moment Q (multiples of barns (10^{-24} cm^2))
9	·F	17*	$\frac{5}{2}$	14.40	—	0.451	4.720	—
9	·F	19	$\frac{1}{2}$	40.0541	100	0.833	2.62727	
9	·F	20*	2	7.977	—	5.26×10^{-2}	2.093	
10	Ne	19*	$(\frac{1}{2})$	28.75	—	0.308	−1.886	
10	Ne	21	$\frac{3}{2}$	3.3611	0.257	2.50×10^{-3}	−0.66140	
11	Na	21*	$\frac{3}{2}$	12.126	—	0.116	2.3861	
11	Na	22*	3	4.436	—	1.81×10^{-2}	1.746	
11	·Na	23	$\frac{3}{2}$	11.262	100	9.25×10^{-2}	2.2161	0.14–0.15
11	Na	24*	4	3.221	—	1.15×10^{-2}	1.690	
12	·Mg	25	$\frac{5}{2}$	2.6054	10.13	2.67×10^{-3}	−0.85449	
13	·Al	27	$\frac{5}{2}$	11.094	100	0.206	3.6385	0.149
14	·Si	29	$\frac{1}{2}$	8.4578	4.70	7.84×10^{-3}	−0.55477	
15	·P	31	$\frac{1}{2}$	17.235	100	6.63×10^{-2}	1.1305	—
15	P	32*	1	1.923	—	2.46×10^{-4}	−0.2523	—
16	·S	33	$\frac{3}{2}$	3.2654	0.76	2.26×10^{-3}	0.64257	-6.4×10^{-2}
16	S	35*	$\frac{3}{2}$	5.08	—	8.50×10^{-3}	1.00	4.5×10^{-2}
17	·Cl	35	$\frac{3}{2}$	4.1717	75.53	4.70×10^{-3}	0.82091	-7.89×10^{-2}
17	Cl	36*	2	4.8931	—	1.21×10^{-2}	1.2838	-1.72×10^{-2}
17	·Cl	37	$\frac{3}{2}$	3.472	24.47	2.71×10^{-3}	0.6833	-6.21×10^{-2}
18	Ar	37*	$\frac{3}{2}$	5.08	—	8.50×10^{-3}	1.0	
19	K	38*	3	3.491	—	8.82×10^{-3}	1.374	
19	·K	39	$\frac{3}{2}$	1.9868	93.10	5.08×10^{-4}	0.39097	0.11
19	K	40*	4	2.470	1.18×10^{-2}	5.21×10^{-3}	−1.296	
19	K	41	$\frac{3}{2}$	1.0905	6.88	8.40×10^{-5}	0.21459	
19	K	42*	2	4.345	—	8.50×10^{-3}	−1.140	
19	K	43*	$\frac{3}{2}$	0.828	—	3.68×10^{-5}	0.163	
20	·Ca	41*	$\frac{7}{2}$	3.4681	—	1.14×10^{-2}	−1.5924	

Z	Element	Isotope	I	Frequency	Abundance (%)	Rel. sensitivity	Magnetic moment	Quadrupole moment
20	·Ca	43	$\tfrac{7}{2}$	2.8646	0.145	6.40×10^{-3}	−1.3153	−0.26
21	Sc	43*	$\tfrac{7}{2}$	10.04	—	0.275	4.61	0.14
21	Sc	44*	2	9.76	—	9.63×10^{-2}	2.56	0.37
21	Sc	44*m	6	5.03	—	9.24×10^{-2}	3.96	−0.22
21	·Sc	45	$\tfrac{7}{2}$	10.343	100	0.301	4.7492	0.12
21	Sc	46*	4	5.77	—	6.65×10^{-2}	3.03	−0.22
21	Sc	47*	$\tfrac{7}{2}$	11.6	—	0.426	5.33	
21	Sc	45*	$\tfrac{7}{2}$	0.207	—	2.40×10^{-6}	0.095	1.5×10^{-2}
22	Ti	47	$\tfrac{5}{2}$	2.4000	7.28	2.09×10^{-3}	−0.78710	
22	·Ti	49	$\tfrac{7}{2}$	2.4005	5.51	3.76×10^{-3}	−1.1022	
23	V	49*	$\tfrac{7}{2}$	9.71	—	0.249	4.46	
23	·V	50	6	4.2450	0.24	5.55×10^{-2}	3.3413	
23	·V	51	$\tfrac{7}{2}$	11.19	99.76	0.382	5.139	
24	·Cr	53	$\tfrac{3}{2}$	2.4065	9.55	9.03×10^{-4}	−0.47354	-4×10^{-2}
25	Mn	52*	6	3.907	—	4.33×10^{-2}	3.075	
25	Mn	52*m	2	0.030	—	2.9×10^{-9}	0.008	
25	Mn	53*	$\tfrac{7}{2}$	11.0	—	0.362	5.05	
25	Mn	54*	(2)	8.4	—	6.11×10^{-2}	(2.2)	
25	Mn	54*m	(3)	6.6	—	5.98×10^{-2}	(2.6)	
25	·Mn	55	$\tfrac{5}{2}$	10.501	100	0.175	3.444	0.55
25	Mn	56*	3	8.233	—	0.116	3.240	
26	·Fe	57	$\tfrac{1}{2}$	1.3758	2.19	3.37×10^{-5}	0.09024	—
27	Co	55*	$\tfrac{7}{2}$	10.0	—	0.274	4.6	
27	Co	56*	4	7.34	—	0.136	3.85	
27	Co	57*	$\tfrac{7}{2}$	10.1	—	0.283	4.65	
27	Co	58*	2	15.4	—	0.381	4.05	
27	·Co	59	$\tfrac{7}{2}$	10.054	100	0.277	4.6163	0.40
27	Co	60*	5	5.793	—	0.101	3.800	
28	·Ni	61	$\tfrac{3}{2}$	3.8047	1.19	3.57×10^{-3}	−0.74868	
29	Cu	61*	$\tfrac{3}{2}$	10.8	—	8.22×10^{-2}	2.13	
29	·Cu	63	$\tfrac{3}{2}$	11.285	69.09	9.31×10^{-2}	2.2206	−0.16
29	Cu	64*	1	3.1	—	9.79×10^{-4}	0.40	

Z	Element	A	Spin I	NMR frequency (MHz for a 10-kG field)	Natural abundance (%)	Relative sensitivity at constant field	Magnetic moment μ (multiples of the nuclear magneton $eh/4\pi Mc$)	Electric quadrupole moment Q (multiples of barns $(10^{-24}\ cm^2)$)
29	·Cu	65	$\frac{3}{2}$	12.089	30.91	0.114	2.3789	−0.15
29	Cu	66*	1	1.65	—	1.54×10^{-4}	−0.216	
30	·Zn	65*	$\frac{5}{2}$	2.345	—	1.95×10^{-3}	0.7692	-2.4×10^{-2}
30	·Zn	67	$\frac{5}{2}$	2.663	4.11	2.85×10^{-3}	0.8733	0.15
31	Ga	68*	1	0.0892	—	2.45×10^{-8}	0.0117	3.1×10^{-2}
31	Ga	69	$\frac{3}{2}$	10.22	60.4	6.91×10^{-2}	2.011	0.178
31	Ga	71	$\frac{3}{2}$	12.984	39.6	0.142	2.5549	0.112
31	Ga	72*	3	0.33591	—	7.80×10^{-6}	−0.13220	0.72
32	Ge	71*	$\frac{1}{2}$	8.4	—	7.64×10^{-3}	0.55	—
32	·Ge	73	$\frac{9}{2}$	1.4852	7.76	1.40×10^{-3}	−0.87679	−0.2
33	·As	75	$\frac{3}{2}$	7.2919	100	2.51×10^{-2}	1.4349	0.3
33	As	76*	2	3.45	—	4.27×10^{-3}	−0.906	
34	·Se	77	$\frac{1}{2}$	8.118	7.58	6.93×10^{-3}	0.5325	
34	Se	79	$\frac{7}{2}$	2.22	—	2.98×10^{-3}	−1.02	0.9
35	Br	76*	1	4.18	—	2.52×10^{-3}	(−)0.548	0.27
35	·Br	79	$\frac{3}{2}$	10.667	50.54	7.86×10^{-2}	2.0990	0.33
35	Br	80*	1	3.92	—	2.08×10^{-3}	0.514	0.20
35	Br	80*m	5	2.008	—	4.20×10^{-3}	1.317	0.76
35	·Br	81	$\frac{3}{2}$	11.498	49.46	9.85×10^{-2}	2.2626	0.28
35	Br	82*	5	2.479	—	7.90×10^{-3}	(+)1.626	(+)0.76
36	·Kr	83	$\frac{9}{2}$	1.638	11.55	1.88×10^{-3}	−0.9671	0.15
36	Kr	85*	$\frac{9}{2}$	1.6956	—	2.08×10^{-3}	−1.001	0.25
37	Rb	81*	$\frac{3}{2}$	10.4	—	7.32×10^{-2}	2.05	
37	Rb	82*m	5	2.29	—	6.20×10^{-3}	1.50	
37	Rb	83*	$\frac{5}{2}$	4.33	—	1.23×10^{-2}	1.42	
37	Rb	84*	2	5.03	—	1.32×10^{-2}	−1.32	

37	·Rb	85	5/2	4.1108	72.15	1.05×10^{-2}	1.3482	0.27
37	Rb	86*	2	6.44	—	2.77×10^{-2}	-1.69	0.13
37	·Rb	87	3/2	13.931	27.85	0.175	2.7414	0.2
38	·Sr	87	9/2	1.8452	7.02	2.69×10^{-3}	-1.0893	0.2
39	Y	89	1/2	2.0859	100	1.18×10^{-4}	-0.13682	-0.16
39	Y	90*	2	6.17	—	2.44×10^{-2}	-1.62	—
39	Y	91*	1/2	2.49	—	1.99×10^{-4}	0.163	—
40	·Zr	91	5/2	3.97249	11.23	9.48×10^{-3}	-1.30284	-0.2
41	·Nb	93	9/2	10.407	100	0.482	6.1435	0.12
42	·Mo	95	5/2	2.774	15.72	3.23×10^{-3}	-0.9097	1.1
42	·Mo	97	5/2	2.832	9.46	3.43×10^{-3}	-0.9289	0.3
43	·Tc	99*	9/2	9.5830	—	0.376	5.6572	—
44	Ru	99	3/2	1.44	12.72	1.95×10^{-4}	-0.284	—
44	Ru	101	5/2	2.1	17.07	1.41×10^{-3}	-0.69	—
45	·Rh	103	1/2	1.3401	100	3.11×10^{-5}	-0.08790	—
46	·Pd	105	5/2	1.95	22.23	1.12×10^{-3}	-0.639	—
47	Ag	104*	5	6.1	—	0.118	4.0	—
47	Ag	104*m	2	14.0	—	0.291	3.7	—
47	Ag	105*	1/2	1.54	—	4.73×10^{-5}	0.101	—
47	·Ag	107	1/2	1.7229	51.82	6.62×10^{-5}	-0.11301	—
47	Ag	108*	1	32.0	—	1.13	4.2	—
47	·Ag	109	1/2	1.9807	48.18	1.01×10^{-4}	-0.12992	—
47	Ag	110*m	6	4.557	—	6.87×10^{-2}	3.587	—
47	Ag	111*	1/2	2.21	—	1.40×10^{-4}	-0.145	—
47	Ag	112*	2	0.2077	—	9.00×10^{-7}	0.0545	—
47	Ag	113*	1/2	2.41	—	1.81×10^{-4}	0.158	—
48	Cd	107*	5/2	1.879	—	1.00×10^{-3}	-0.6162	0.8
48	Cd	109*	5/2	2.529	—	2.44×10^{-3}	-0.8293	0.8
48	·Cd	111	1/2	9.028	12.75	9.54×10^{-3}	-0.5922	—
48	Cd	113	1/2	9.445	12.26	1.09×10^{-2}	-0.6195	—
48	Cd	113*m	11/2	1.51	—	2.13×10^{-3}	-1.09	-0.79
48	Cd	115*	1/2	9.862	—	1.24×10^{-2}	-0.6469	—

Z	Isotope Element	Isotope A	Spin I	NMR frequency (MHz for a 10-kG field)	Natural abundance (%)	Relative sensitivity at constant field	Magnetic moment μ (multiples of the nuclear magneton $eh/4\pi Mc$)	Electric quadrupole moment Q (multiples of barns (10^{-24} cm^2))
48	Cd	115*m	$\frac{11}{2}$	1.447	—	1.87×10^{-3}	−1.044	−0.61
49	·In	113	$\frac{9}{2}$	9.3099	4.28	0.345	5.4960	1.14
49	In	113*m	$\frac{1}{2}$	3.209	—	4.28×10^{-4}	−0.2105	—
49	In	114*m	5	7.2	—	0.191	4.7	—
49	·In	115*	$\frac{9}{2}$	9.3301	95.72	0.347	5.5079	1.16
49	In	115*m	$\frac{1}{2}$	3.715	—	6.64×10^{-4}	−0.2437	—
49	In	116*	5	6.42	—	0.137	4.21	—
49	In	116*m	5	6.7	—	0.156	4.4	—
50	·Sn	115	$\frac{1}{2}$	13.922	0.35	3.50×10^{-2}	−0.91320	—
50	·Sn	117	$\frac{1}{2}$	15.168	7.61	4.52×10^{-2}	−0.99490	—
50	·Sn	119	$\frac{1}{2}$	15.869	8.58	5.18×10^{-2}	−1.0409	—
51	·Sb	121	$\frac{5}{2}$	10.189	57.25	0.160	3.3415	−0.5
51	Sb	122*	2	7.24	—	3.94×10^{-2}	−1.90	—
51	·Sb	123	$\frac{7}{2}$	5.5176	42.75	4.57×10^{-2}	2.5334	−0.7
52	Te	119*	$\frac{1}{2}$	4.12	—	9.04×10^{-4}	0.27	—
52	·Te	123	$\frac{1}{2}$	11.16	0.87	1.8×10^{-2}	−0.7319	—
52	·Te	125	$\frac{1}{2}$	13.45	6.99	3.15×10^{-2}	−0.8824	—
53	I	125*	$\frac{5}{2}$	9.0	—	0.116	3	−0.66
53	·I	127	$\frac{5}{2}$	8.5183	100	9.34×10^{-2}	2.7937	−0.69
53	·I	129*	$\frac{7}{2}$	5.6694	—	4.96×10^{-2}	2.6031	−0.48
53	I	131*	$\frac{7}{2}$	5.963	—	5.77×10^{-2}	2.738	−0.41
54	·Xe	129	$\frac{1}{2}$	11.777	26.44	2.12×10^{-2}	−0.77247	—
54	·Xe	131	$\frac{3}{2}$	3.4911	21.18	2.76×10^{-3}	0.68697	−0.12
55	Cs	127*	$\frac{1}{2}$	21.8	—	0.134	1.43	—
55	Cs	129*	$\frac{1}{2}$	22.4	—	0.146	1.47	—
55	Cs	130*	1	10.7	—	4.20×10^{-2}	1.4	—
55	Cs	131*	$\frac{5}{2}$	10.72	—	0.186	3.517	—

55	Cs	132*	2	8.46	—	6.28×10^{-2}	2.22	-3×10^{-3}
55	·Cs	133	7/2	5.58469	100	4.74×10^{-2}	2.56422	0.43
55	Cs	134*	4	5.666	—	6.28×10^{-2}	2.973	
55	Cs	134*m	8	1.0447	—	1.42×10^{-3}	1.0964	
55	Cs	135*	7/2	5.9096	—	5.62×10^{-2}	2.7134	
55	Cs	137*	7/2	6.1459	—	6.32×10^{-2}	2.8219	
56	·Ba	135	3/2	4.2296	6.59	4.90×10^{-3}	0.83229	0.25
56	·Ba	137	3/2	4.7315	11.32	6.86×10^{-3}	0.93107	0.2
57	·La	138	5*	5.6171	0.089	9.19×10^{-2}	3.6844	2.7
57	·La	139	7/2	6.0144	99.911	5.92×10^{-2}	2.7615	0.21
58	Ce	137*	3/2	4.6	—	6.20×10^{-3}	0.9	
58	Ce	137*m	11/2	0.96	—	5.40×10^{-4}	0.69	
58	Ce	139*	3/2	5.1	—	8.50×10^{-3}	1.0	
58	Ce	141*	7/2	2.1	—	2.57×10^{-3}	0.97	
58	Ce	143*	5/2	2.2	—	2.81×10^{-3}	1.0	
59	·Pr	141	5/2	12.5	100	0.293	4.09	-5.9×10^{-2}
59	Pr	142*	2	1.1	—	1.55×10^{-4}	0.30	4×10^{-2}
60	Nd	143	7/2	2.315	12.17	3.38×10^{-3}	−1.063	−0.48
60	Nd	145	7/2	1.42	8.30	7.86×10^{-4}	−0.654	−0.25
60	Nd	147*	5/2	1.77	—	8.32×10^{-4}	0.579	
61	Pm	143*	(5/2)	11.6	—	0.235	(3.8)	
61	Pm	143*	(7/2)	8.5	—	0.167	(3.9)	
61	Pm	144*	(5)	2.6	—	9.02×10^{-3}	(1.7)	
61	Pm	144*	(6)	2.3	—	8.68×10^{-3}	(1.8)	
61	Pm	147*	7/2	5.62	—	4.83×10^{-2}	2.58	0.7
61	Pm	148*	1	16	—	0.142	2.1	0.2
61	Pm	148*m	6	2.3	—	8.68×10^{-3}	1.8	
61	Pm	149*	7/2	7.2	—	0.101	3.3	
61	Pm	151*	5/2	5.5	—	2.50×10^{-2}	1.8	
62	Sm	147	7/2	1.76	14.97	1.48×10^{-3}	−0.807	1.9
62	Sm	149	7/2	1.40	13.83	7.47×10^{-4}	−0.643	−0.208
63	·Eu	151	5/2	10.559	47.82	0.178	3.4630	6.0×10^{-2}
63	Eu	152*	3	4.858	—	2.38×10^{-2}	1.912	1.16

Isotope Element	A	Z	Spin I	NMR frequency (MHz for a 10-kG field)	Natural abundance (%)	Relative sensitivity at constant field	Magnetic moment μ (multiples of the nuclear magneton $eh/4\pi Mc$)	Electric quadrupole moment Q (multiples of barns (10^{-24} cm²))
·Eu	153	63	$\frac{5}{2}$	4.6627	52.18	1.53×10^{-2}	1.5292	2.9
Eu	154*	63	3	5.084	—	2.72×10^{-2}	2.001	
·Gd	155	64	$\frac{3}{2}$	1.6	14.73	2.79×10^{-4}	-0.32	1.6
·Gd	157	64	$\frac{3}{2}$	2.0	15.68	5.44×10^{-4}	-0.40	2
Tb	156*	65	3	3.8	—	1.15×10^{-2}	1.5	1.4
Tb	159	65	$\frac{3}{2}$	9.66	100	5.83×10^{-2}	1.90	1.3
Tb	160*	65	3	4.1	—	1.39×10^{-2}	1.6	1.9
Dy	155*	66	$(\frac{3}{2})$	1.1	—	7.87×10^{-5}	0.21	
Dy	157*	66	$(\frac{3}{2})$	1.6	—	2.79×10^{-4}	0.32	
Dy	161	66	$\frac{5}{2}$	1.4	18.88	4.17×10^{-4}	-0.46	1.4
Dy	163	66	$\frac{5}{2}$	2.0	24.97	1.12×10^{-3}	0.64	1.6
Ho	165	67	$\frac{7}{2}$	8.73	100	0.181	4.01	2.82
Er	165*	68	$\frac{5}{2}$	2.0	—	1.18×10^{-3}	0.65	2.2
Er	167	68	$\frac{7}{2}$	1.23	22.94	5.07×10^{-4}	-0.565	2.83
Er	169*	68	$\frac{1}{2}$	7.8	—	6.09×10^{-3}	0.51	—
Er	171*	68	$\frac{5}{2}$	2.1	—	1.47×10^{-3}	0.70	
Tm	166*	69	2	0.19	—	7.17×10^{-7}	0.05	4.6
·Tm	169	69	$\frac{1}{2}$	3.52	100	5.66×10^{-4}	-0.231	—
Tm	170*	69	1	2.0	—	2.69×10^{-4}	0.26	0.61
Tm	171*	69	$\frac{1}{2}$	3.46	—	5.37×10^{-4}	0.227	—
·Yb	171	70	$\frac{1}{2}$	7.4990	14.31	5.46×10^{-3}	0.49188	—
Yb	173	70	$\frac{5}{2}$	2.0659	16.13	1.33×10^{-3}	-0.67755	2.8
Yb	175*	70	$(\frac{7}{2})$	0.33	—	9.40×10^{-6}	-0.15	
·Lu	175	71	$\frac{7}{2}$	4.86	97.41	3.12×10^{-2}	2.23	5.68
Lu	176*	71	7	3.4	2.59	3.72×10^{-2}	3.1	8.0
Lu	177*	71	$\frac{7}{2}$	4.84	—	3.08×10^{-2}	2.22	5.51
Hf	177	72	$\frac{7}{2}$	1.3	18.50	6.38×10^{-4}	0.61	3

Z		A	I		%			
72	Hf	179	9/2	0.80	13.75	2.16×10^{-4}	-0.47	3
73	·Ta	181	7/2	5.096	99.988	3.60×10^{-2}	2.340	3
74	·W	183	1/2	1.7716	14.40	7.20×10^{-5}	0.116205	—
75	·Re	185	5/2	9.5855	37.07	0.133	3.1437	2.8
75	Re	186*	1	13.17	—	7.90×10^{-2}	1.728	
75	·Re	187*	5/2	9.6837	62.93	0.137	3.1759	2.6
75	Re	188*	1	13.55	—	8.59×10^{-2}	1.777	
76	·Os	187	1/2	0.98059	1.64	1.22×10^{-5}	0.06432	—
76	·Os	189	3/2	3.3034	16.1	2.34×10^{-3}	0.65004	0.8
77	·Ir	191	3/2	0.7318	37.3	2.53×10^{-5}	0.1440	1.5
77	·Ir	193	3/2	0.7968	62.7	3.27×10^{-5}	0.1568	1.5
78	·Pt	195	1/2	9.153	33.8	9.94×10^{-3}	0.6004	—
79	Au	190*	1	0.496	—	4.20×10^{-6}	0.065	
79	Au	194*	1	0.56	—	5.95×10^{-6}	0.073	
79	Au	195*	3/2	0.742	—	2.65×10^{-5}	0.146	
79	·Au	196*	2	2.3	—	1.24×10^{-3}	0.6	
79	·Au	197	3/2	0.729188	100	2.51×10^{-5}	0.143489	0.59
79	·Au	198*	2	2.227	—	1.14×10^{-3}	0.5842	
79	·Au	199*	3/2	1.358	—	1.62×10^{-4}	0.2673	
80	Hg	193*m	13/2	3.1	—	1.93×10^{-3}	-0.61	1.37
80	Hg	195*	1/2	1.23	—	1.57×10^{-3}	-1.05	
80	Hg	195*m	13/2	8.1	—	6.84×10^{-3}	0.53	1.41
80	Hg	197*	1/2	1.22	—	1.53×10^{-3}	-1.04	
80	Hg	197*m	13/2	7.9	—	6.46×10^{-3}	0.52	0.50
80	·Hg	199	1/2	7.59012	16.84	5.67×10^{-3}	0.497859	
80	·Hg	201	3/2	2.8099	13.22	1.44×10^{-3}	-0.55293	0.5
80	Hg	203*	5/2	2.5	—	2.45×10^{-3}	0.83	
81	Tl	197*	1/2	23.6	—	0.171	1.55	—
81	Tl	199*	1/2	23.9	—	0.178	1.57	—
81	Tl	200*	2	0.57	—	1.94×10^{-5}	(0.15)	—
81	Tl	201*	1/2	24.1	—	0.181	1.58	
81	Tl	202*	2	0.57	—	1.94×10^{-5}	(0.15)	
81	·Tl	203	1/2	24.332	29.50	0.187	1.5960	—

Z	Isotope Element	A	Spin I	NMR frequency (MHz for a 10-kG field)	Natural abundance (%)	Relative sensitivity at constant field	Magnetic moment μ (multiples of the nuclear magneton $eh/4\pi Mc$)	Electric quadrupole moment Q (multiples of barns $(10^{-24}\ \text{cm}^2)$)
81	Tl	204*	2	0.34	—	4.05×10^{-6}	0.089	—
81	·Tl	205	$\frac{1}{2}$	24.570	70.50	0.192	1.6116	—
82	·Pb	207	$\frac{1}{2}$	8.90771	22.6	9.16×10^{-3}	0.584284	—
83	Bi	203*	$\frac{9}{2}$	7.78	—	0.201	4.59	−0.64
83	Bi	204*	6	5.40	—	0.114	4.25	−0.41
83	Bi	205*	$\frac{9}{2}$	9.3	—	0.346	(5.5)	
83	Bi	206*	6	5.79	—	0.141	4.56	−0.19
83	Bi	209*	$\frac{9}{2}$	6.84178	100	0.137	4.03896	−0.4
83	Bi	210*	1	0.337	—	1.32×10^{-6}	0.0442	0.13
84	Po	205*	$\frac{5}{2}$	0.79	—	7.55×10^{-5}	0.26	0.17
84	Po	207*	$\frac{5}{2}$	0.82	—	8.43×10^{-5}	0.27	0.28
89	Ac	227*	$\frac{3}{2}$	5.6	—	1.13×10^{-2}	1.1	−1.7
90	Th	229*	$\frac{5}{2}$	1.2	—	2.74×10^{-4}	0.4	4.6
91	Pa	231*	$\frac{3}{2}$	9.96	—	6.40×10^{-2}	1.96	
91	Pa	233*	$\frac{3}{2}$	17	—	0.334	3.4	−3.0
92	U	233*	$\frac{5}{2}$	1.6	—	6.75×10^{-4}	0.54	3.5
92	U	235*	$\frac{7}{2}$	0.76	0.72	1.21×10^{-4}	0.35	4.1
93	Np	237*	$\frac{5}{2}$	18	—	0.926	(6)	

Z		A	I					
94	Pu	239*	$\frac{1}{2}$	3.05	3.67×10^{-4}	0.200	—	—
94	Pu	241*	$\frac{5}{2}$	2.09	1.38×10^{-3}	−0.686	—	4.9
95	Am	241*	$\frac{5}{2}$	4.82	1.69×10^{-2}	1.58	—	−2.8
95	Am	242*	1	2.90	8.46×10^{-4}	0.381	—	4.9
95	Am	243*	$\frac{5}{2}$	4.79	1.66×10^{-2}	1.57	—	4.9
Free electron with $g = 2.00232$			$\frac{1}{2}$	2.80246×10^{4}	2.84×10^{8}	−1836.09	—	—

[a] Compiled by K. Lee and W. A. Anderson, October, 1967. Z, atomic number; A, atomic weight (mass number); I, nuclear spin in units of $h/2\pi$; μ, magnetic moments in units of the nuclear magneton $eh/4\pi Mc$; Q, quadrupole moment in units of barns (10^{-24} cm²); *, magnetic moment observed by NMR; *, radioactive isotope; ( ), assumed or estimated values; m, metastable excited state.

Assuming a nuclear magneton value of 5.0505×10^{-24} erg/gauss, the NMR frequency was calculated for a total field of 10^{4} gauss. The sensitivities, relative to the proton, are calculated from:

$$\text{Sensitivity at constant field} = 7.652 \times 10^{-3} \, \mu^{3} \, (I + 1)/I^{2}.$$

This expression assumes an equal number of nuclei, a constant temperature, and that $T_1 = T_2$ (the longitudinal relaxation time equals the transverse relaxation time). This sensitivity represents the ideal induced voltage in the receiver coil at saturation and with a constant noise source. The calculated values are therefore determined under complete optimum conditions and should be regarded as such.

Further details and literature references are given in the original tabulation, printed and distributed by Varian Associates, Palo Alto, California.

Proton and Carbon-13 NMR Spectra of "Unknowns"

All proton spectra in this section were obtained at 60 MHz except where specified. Carbon-13 spectra were run at 15.1 MHz. Tetramethylsilane was used as an internal reference for all spectra except No. 39. Except where noted, samples were solutions in chloroform-d. Chemicals were reagent grade but were not purified. Some weak lines due to impurities appear in the spectra but are not usually identified in order to simulate the situation encountered in practice.

Spectra have been selected to illustrate points covered in the text. Appropriate spectra are assigned in the problems at the ends of the various chapters. Answers to the *odd-numbered* spectra are given in Appendix D.

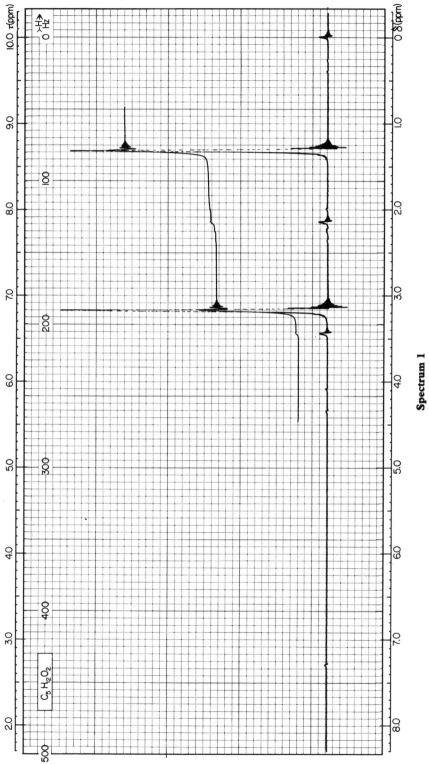

Spectrum 1

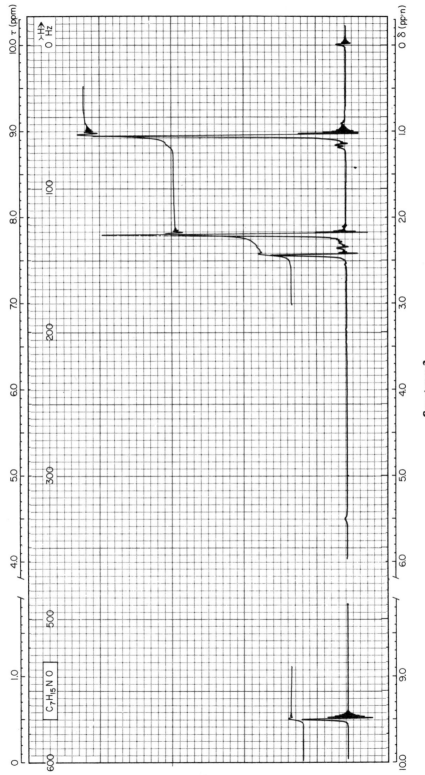

C₇H₁₅NO

Spectrum 2

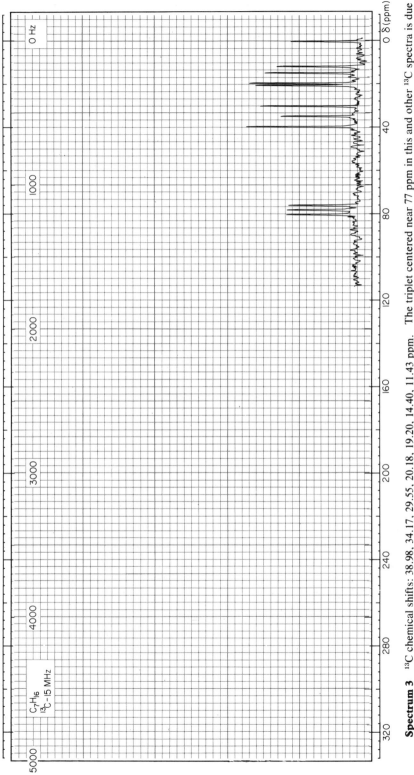

Spectrum 3 ^{13}C chemical shifts: 38.98, 34.17, 29.55, 20.18, 19.20, 14.40, 11.43 ppm. The triplet centered near 77 ppm in this and other ^{13}C spectra is due to the solvent $CDCl_3$.

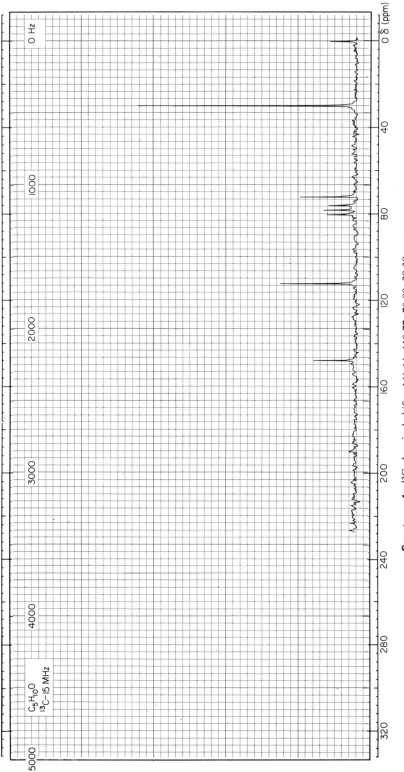

O Hz

0 δ (ppm)

1000

2000

3000

4000

5000

$C_5H_{10}O$
^{13}C–15 MHz

Spectrum 4 ^{13}C chemical shifts: 146.14, 110.77, 70.99, 29.38 ppm.

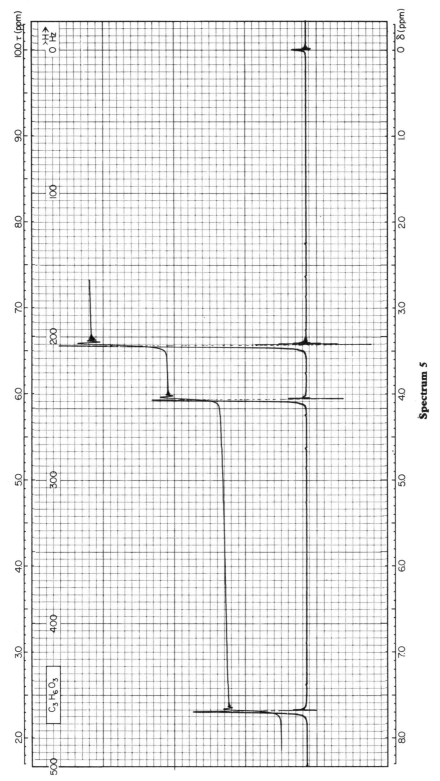

Spectrum 5

C_3 H_6 O_3

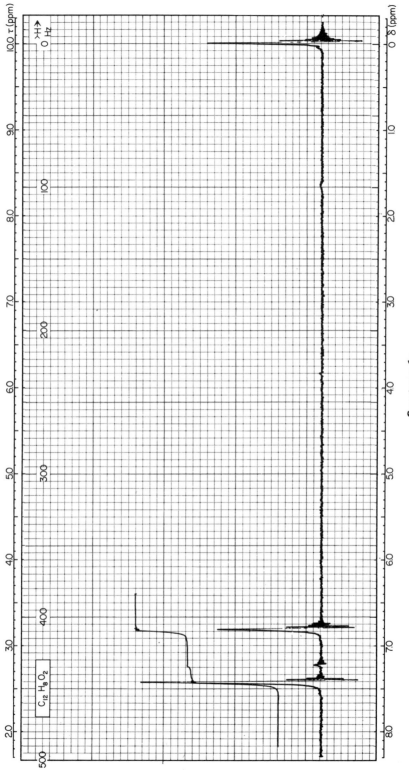

Spectrum 6a

$C_{12}H_8O_2$

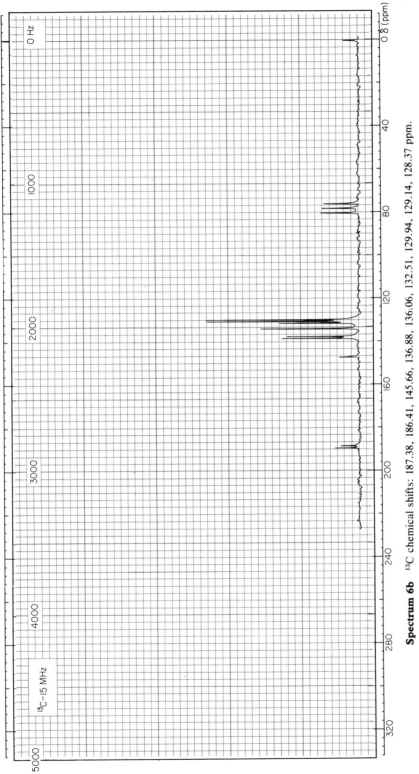

Spectrum 6b 13C chemical shifts: 187.38, 186.41, 145.66, 136.88, 136:06, 132.51, 129.94, 129.14, 128.37 ppm.

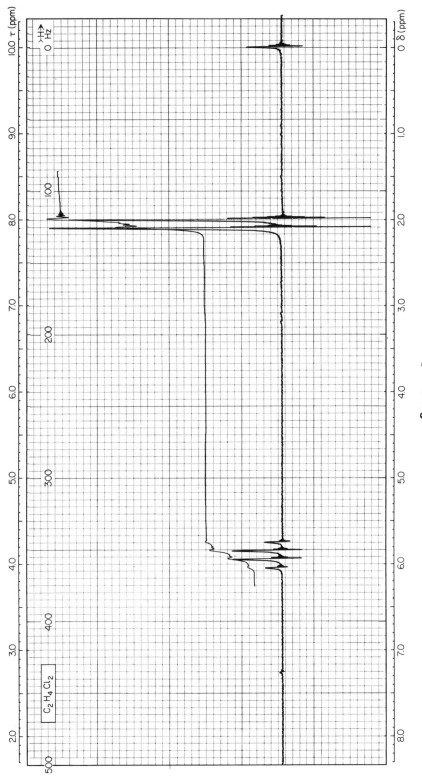

Spectrum 7

$C_2H_4Cl_2$

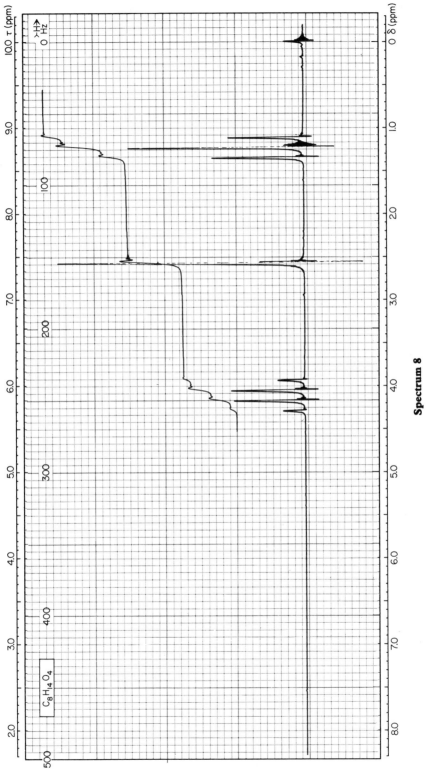

Spectrum 8

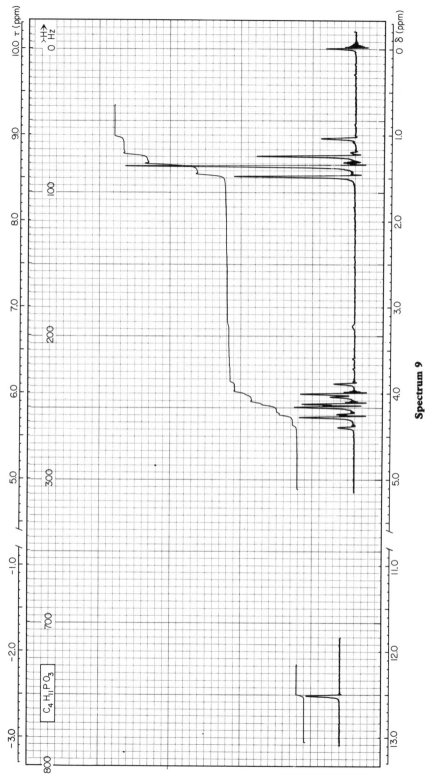

Spectrum 9

$C_4H_{11}PO_3$

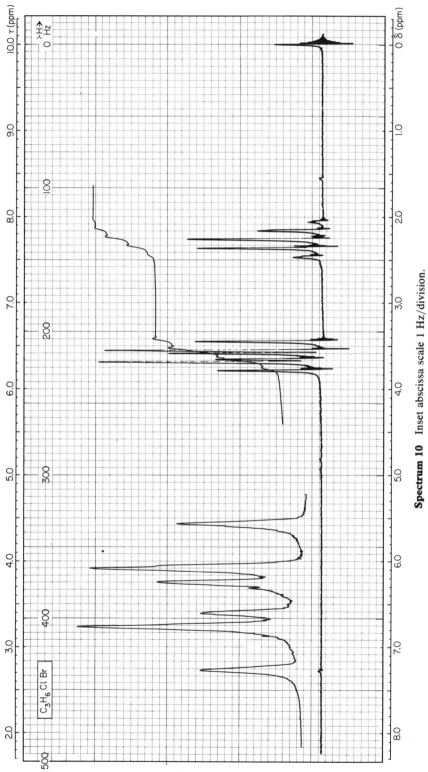

Spectrum 10 Inset abscissa scale 1 Hz/division.

C_3H_6ClBr

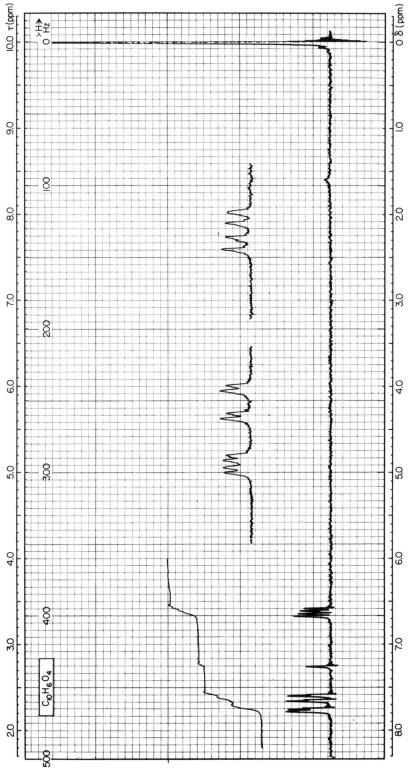

Spectrum 11 Inset abscissa scale 1 Hz/division.

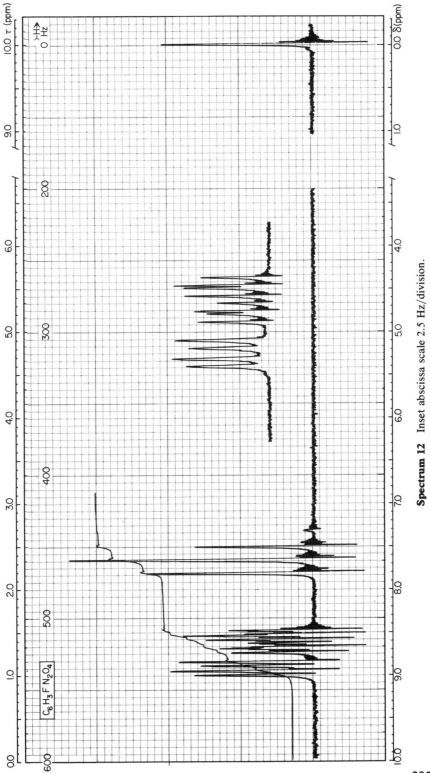

Spectrum 12 Inset abscissa scale 2.5 Hz/division.

$C_6H_3FN_2O_4$

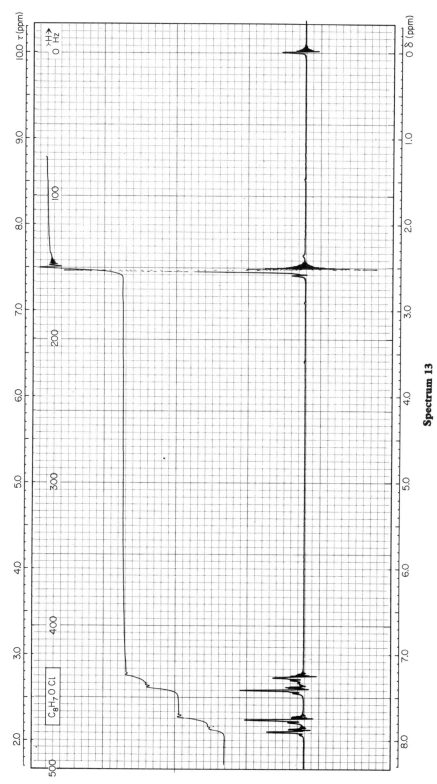

C₈H₇OCl

Spectrum 13

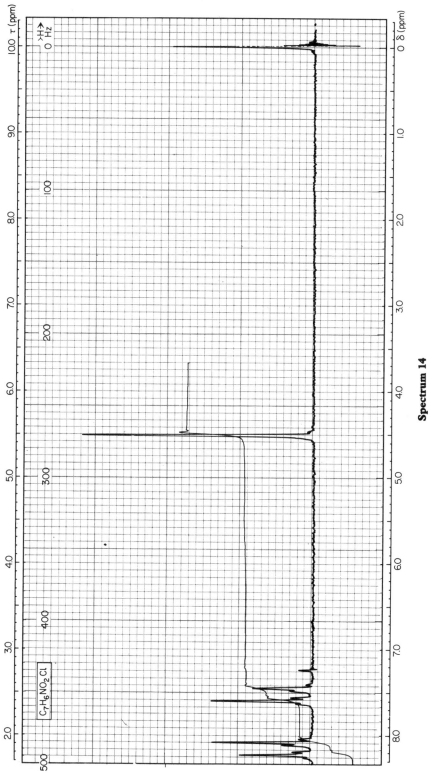

Spectrum 14

$C_7H_6NO_2Cl$

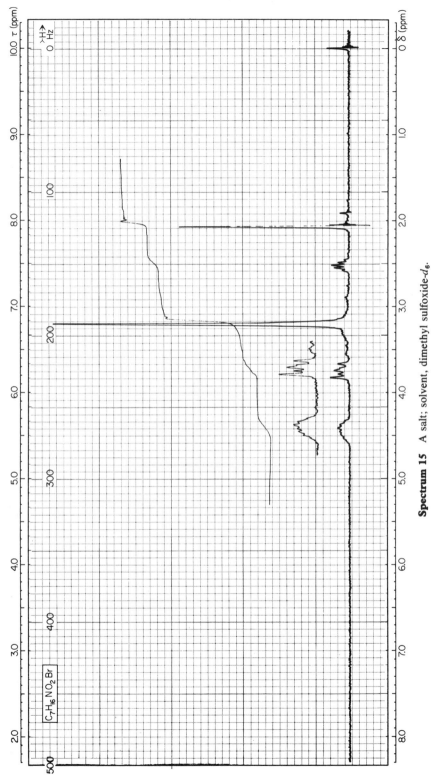

Spectrum 15 A salt; solvent, dimethyl sulfoxide-d_6.

C₇H₁₆NO₂Br

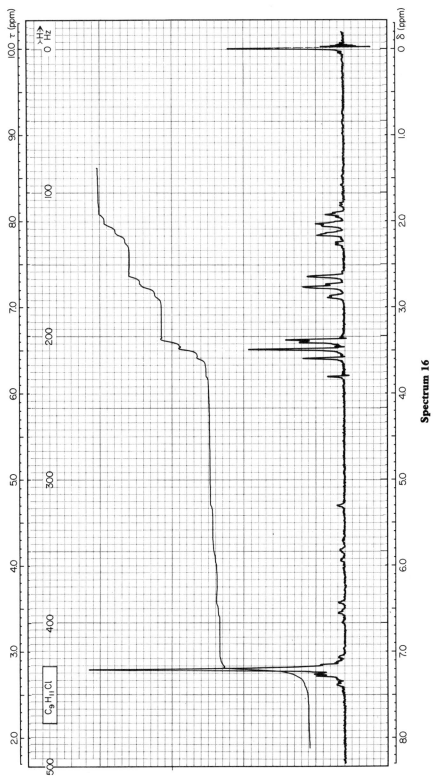

Spectrum 16

C$_9$H$_{11}$Cl

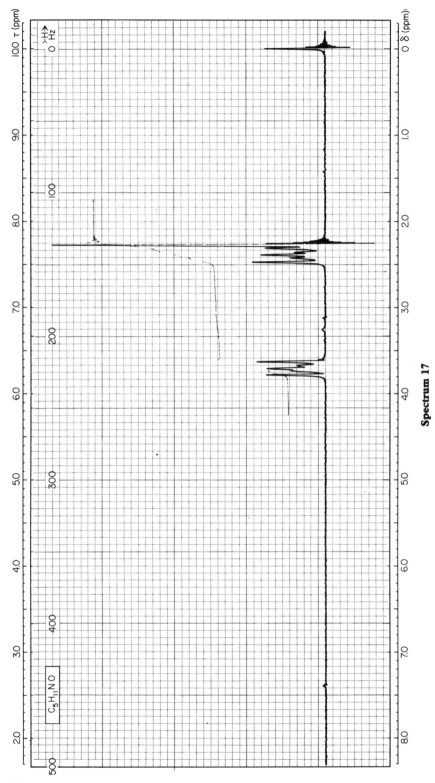

Spectrum 17

C$_5$H$_{11}$NO

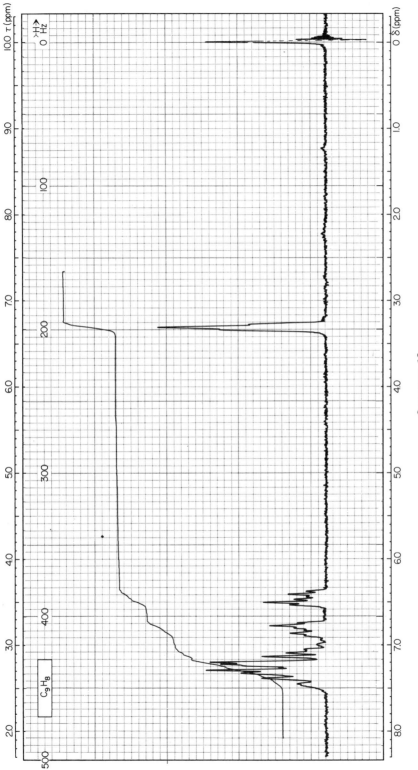

C₉H₈

Spectrum 18

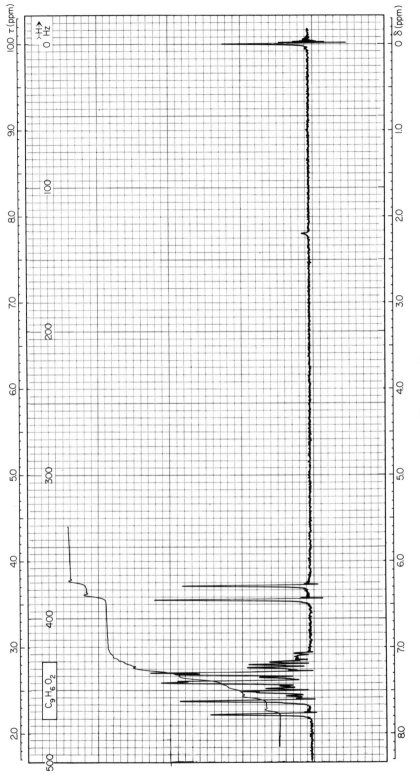

Spectrum 19

$C_9H_6O_2$

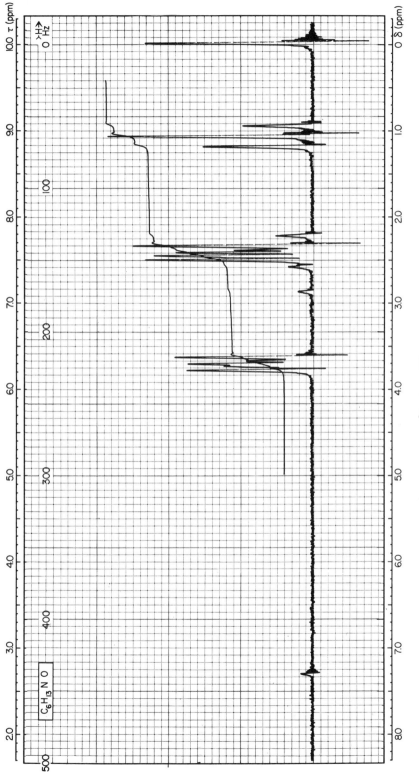

$C_6H_{13}NO$

Spectrum 20

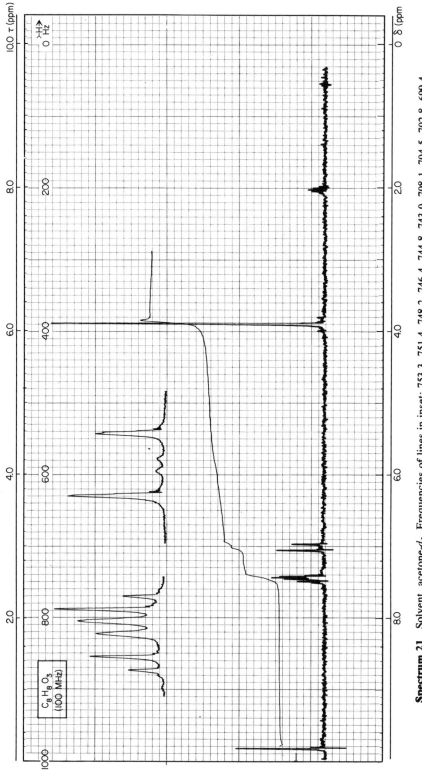

Spectrum 21 Solvent, acetone-d_6. Frequencies of lines in inset: 753.3, 751.4, 748.2, 746.4, 744.8, 743.0, 708.1, 704.5, 702.8, 699.4.

$C_8H_8O_3$
(100 MHz)

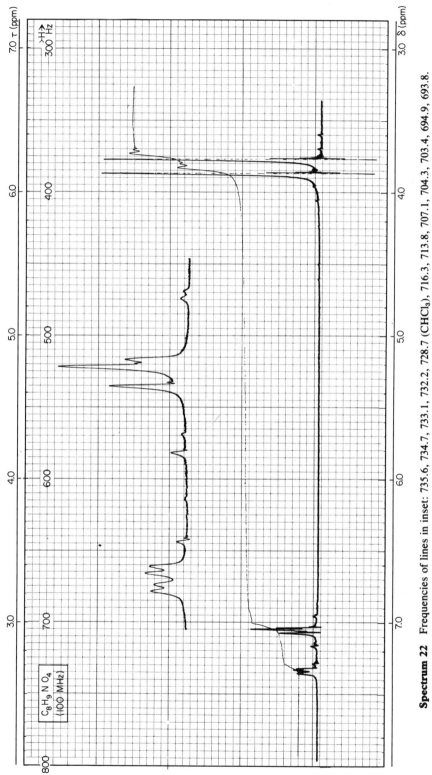

Spectrum 22 Frequencies of lines in inset: 735.6, 734.7, 733.1, 732.2, 728.7 (CHCl₃), 716.3, 713.8, 707.1, 704.3, 703.4, 694.9, 693.8.

C₈H₉NO₄
(100 MHz)

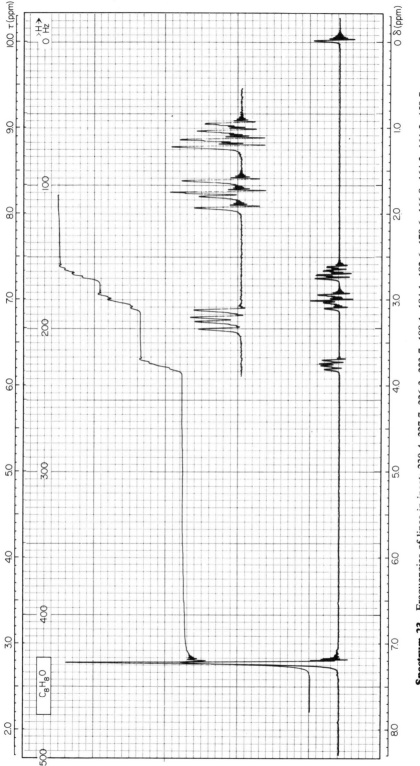

Spectrum 23 Frequencies of lines in inset: 230.4, 227.7, 226.3, 223.7, 188.1, 184.1, 182.6, 178.6, 166.8, 164.3, 161.2, 158.7.

C_8H_8O

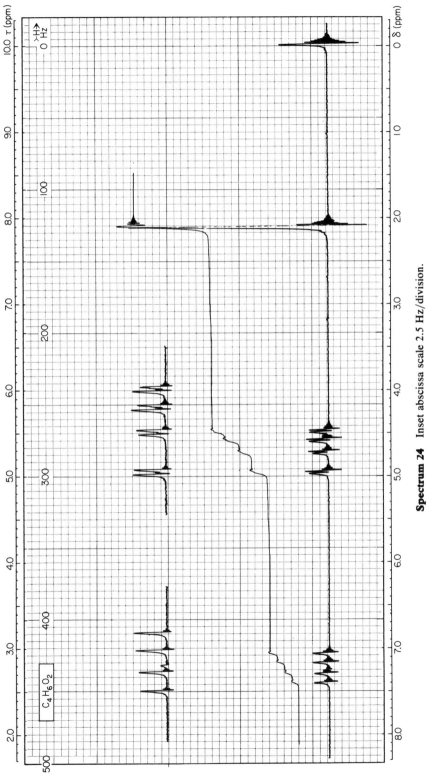

$C_4H_6O_2$

Spectrum 24 Inset abscissa scale 2.5 Hz/division.

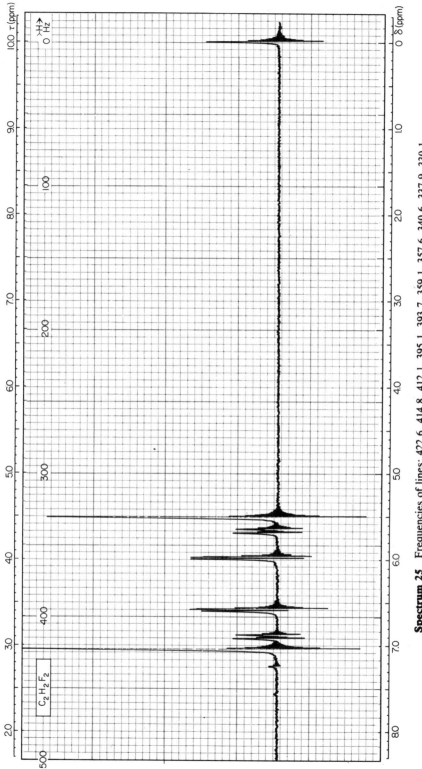

Spectrum 25 Frequencies of lines: 422.6, 414.8, 412.1, 395.1, 393.7, 359.1, 357.6, 340.6, 337.9, 330.1.

$C_2H_2F_2$

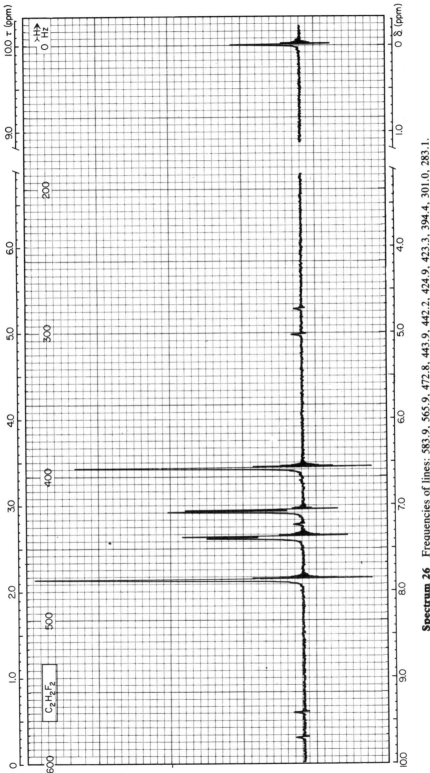

Spectrum 26 Frequencies of lines: 583.9, 565.9, 472.8, 443.9, 442.2, 424.9, 423.3, 394.4, 301.0, 283.1.

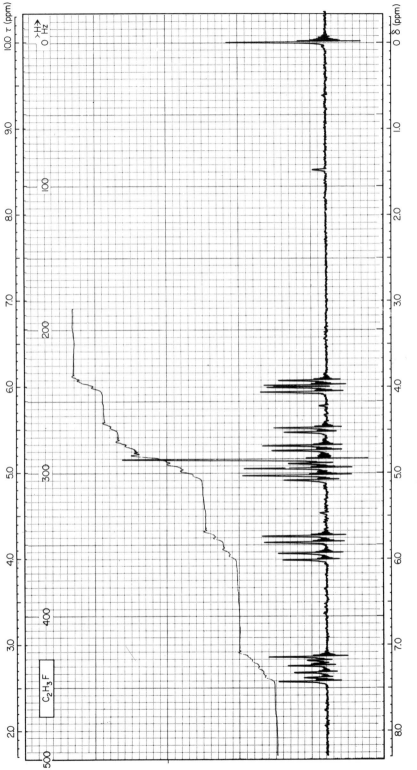

Spectrum 27a See 100 MHz spectrum below.

C₂H₃F

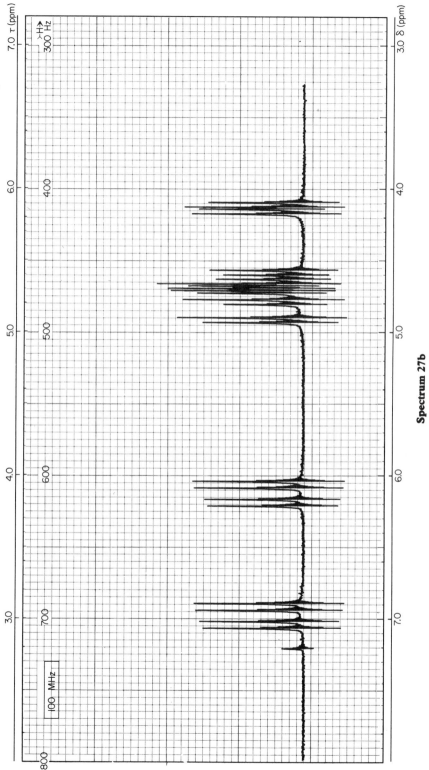

Spectrum 27b

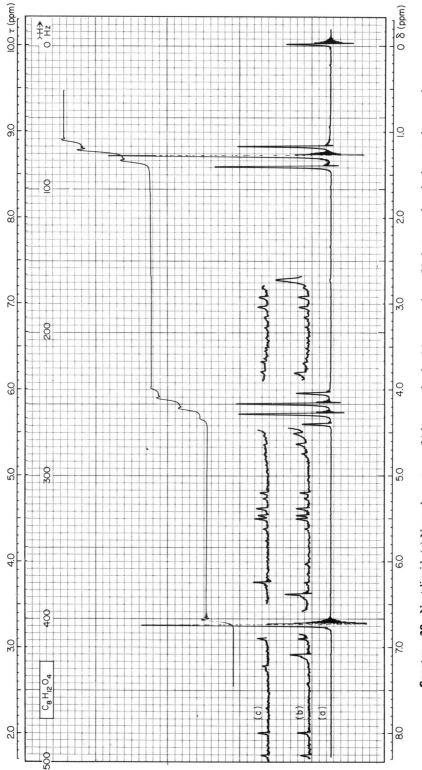

Spectrum 28 Neat liquid. (a) Normal spectrum; (b) increased gain; (c) same gain as (b), but sample spinning rate increased.

322

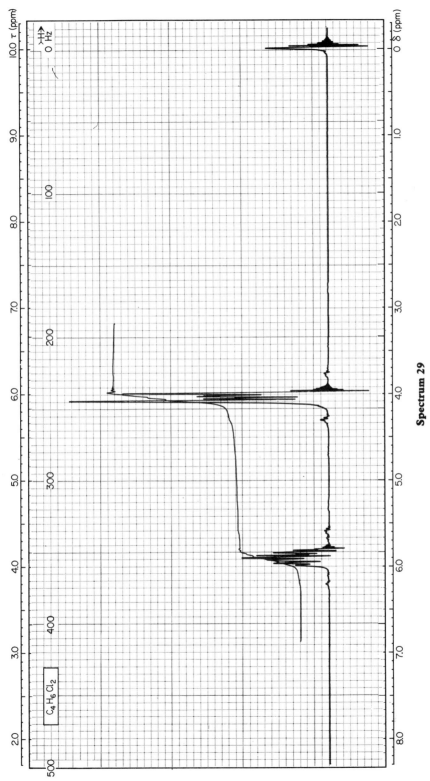

C₄H₆Cl₂

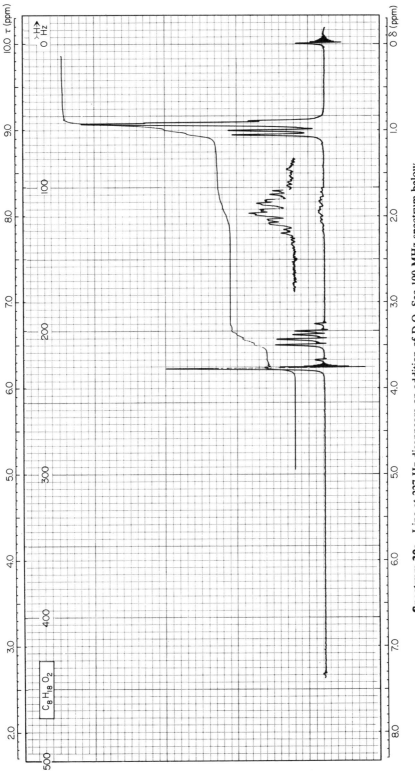

$C_8H_{18}O_2$

Spectrum 30a Line at 227 Hz disappears on addition of D_2O. See 100 MHz spectrum below.

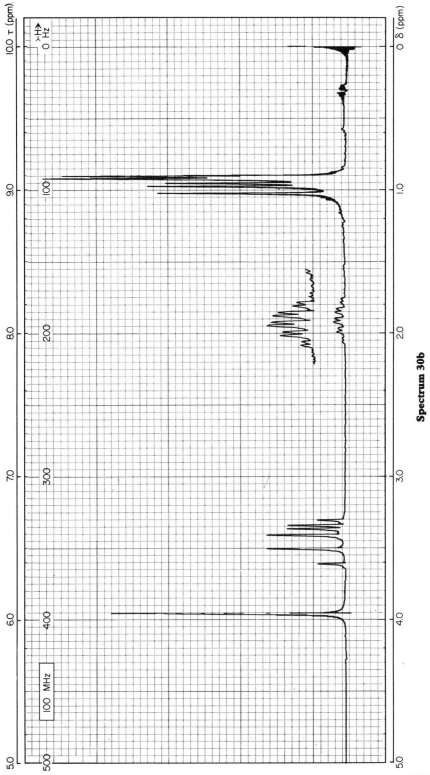

Spectrum 30b

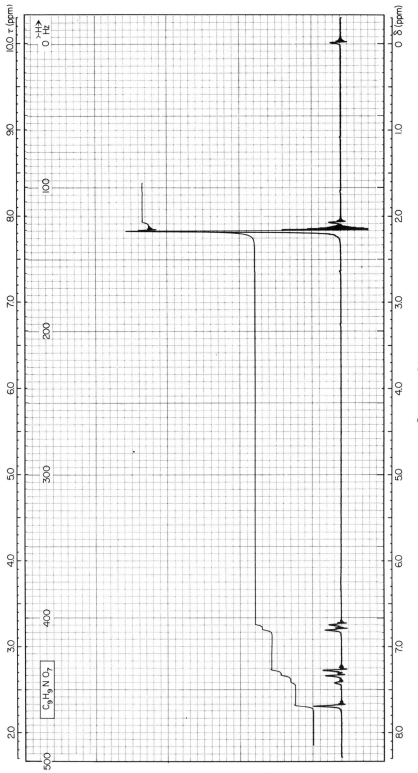

Spectrum 31a

$C_9H_9NO_7$

326

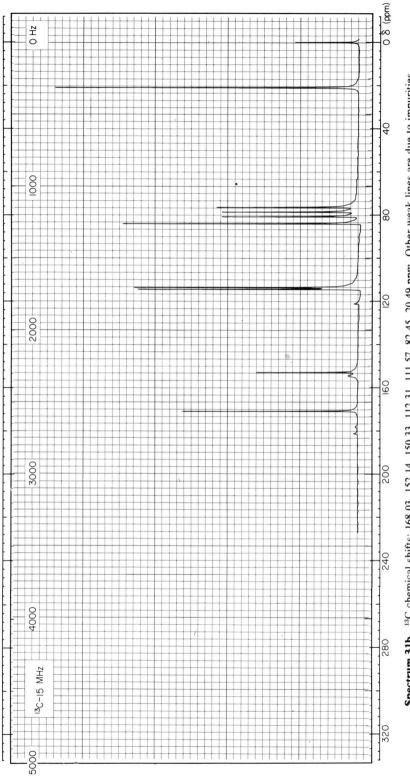

0 Hz

1000

2000

3000

4000

5000

13C–15 MHz

0 δ (ppm)

40

80

120

160

200

240

280

320

Spectrum 31b ^{13}C chemical shifts: 168.03, 152.14, 150.33, 112.31, 111.57, 82.45, 20.49 ppm. Other weak lines are due to impurities.

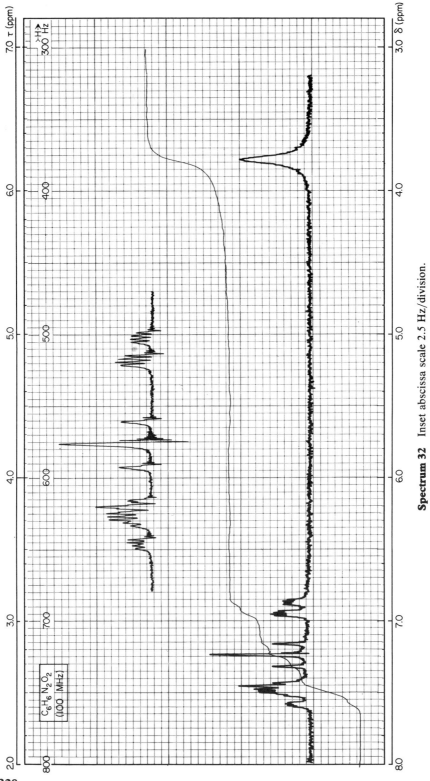

Spectrum 32 Inset abscissa scale 2.5 Hz/division.

7.0 τ (ppm)

→H→
300 Hz

400

500

600

700

800

C₆H₆N₂O₂
(100 MHz)

3.0 δ (ppm)

4.0

5.0

6.0

7.0

8.0

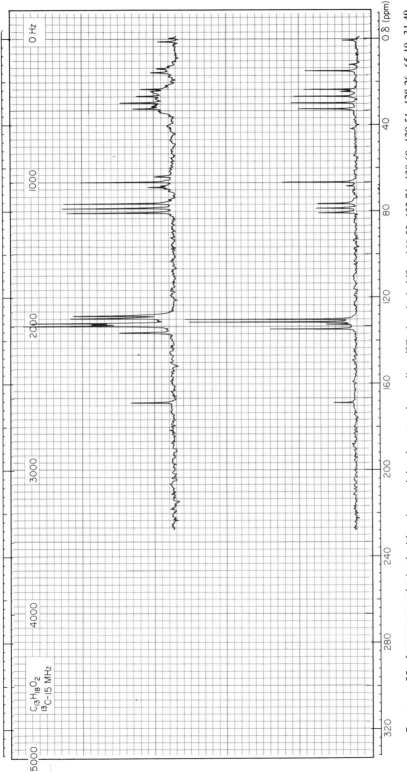

Spectrum 33 Lower trace obtained with noise-modulated proton decoupling. ^{13}C chemical shifts: 166.55, 132.71, 130.60, 129.51, 128.26, 65.10, 31.49, 28.75, 25.78, 22.58, 14.00 ppm. Upper trace obtained with off-resonance proton decoupling.

$C_{13}H_{18}O_2$
^{13}C–15 MHz

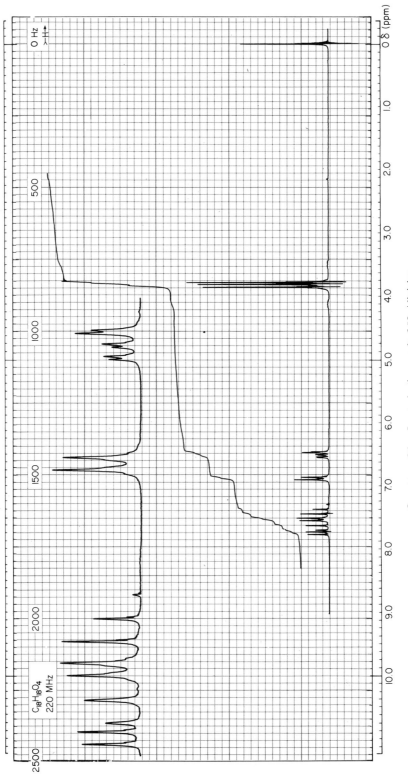

Spectrum 34a Inset abscissa scale 5 Hz/division.

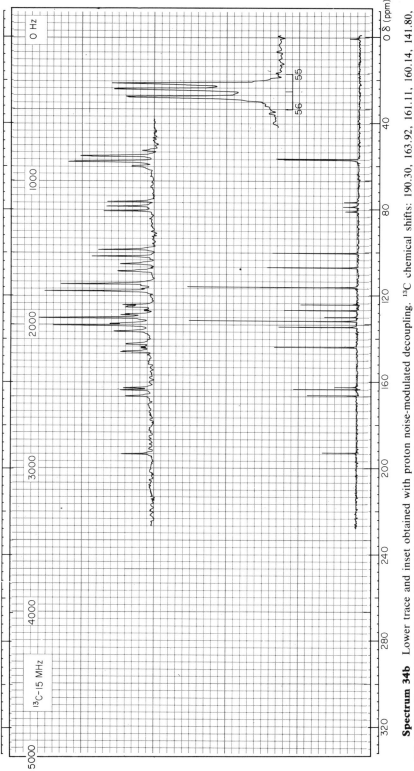

Spectrum 34b Lower trace and inset obtained with proton noise-modulated decoupling. ^{13}C chemical shifts: 190.30, 163.92, 161.11, 160.14, 141.80, 132.57, 129.83, 127.97, 124.91, 122.25, 114.20, 105.13, 98.50, 55.61, 55.41, 55.24 ppm. Upper trace obtained with off-resonance proton decoupling.

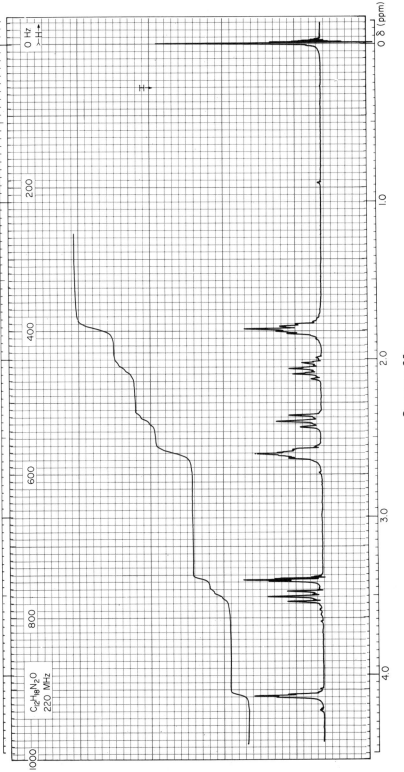

0 Hz
>—H→

H→

200

400

600

800

1000

$C_{12}H_{18}N_2O$
220 MHz

0 δ (ppm)

1.0

2.0

3.0

4.0

Spectrum 35a

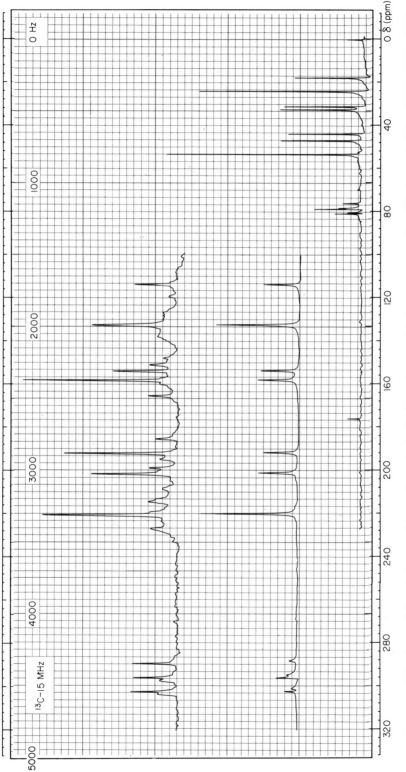

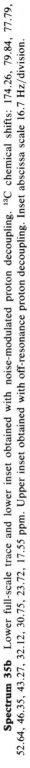

Spectrum 35b Lower full-scale trace and lower inset obtained with noise-modulated proton decoupling. ^{13}C chemical shifts: 174.26, 79.84, 77.79, 52.64, 46.35, 43.27, 32.12, 30.75, 23.72, 17.55 ppm. Upper inset obtained with off-resonance proton decoupling. Inset abscissa scale 16.7 Hz/division.

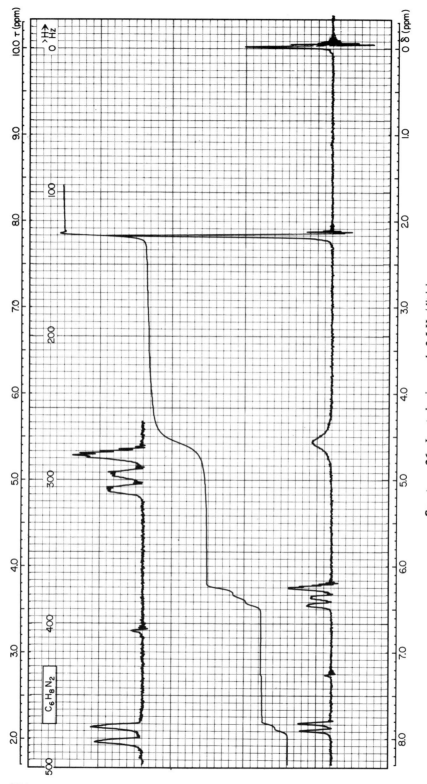

Spectrum 36 Inset abscissa scale 2.5 Hz/division.

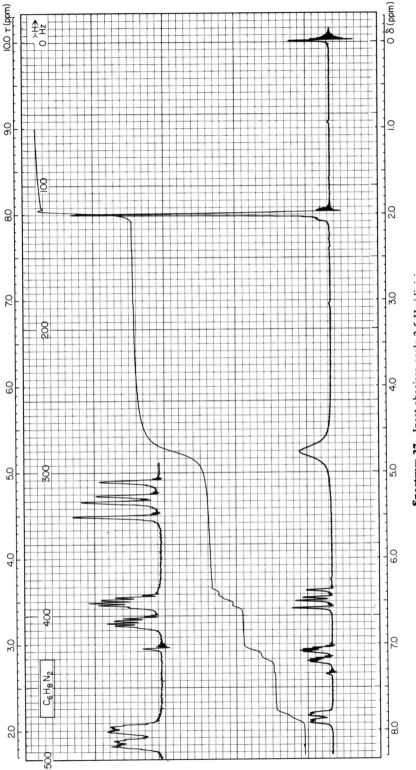

Spectrum 37 Inset abscissa scale 2.5 Hz/division.

$C_6H_8N_2$

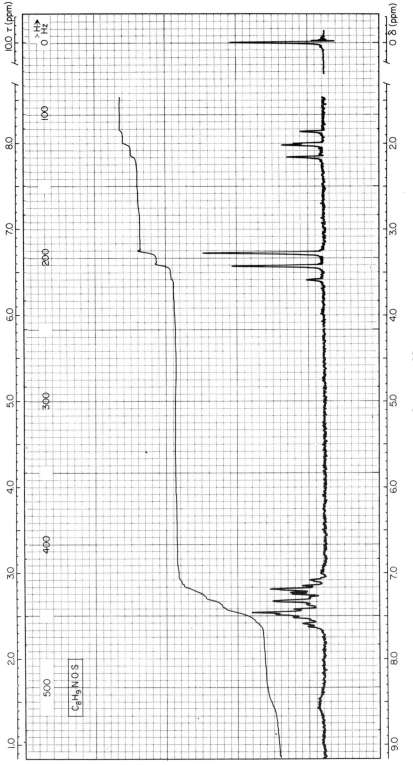

10.0 τ (ppm)

H

0 Hz

100

200

300

400

500

C_8H_9NOS

0 δ (ppm)

Spectrum 38

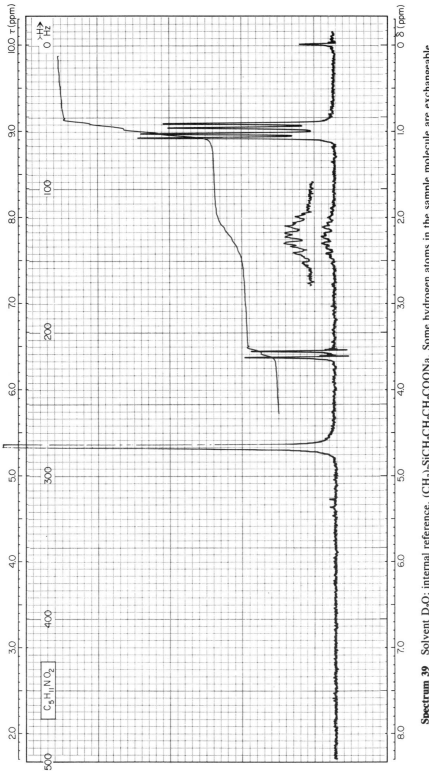

$C_5H_{11}NO_2$

Spectrum 39 Solvent D_2O; internal reference, $(CH_3)_3SiCH_2CH_2CH_2COONa$. Some hydrogen atoms in the sample molecule are exchangeable.

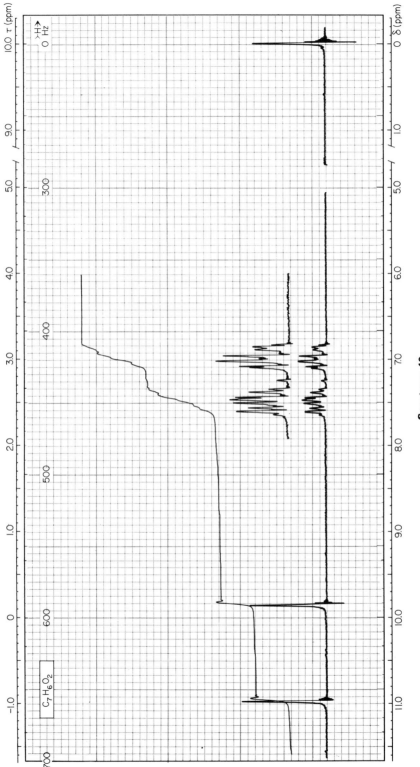

Spectrum 40

$C_7H_6O_2$

Answers to Selected Problems

Chapter 2

4. Since signal strength is proportional to n, the difference in population of the two energy levels, Eq. (2.39) gives the desired relation. It is convenient to cast it into the form

$$\ln(n - n_{eq}) = -\frac{1}{T_1} t + \ln(n - n_{eq})_{t=0}$$

or

$$\log(n - n_{eq}) = -\frac{1}{2.303 \ T_1} t + \log(n - n_{eq})_{t=0}.$$

Thus a plot against time of the difference between the signal height at time t and the asymptotic value it assumes after a sufficiently long time should give a straight line, from the slope of which we obtain T_1.

7. For a Gaussian line the shape function is

$$g(\nu) = A \ \exp[-\alpha(\nu - \nu_0)^2].$$

From Eq. (2.42),

$$A = 2T_2.$$

At the half-maximum point, where $\nu = \nu_0 + \frac{1}{2}\nu_{1/2}$, we obtain

$$\exp\left(-\frac{\alpha\nu_{1/2}^2}{4}\right) = \frac{1}{2}$$

or

$$\alpha = \frac{4 \ln 2}{\nu_{1/2}^2}.$$

From Eq. (2.19),

$$\int_0^\infty g(v)\,dv = 1.$$

A table of definite integrals shows that

$$\int_{-\infty}^\infty \exp[-ax^2]\,dx = \left(\frac{\pi}{a}\right)^{1/2}.$$

Making the substitution $x = v - v_0$,

$$\int_{-v_0}^\infty 2T_2 \exp\left(-\frac{4\ln 2}{v_{1/2}^2}x^2\right) dx = 1,$$

$$2\left(\frac{\pi}{4\ln 2}\right)^{1/2} v_{1/2}T_2 = 1,$$

$$v_{1/2} = \left(\frac{\ln 2}{\pi}\right)^{1/2}\frac{1}{T_2} = \frac{1}{2.14T_2}.$$

(Note that the limit of integration, $-v_0$, is replaced by $-\infty$. Why is this change valid?)

Chapter 3

2. The expression for the amplitude modulated rf wave is

$$AB\cos(2\pi v_0 t)\cos(2\pi v_m t).$$

From the well-known trigonometric identities for the sum and difference of two angles we have

$$\cos(a + b) = \cos a \cos b - \sin a \sin b,$$
$$\cos(a - b) = \cos a \cos b + \sin a \sin b.$$

The sum of these identities is

$$\cos(a + b) + \cos(a - b) = 2\cos a \cos b.$$

Thus the expression for the modulated wave can be written

$$\frac{AB}{2}\{\cos[2\pi(v_0 + v_m)t] + \cos[2\pi(v_0 - v_m)t]\}.$$

In this form we see that there are components oscillating at $(v_0 + v_m)$ and $(v_0 - v_m)$.

Chapter 4

3. We are given that dioxane in CCl_4 has a chemical shift $\tau = 6.43$ ppm, or that $\delta = 3.57$ ppm downfield from TMS as an internal reference. From Fig. 4.1 we see that liquid dioxane is 3.37 ppm lower in field than TMS dissolved in CCl_4. Thus dioxane dissolved in CCl_4 must be 0.20 ppm lower in field than liquid dioxane.

5. Using the same reasoning as in Problem 3, we find that the chemical shift of benzene in CCl_4 is 0.75 ppm lower in field than that of liquid benzene. We can use Eq. (4.13) to correct for the effect of magnetic susceptibility:

$$\delta \text{ (true)} = 0.75 - \tfrac{2}{3}\pi \times 10^6[-0.611 - (-0.691)] \times 10^{-6}$$
$$= 0.58 \text{ ppm.}$$

Thus of the measured difference in chemical shift between benzene in CCl_4 and in the neat liquid, 23% is due to susceptibility effects. The remainder represents a real change in shielding due to the solvent environment (see Chapter 12).

Chapter 5

2. The splitting is measured in CH_3D, and Eq. (5.2) is applied.

5. The ^1H spectrum consists of a doublet with equal intensities and a separation of 12 Hz. The ^{31}P spectrum contains 10 lines, each separated from the next by 12 Hz. The relative intensities, by extension of Table 5.1, are $1:9:36:84:126:126:84:36:9:1$. Experimentally the spectrum often appears as an octet since the outermost lines are so weak.

9. A nucleus with $I = 1$ may be oriented three ways relative to H_0, with a projection, m_i, on H_0 of $+1$, 0, or -1. We must consider what arrangements of n spins are possible to give the same total projection, M. For $n = 2$, the total projection M can equal 2 only if both spins have projections of $+1$; i.e., $m_1 = m_2 = 1$. But $M = 1$ can be obtained with $m_1 = 1$ and $m_2 = 0$, or with $m_1 = 0$ and $m_2 = 1$. Thus the state with $M = 1$ is twice as likely to occur as that with $M = 2$. $M = 0$ can occur three ways: $m_1 = 1$ and $m_2 = -1$; $m_1 = -1$ and $m_2 = 1$; or $m_1 = 0$ and $m_2 = 0$. Other cases can be treated in the same way. A table analogous to Table 5.1 for the $I = 1$ case is

n	Relative intensity
0	1
1	1 1 1
2	1 2 3 2 1
3	1 3 6 7 6 3 1

(The table may be extended by noting that each entry is the sum of the one immediately above and its two neighbors.)

Chapter 7

1. $CH_2{=}CHF$, ABCX or ABMX; PF_3, A_3X; cubane, A_8; $CH_3CHOHCH_3$, A_6MX or A_6XY if OH is not exchanging, or A_6X with exchange; chlorobenzene, AA'BB'C; $CH_3CH_2CH_3$, A_6B_2 (assuming long-range coupling can be neglected; otherwise, $A_3A_3'B_2$).

4. (a) and (d) are AB spectra. (b) has incorrect line spacings, and (c) has the wrong intensity ratio.

20. (a) A_3X_2; (b) AA'XX'; (c) A_3; (d) ABX; (e) AX.

21. (a) ABB'XX'; (b) AA'XX' and ABXY, with intensity ratio of $1:2$ if all three conformations are equally populated; (c) ABC; (d) three different ABX spectra; (e) three different AX spectra.

Chapter 8

1. (a) From Eq. (8.8),

$$R_1 = \frac{4\,\gamma_C^2\,\gamma_H^2\,\hbar^2\,S(S+1)}{3r_{CH}^6}\,\tau_c$$

This relation is applicable to each of the three methyl protons; hence the overall contribution is

$$R_1 = \frac{(4)(6723)^2(26750)^2(1.054 \times 10^{-27})^2(1/2)(3/2)}{(2.14 \times 10^{-8})^6}\,\tau_c$$

$$= 1.12 \times 10^9\,\tau_c.$$

(b) From Eq. (8.8),

$$R_1 = \frac{(4)(6723)^2(1929)^2(1.054 \times 10^{-27})^2(1)(2)}{(3)(1.14 \times 10^{-8})^6} \tau_c$$

$$= 0.227 \times 10^9 \, \tau_c.$$

(c) From Eq. (8.10),

$$R_1 = (2/15)\gamma_C^2 H_0^2 (\sigma^\| - \sigma^\perp)^2 \, \tau_c$$
$$= (0.133)(2\pi)^2 \, (68 \times 10^6)^2 \, (300 \times 10^{-6})^2 \, \tau_c$$
$$= 2.19 \times 10^9 \, \tau_c.$$

$R_1(a + b + c) = 3.54 \times 10^9 \, \tau_c$, with csa accounting for about 62% and ^{1}H dipolar relaxation 32% of the total for the three processes considered. What other relaxation mechanisms might be important?

2. From Eq. (8.17),

$$R_2 = \frac{4\pi^2 J^2}{3} S(S + 1) \left[T_1^S + \frac{T_1^S}{1 + (\omega_I - \omega_S)^2 \, (T_1^S)^2} \right]$$

$$= \frac{4\pi^2 (12.1)^2 (1)(2)}{3} \left[1.8 \times 10^{-3} + \frac{1.8 \times 10^{-3}}{1 + 4\pi^2 (48 \times 10^6)^2 (1.8 \times 10^{-3})^2} \right]$$

$$= (3853)[1.8 \times 10^{-3} + 6.1 \times 10^{-15}]$$
$$= 6.9 \, \text{sec}^{-1}.$$

$T_2 = 0.14$ sec, and the minimum linewidth is 2.3 Hz. From Eq. (8.16), $R_1 \approx 0$, since only the second term in the brackets contributes.

Chapter 9

1. (a) $\gamma H_2/2\pi \geqslant 20$ Hz. (b) From Eq. (9.5),

$$(173 - 243) = (173 - \nu_B) - \frac{(20)^2}{2(173 - \nu_B)}.$$

To a high degree of approximation, ν_B in the last term can be replaced by ν_2. Then solution of the equation gives $\nu_B = 245.9$ Hz.

6. The labeling of the states would change, but the experimental result would appear to be the same. [For a discussion of this point, see R. Johannesen, *J. Chem. Phys.* **48**, 1414 (1968).]

Chapter 10

1. $H_1 = 11.6$ G

2. $M_z = M_0(1 - e^{-t/T_1})$

Appendix C

1. $(CH_3)_2C(OCH_3)_2$.

3. 3-methylhexane: $CH_3CH_2CHCH_2CH_2CH_3$. From the molecular for-
                                    $|$
                                  CH_3
   mula, this is obviously an alkane. There are nine possible isomers,
   but all except 3-methylhexane have some symmetry, so that less than
   seven lines would be observed. In instances where the number of observed lines is not definitive, Eq. (4.23) can be used to predict the
   chemical shifts. It is instructive to apply it to the present case:

   $\delta_1 = 6.80 + (2)(-2.99) + (1)(0.49) + (1)(9.56) = 10.87$,
   $\delta_2 = 15.34 + (1)(-2.69) + (1)(0.25) + (1)(16.70) = 29.60$,
   $\delta_3 = 23.46 + (1)(-2.07) + (2)(6.60) = 34.59$,
   $\delta_4 = 15.34 + (1)(-2.69) + (1)(9.75) + (1)(16.70) = 39.10$,
   $\delta_5 = 15.34 + (2)(-2.69) + (1)(0.25) + (1)(9.75) = 19.96$,
   $\delta_6 = 6.80 + (1)(-2.99) + (2)(0.49) + (1)(9.56) = 14.35$,
   $\delta_7 = 6.80 + (2)(-2.99) + (1)(0.49) + (1)(17.83) = 19.14$.

   Comparison with the spectrum shows that the calculated values are in
   excellent agreement with the observed chemical shifts and permit
   clear assignments of all lines.

5. CH_3OCH_2COOH. Note that the integral of the carboxyl proton peak
   is too large. This discrepancy is due to the fact that this hygroscopic
   sample has absorbed water. The line at 462 Hz is actually due to the
   rapidly exchanging water and carboxyl protons. See Chapter 11 for a
   discussion of exchange phenomena.

7. CH_3CHCl_2. The symmetric isomer, CH_2ClCH_2Cl, would have only a
   single line in its spectrum.

9. $(C_2H_5O)_2PH$. $^1J_{PH} = 688$ Hz, while $^3J_{PH} = 9$ Hz.
            $\|$
            O

11.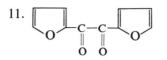

13. 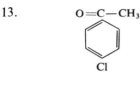 Note the pattern typical of a para-substituted aromatic.

15. $(CH_3)_3N^+CH_2CH_2OCCH_3Br^-$.

17. 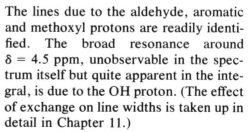 Note the complex but symmetric pattern of lines due to the ring protons, which results from the magnetic nonequivalence of the two protons attached to the same carbon atom.

19. An α,β-unsaturated ketone usually displays the resonance of the β proton at lower field than most olefinic protons. As seen in this spectrum it can be downfield as far as or farther than the aromatic proton lines.

21. The lines due to the aldehyde, aromatic and methoxyl protons are readily identified. The broad resonance around $\delta = 4.5$ ppm, unobservable in the spectrum itself but quite apparent in the integral, is due to the OH proton. (The effect of exchange on line widths is taken up in detail in Chapter 11.)

So far as the positions of the substituents on the aromatic ring are concerned, it is nearly impossible to distinguish between the OH and OCH$_3$ since both have almost the same effect on the chemical shifts of nearby aromatic protons. Treating these two substituents as equivalent, we can still distinguish six position isomers:

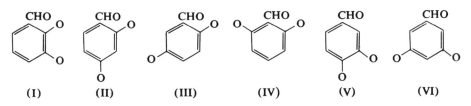

(I)   (II)   (III)   (IV)   (V)   (VI)

Isomers (IV) and (VI) can be excluded since they would give essentially A_2X spectra, which would differ from the experimental spectrum. The observed spectrum should be analyzed as ABX (see next paragraph), but for our present purposes we can use first-order analysis. The single proton responsible for the resonance near 700 Hz is evidently ortho to exactly one other proton and meta to no protons. Since it is shielded, it is probably ortho to the OH or OCH_3 rather than to the CHO group. These considerations rule out (I) and (II). Isomer (III) would be expected to have the chemical shifts of protons 3 and 4 nearly equal and at high field, whereas it is actually the two low-field chemical shifts that are almost equal. Thus (V) is preferred. The correct formula, given above, is vanillin; iso-vanillin, which differs only in the interchange of OH and OCH_3, gives a very similar NMR spectrum.

ABX analysis: The AB portion shows only six lines, but the two central lines are obviously broader than the others and must each consist of two almost coincident lines. Thus the two ab subspectra may be identified, and J_{AB} is found to be 1.8 Hz. From the distance between the centers of the subspectra $|J_{AX} + J_{BX}| = 8.4$ Hz, while the X portion shows that this quantity is 8.7 Hz (reasonable agreement for a single spectral trace with some ambiguity in selecting certain line positions). From the AB portion we obtain $2D_+ = 5.1$ and $2D_- = 3.4$, while from the X portion we get 5.2 and 3.5 Hz. Using average values for these quantities and following the procedure of Table 7.6, we obtain the two solutions (1) $\nu_A - \nu_B = 3.9$ and $J_{AX} - J_{BX} = 0.9$, or (2) $\nu_A - \nu_B = 0.9$ and $J_{AX} - J_{BX} = 3.9$. The former leads to the result $J_{AX} = 5.2$ Hz and $J_{BX} = 3.4$ Hz, which would be highly unusual for an aromatic system of this sort. The latter solution gives the perfectly acceptable values $J_{AX} = 8.2$ Hz and $J_{BX} = 0.4$ Hz, with $\nu_A = 748.2$ Hz and $\nu_B = 747.3$ Hz. We did not need to consider the relative intensities of the X lines in this case since we could reject one possible solution on the basis of our prior knowledge of the magnitudes of certain coupling constants.

23.   Within the accuracy of the data given, this spectrum can be analyzed by the first-order procedure. Alternatively the two

ab subspectra can be identified, and an ABX analysis carried out. There is ambiguity in the association of the left and right halves of the ab subspectra, but that simply means that the relative signs cannot be determined.

25. From the empirical formula this compound must be one of the three isomers of difluoroethylene. The analysis of the AA'XX' spectrum follows the procedure outlined in Section 7.22. With the notation of Table 7.7, $N = 92.5$, $K = 21.2$, $M = 17.0$, and $L = 51.7$ Hz, provided we take one of the ab subspectra as lines 2, 5, 6, and 9 (reading lines in order across Spectrum 25) and the other as lines 3, 4, 7, and 8. This gives $J_{AA'} = 19.1$, $J_{XX'} = 2.1$, $J_{AX} = 72.0$, and $J_{AX'} = 20.5$ Hz. The association of these values with the two geminal, the cis, and the trans couplings, and the determination of the correct geometric isomers is best done in conjunction with Spectrum 26 and Figure 5.3.

It might appear that the intensity relations would permit the ab subspectra to be chosen as follows: lines 2, 4, 7, and 9 as one subspectrum and lines 3, 5, 6, and 8 as the other. However, this choice is not valid, for it would lead to different values of L from the two subspectra. Can you show that this is a general result for all AA'XX' spectra?

27. The compound is vinyl fluoride, $CH_2=CHF$. The 60 MHz spectrum departs considerably from first order due to the fortuitous coincidence of several lines. The 100 MHz spectrum, on the other hand, shows the 24 lines predicted by the first-order rules and may be analyzed accordingly to give the values

$$J_{HF}(gem) = 85, \quad J_{HF}(cis) = 20.5, \quad J_{HF}(trans) = 53.5,$$
$$J_{HH}(gem) = 3, \quad J_{HH}(cis) = 5, \quad and \quad J_{HH}(trans) = 12.5.$$

The exact treatment of this molecule as an ABCX system shows that the proton resonance spectrum consists of two overlapping abc (approximately abx) subspectra, which may be analyzed to give the proton chemical shifts and all six coupling constants (signs, as well as magnitudes). For details see Emsley, Feeney, and Sutcliffe,[43] p. 423.

29. $CH_2ClCH=CHCH_2Cl(trans)$. Note the effect of virtual coupling.

31.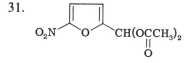

A chemical shift of 7.7 ppm is highly unusual for an aliphatic proton, but arises from the concerted electron-withdrawing action of three electronegative groups. A similar situation occurs with the ^{13}C line at 82 ppm. Another remarkable feature of the ^{13}C spectrum in the broad line at 152 ppm. This line is most probably assigned to C-5, with the broadening attributed to T_2 scalar relaxation as a result of coupling to the rapidly relaxing ^{14}N.

33. 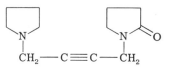 Off-resonance decoupling provides useful information on the multiplicity of lines, hence on the number of protons directly bonded to each carbon. The C $=$ O line remains a singlet, while several of the aromatic carbon lines become doublets. Only one of the CH_2 groups shows a clear triplet, but the others display a central peak in each multiplet that coincides in frequency with the noise decoupled line—a situation compatible with a CH_2 or a quaternary carbon. However, the latter would be a clean singlet, as in C $=$ O. Likewise, the CH_3 quartet is not clear, but the two most intense peaks of the quartet are evident, with additional unresolved lines.

35. Oxotremorine

The pyrrolidone ring gives rise to the clearly defined multiplets in the ^{1}H spectrum suggestive of a $-CH_2-CH_2-CH_2-$ fragment, while the pyrrolidine ring gives two complex multiplets that indicate more extensive coupling, along with "virtual coupling." The ^{13}C spectrum indicates the presence of a C=O, and the off-resonance decoupling results point to the presence of CH_2 groups and two unprotonated carbons (nearly hidden by the $CDCl_3$ solvent lines).

37. The chemical shifts of the three ring protons suggest that the substituents are located on the 2 and 3 positions, and this conclusion is confirmed by the observed splittings. The long-range coupling of 4-H to the CH_3 establishes the methyl group in the 3 position. Note the broadening of the amino protons and the proton α to the ring nitrogen as a result of ^{14}N relaxation.

39. $(CH_3)_2CHCHCOOH.$ The two methyl groups are chemically nonequivalent because of the presence of the asymmetric center.
    $\quad\quad\quad\quad |$
    $\quad\quad\quad NH_2$

Index